PID CONTROLLER DESIGN APPROACHES – THEORY, TUNING AND APPLICATION TO FRONTIER AREAS

Edited by **Marialena Vagia**

PID Controller Design Approaches – Theory, Tuning and Application to Frontier Areas
Edited by Marialena Vagia

Published by InTech

Janeza Trdine 9, 51000 Rijeka, Croatia

Publishing Process Manager Irena Voric

Technical Editor Teodora Smiljanic

Cover Designer InTech Design Team

Additional hard copies can be obtained from orders@intechopen.com

PID Controller Design Approaches – Theory, Tuning and Application to Frontier Areas,
Edited by Marialena Vagia
p. cm.
ISBN 978-953-51-0405-6

For EU product safety concerns:
IN TECH d.o.o., Prolaz Marije Krucifikse Kozulić 3, 51000 Rijeka, Croatia,
info@intechopen.com or visit our website at intechopen.com.

Meet the editor

Dr. Marialena Vagia (BSc. 2005, PhD. 2009) is currently a Marie Curie postdoctoral fellow funded by ERCIM Alain Bensoussan Network at the Norwegian University of Science and Technology Trondheim, working in the area of control and nanopositioning. Her research interests include the areas of Robust, Switching and Nonlinear Control, LMIs, Electrostatic MicroActuators, MicroCantilever Beams, Nanostructures, PiezoActuators and Nanopositioning. In the past, she received the C. Karatheodoris fellowship funded by University of Patras for PhD research, a postdoctoral scholarship from "Greek State Scholarships Foundation", and also a scholarship for carrying post-doctoral research from the General Secretariat Research and Technology, Greece. She has published in 7 journals and has 20 conference papers, and received the BEST Student paper award in IEEE MED Control Conference in 2007.

Contents

Preface

First placed on the market in 1939, the PID controller is one of the most dominating forms of feedback used until the recent years. Although PID controllers are used for a wide range of technology applications, many important issues regarding their tuning and performance remain open. Numerous everyday applications are incorporating PID control structures into their design, such as process control, automotive applications, magnetic and optic memories, flight control, instrumentation, microsystems technology, pneumatic mechanisms, dc motors, etc. The augmented complexity of modern applications requires that the controllers implemented would incorporate into their design important characteristics of the systems. These characteristics include but are not limited to: model uncertainties, system's nonlinearities, time delays, disturbance rejection requirements and performance criteria.

The scope of this book is to propose different PID control design strategies for numerous modern technology applications. The book is divided into four sections, in order to categorize the different applications and control methodologies described in each one of them. The first section is entitled "Tuning Methods for 3 Terms Controllers - Classical Approach ". The first chapter of this section describes the idea of a multiresolution PID controller using the wavelet theory for the decomposition of the tracking error signal. Innovative methods of tuning three-term controllers for integrating processes incorporating both time-delay and a non-minimum phase zero are presented in the second chapter. The next chapter describes a novel test bed for evaluation of a golf putting pneumatic mechanism and the last chapter deals with the development of a controller for the active control of a suspension system, which improves the inherent tradeoff among ride comfort, suspension travel and road-holding ability.

The second section is called "Intelligent Control and Genetic Algorithms Approach". In this section, fuzzy controllers and controllers tuned with the methods of genetic algorithms are presented. In the first place, a fuzzy PID supervision scheme associated to a PID control is designed, and an automotive application is chosen for illustration and validation of the proposed approach. In addition, a conceptual model for a knowledge-based system for the achievement of the best parameters based in Closed-Loop design of PID controllers is presented. In the sequel, a nominal characteristic trajectory following controller is proposed as a practical approach for point to point positioning systems. At the end of this section another strategy is proposed, where the ranges of proportional, derivative and integral gains are not set arbitrary but they are set within the stabilizing regions determined, so that we are searching among the stabilizing values of the PID controller.

Third section "Robust PID Controller Design" deals with robustness analysis for PID controller design. The first chapter deals with the famous Monte-Carlo method, and a performance robustness criterion is proposed. The performance robustness criterion would give a new view to study the important issue of how the PID controller performs while the parameters of model are uncertain. In the last chapter, a robust PID controller tuned with the use of the Linear Matrix Inequalities is designed, and applied on an electrostatic micro-beams structure.

Finally the last section of this book, entitled "Disturbance Rejection for PID Controller Design" proposes tuning techniques in order to minimize the undesired effects of disturbances, which are present in many real control application problems.

Book "PID Controller Design Approaches - Theory, Tuning and Application to Frontier Areas" proposes different PID control strategies for numerous modern technology applications in order to cover the needs of an audience including researchers, scholars and professionals who are interested in advances in PID controllers and related topics.

Last but not least, at this point I would like to thank all the authors of this book for the efforts they made in order to submit their chapters and to the InTech publishing working team for their support during all the stages of the publishing process.

Dr. Marialena Vagia
Electrical & Computer Engineering Department,
University of Patras,
Greece

Part 1

Tuning Methods for 3 Terms Controllers – Classical Approach

1

Wavelet PID and Wavenet PID: Theory and Applications

José Alberto Cruz Tolentino[1], Alejandro Jarillo Silva[1], Luis Enrique Ramos Velasco[2] and Omar Arturo Domínguez Ramírez[2]
[1]*Universidad de la Sierra Sur*
[2]*Univerisidad Politécnica de Pachuca*
Universidad Autónoma del Estado de Hidalgo
México

1. Introduction

We introduce in this chapter a new area in PID controllers, which is called multiresolution PID (MRPID). Basically, a MRPID controller uses wavelet theory for the decomposition of the tracking error signal. We present a general error function in terms of partial errors which gives us the various frequencies appearing in the general errors. Once we obtain the spectrum of the error signal, we divide the error at frequencies that are weighted by gains proposed by the designer. We can say that the MRPID is a generalization of conventional PID controller in the sense that the error decomposition is not only limited to three terms.

The PID is the main controller used in the control process. However, the linear PID algorithm might be difficult to be used with processes with complex dynamics such as those with large dead time and highly nonlinear characteristics. The PID controller operation is based on acting proportionally, integrally and derivative way over the error signal $e(t)$, defined it as the difference between the reference signal y_{ref} and the process output signal $y(t)$, for generating the control signal $u(t)$ that manipulates the output of the process as desired, as shown in the Fig. 2, where the constants k_P k_I and k_D are the controller gains. There are several analytical and experimental techniques to tune these gains (Aström & Hägglund, 2007). One alternative is auto-tuning online the gains as in (Cruz et al., 2010; O. Islas Gómez, 2011a; Sedighizadeh & Rezazadeh, 2008a) where they use a wavelet neural networks to identify the plant and compute these gain values, this approach has been applied in this chapter.

The chapter is organized as follows: a general overview of the wavelets and multiresolution decomposition is given in Section 2. In Section 3 we preset some experimental results of the close-loop system with the MRPID controller. The PID controller based on wavelet neural network and experimental is given in Section 4, while the experimental results are given in Section 5. Finally, the conclusions of the contribution about wavelet PID and wavenet controllers are presented in Section 6.

2. PID controller based on wavelet theory and multiresolution analysis

2.1 Wavelet theory and multiresolution analysis

Here, we briefly summarize some results from the wavelet theory that are relevant to this work, for it we use the notation presented in the Table 1. For more comprehensive discussion of wavelets and their applications in control, signal processing, see e.g., (Daubechies, 1992; Hans, 2005; Mallat, 1989a;b; Parvez, 2003; Parvez & Gao, 2005; Vetterli & Kovačević, 1995).

$\psi(t)$	Mother wavelet function
$\psi_{a,b}$	Daughter wavelet function
$W_f(a,b)$	Continuous wavelet transform
$W_f[a,b]$	Discrete wavelet transform
$< f,g >$	Inner product between f and g
$\bigoplus$	Direct sum of subspaces
$V \perp W$	V is orthogonal to W
$L^2(\mathbb{R})$	Vector space of all measurable, square integrable functions
$\mathbb{R}$	Vector space of the real numbers
$\mathbb{Z}$	Set of all integers

Table 1. Notation

A wavelet is defined as an oscillatory wave ψ of very short duration and satisfy the admissibility condition (Daubechies, 1992), given by

$$\Psi(0) = \int_{-\infty}^{\infty} \psi(t)dt = 0, \tag{1}$$

where Ψ is the Fourier transform of wavelet function ψ, the latter also called wavelet mother function, the mathematical representation of some mother wavelet are shown in Table 2 and their graphs are plotted in Fig. 1. Wavelet function ψ is called the "mother wavelet" because different wavelets generated from the expansion or contraction, and translation, they are called "daughter wavelets", which have the mathematical representation given by:

$$\psi_{a,b}(t) = \frac{1}{\sqrt{a}}\psi\left(\frac{t-b}{a}\right), \quad a \neq 0;\ a,b \in \mathbb{R}, \tag{2}$$

where a is the dilation variable that allows for the expansions and contractions of the ψ and b is the translation variable and allows translate in time.

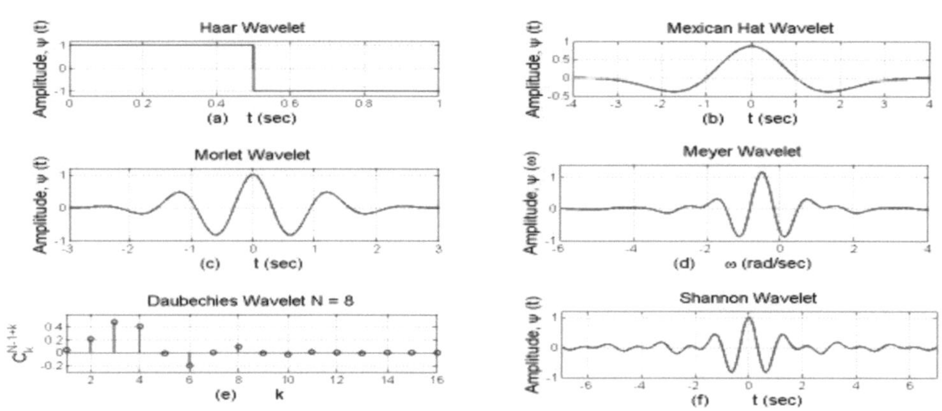

Fig. 1. Graphics of the mother wavelets showed in Table 2.

Haar	$\psi(t) = \begin{cases} 1, \text{ if } t \in [0, \frac{1}{2}] \\ -1, \text{ if } t \in (\frac{1}{2}, 1] \\ 0, \text{ otherwise} \end{cases}$
Mexican hat	$\psi(t) = \frac{2}{\sqrt{3}} \pi^{-\frac{1}{4}} (1 - t^2) e^{(-\frac{1}{2}t^2)}$
Morlet	$\psi(t) = e^{-\frac{t^2}{2}} cos(5t)$
Shannon	$\psi(t) = \frac{sin(\frac{\pi}{2}t)}{\frac{\pi}{2}t} cos(\frac{3\pi}{2})t$
Daubechies	$P(y) = \sum_{k=0}^{N-1} C_k^{N-1+k} y^k$; C_k^{N-1+k} are binomial coefficients, N is the order of the wavelet
Meyer	$\psi(\omega) = \begin{cases} \frac{e^{\frac{i\omega}{2}}}{\sqrt{2\pi}} \sin(\frac{\pi}{2} v (\frac{3}{2\pi}\lvert\omega\rvert - 1)), \\ \text{if } \frac{2\pi}{3} \le \lvert\omega\rvert \le \frac{4\pi}{3} \\ \frac{e^{\frac{i\omega}{2}}}{\sqrt{2\pi}} \cos(\frac{\pi}{2} v (\frac{3}{4\pi}\lvert\omega\rvert - 1)), \\ \text{if } \frac{4\pi}{3} \le \lvert\omega\rvert \le \frac{8\pi}{3} \\ 0, \text{ otherwise} \end{cases}$ $v = a^4(35 - 84a + 70a^2 - 20a^3)$, $a \in [0, 1]$

Table 2. Some examples of common mother wavelets

There are two types of wavelet transform: continuous wavelet transform (CWT) and discrete wavelet transform (DWT), whose mathematical definition are given by (3) and (4), respectively (Daubechies, 1992):

$$W_f(a,b) = \langle f, \psi_{a,b} \rangle = \frac{1}{\sqrt{a}} \int_{-\infty}^{\infty} f(t) \psi\left(\frac{t-b}{a}\right) dt, \tag{3}$$

$$W_f[a,b] = \frac{1}{\sqrt{a_0^m}} \int_{-\infty}^{\infty} f(t)\, \psi\left(\frac{t}{a_0^m} - kb_0\right) dt, \tag{4}$$

for CWT, the expansion parameters a and translation b vary continuously on $\mathbb{R}$, with the restriction $a > 0$. For DWT, the parameters a and b are only discrete values: $a = a_0^m$, $b = kb_0a_0^m$, where $a_0 > 1$, b_0 and are fixed values. In both cases $f \in L^2(\mathbb{R})$, i.e., a function that belongs to the space of all square integrable functions.

In DWT, one of the most important feature is the multiresolution analysis (Mallat, 1989a;b). Multiresolution analysis with a function $f \in L^2(\mathbb{R})$, can be decomposed in the form of successive approximations, using wavelet basis functions. The multiresolution analysis consists of a sequence successive approximations of enclosed spaces, nested spaces $\{V_N : N \in \mathbb{Z}\}$ with the following properties (Daubechies, 1992):

1. Nesting: $V_N \subset V_{N+1}, \ \forall N \in \mathbb{Z}$.
2. Closure: $clos\left(\bigcup_{N\in\mathbb{Z}} V_N\right) = L^2(\mathbb{R})$.
3. Shrinking: $\bigcap_{N\in\mathbb{Z}} V_N = \{0\}$.
4. Multiresolution: $f[n] \in V_N \Longleftrightarrow f[2n] \in V_{N+1} \ \forall N \in \mathbb{Z}$.
5. Shifting: $f[n] \in V_N \Longleftrightarrow f[n - 2^{-N}k] \in V_N \ \forall N \in \mathbb{Z}$.

6. There exists a *scaling function* $\phi \in V_0$ such that the integer shifts of ϕ form an orthonormal basis for V_0, i.e.,

$$V_0 = span\{\phi_{N,k}[n], \ N, k \in \mathbb{Z}\},$$

where

$$\phi_{N,k}[n] = 2^{-\frac{N}{2}} \phi[2^{-N} n - k], \tag{5}$$

forming an orthogonal basis of V_0. Then for each V_N exists additional space W_N that meets the following conditions (Daubechies, 1992)

$$V_{N+1} = V_N \oplus W_N, \tag{6}$$

$$V_N \perp W_N = 0, \ \ \forall \, N \in \mathbb{Z}, \tag{7}$$

and is

$$\psi_{N,k}[n] = 2^{-\frac{N}{2}} \psi[2^{-N} n - k], \ \ \forall \, N, k \in \mathbb{Z}, \tag{8}$$

forming an orthogonal basis for W_N, i.e. at $\psi[n]$ can generate the space W_N.

From the above we can say that the purpose of analysis multiresolution is to determine a function $f[n]$ by successive approximations, as

$$f[n] = \sum_{k=-\infty}^{\infty} c_{N,k} \phi_{N,k}[n] + \sum_{m=1}^{N} \sum_{k=-\infty}^{\infty} d_{m,k} \psi_{m,k}[n], \tag{9}$$

with

$$c_{m,k} = \sum_{k=-\infty}^{\infty} f[n] \overline{\phi_{m,k}[n]},$$

$$d_{m,k} = \sum_{k=-\infty}^{\infty} f[n] \overline{\psi_{m,k}[n]}. \tag{10}$$

Where N is the level at which decomposes $f[n]$ and $\overline{\phi[n]}$, $\overline{\psi[n]}$ are conjugate functions for $\phi[n]$ and $\psi[n]$, respectively. Multiresolution analysis, in addition to being intuitive and useful in practice, form the basis of a mathematical framework for wavelets. One can decompose a function a soft version and a residual, as we can see from (9), where the wavelet transform decomposes a signal $f[n]$ in one approach or trend coefficients c and detail coefficients d which, together with $\phi[n]$ and $\psi[n]$, are the smoothed version and the residue, respectively.

The important thing here is that the decomposition of the $f[n]$ for large enough value of N can be approximated arbitrarily close to V_N. This is that $\exists$ some $\epsilon > 0$ such that

$$\| f[n] - \sum_{k=-\infty}^{\infty} c_{N,k} \phi_{N,k}[n] \| < \epsilon. \tag{11}$$

The approach by the truncation of the wavelet decomposition can be approximated as:

$$f[n] \approx \sum_{k=-\infty}^{\infty} c_{N,k} \phi_{N,k}[n]. \tag{12}$$

This expression indicates that some fine components (high frequency) belonging to the wavelet space W_N for the $f[n]$ are removed and the components belonging to the coarse scale space V_N are preserved to approximate the original function at a scale N. Then (12) tells us that any function $f[n] \in L^2(\mathbb{R})$ can be approximated by a finite linear combination.

2.2 Wavelet PID controller design

A classic control scheme consists of three basic blocks as shown in Fig. 2: the plant can be affected by external perturbation P, the sensor measures, the variable of interest y, and finally the controller makes the plant behaves in a predetermined manner, y_{ref}. One of the most employed controller in the modern industry is a classical control Proportional, Integral and Derivative, PID because its easy of implementation, requiring only basics testing for tuning gains k_P k_I and k_D (Aström & Hägglund, 2007).

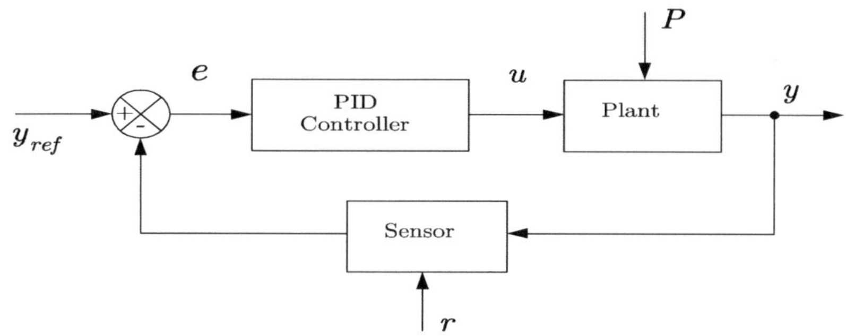

Fig. 2. Scheme of a SISO system with a PID controller.

In general, a PID controller takes as input the error signal e and acts on it to generate an output control signal u, as

$$u = k_P e + k_I \int_0^t e dt + k_D \frac{de}{dt}, \tag{13}$$

where k_P, k_I y k_D are the PID gains to be tuned, and e is the error signal which is defined as

$$e = y_{ref} - y, \tag{14}$$

The form of a discrete PID is (Visioli, 2006):

$$u(k) = u(k-1) + k_P\left[e(k) - e(k-1)\right] + k_I e(k) + k_D\left[e(k) - 2e(k-1) + e(k-2)\right], \tag{15}$$

whose transfer function is given by

$$\frac{u(z)}{e(z)} = k_P + k_I \frac{T}{2}\frac{z+1}{(z-1)} + k_D \frac{1}{T}\frac{(z-1)}{z}, \tag{16}$$

and its operation is the same way that the continuous PID.

Taking the parameters k_P, k_I and k_D of the PID, as adjustment variables, then (15) can be described as

$$u(k) = u(k-1) + \sum_{i=0}^{2} k_i e(k-i), \tag{17}$$

or equivalently

$$\Delta u(k) = \sum_{i=0}^{2} k_i e(k-i), \tag{18}$$

where $k_0 = k_P + k_I + k_D$, $k_1 = -k_P - 2k_D$ y $k_2 = k_D$. From (18), we see that the control law of a classic PID is a linear decomposition of the error, only that this decomposition is fixed, that is, always has three terms, this makes the difference between the classic PID and the MRPID, where here the number of decompositions can be infinite and even more than each one is different scales of time-frequency, this means that the MRPID controller decomposes the signal error e for high, low and intermediate frequencies, making use of multiresolution analysis for the decomposition. Where the components of the error signal are computed using (9) through a scheme of sub-band coding, as shown in Fig. 3.

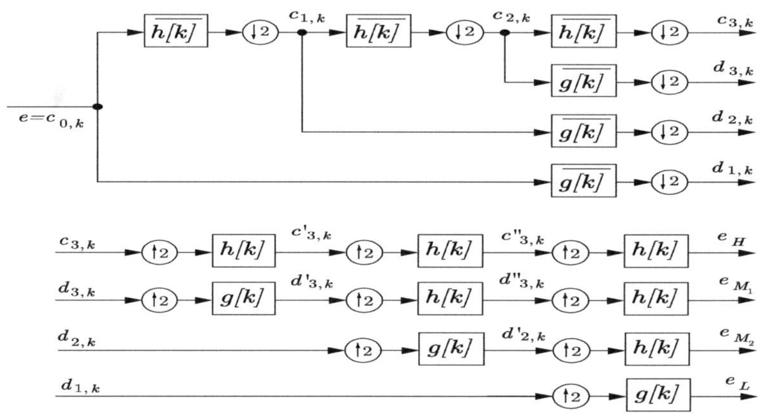

Fig. 3. Sub-band coding scheme for decomposition of the error signal e for N=3.

Thus each of these components are scaled with their respective gains and added together to generate the control signal u, as follows:

$$u = K_H e_H + K_{M_1} e_{M_1} + \cdots + k_i e_i + \cdots + K_{M_{N-1}} e_{M_{N-1}} + K_L e_L, \tag{19}$$

$$u(k) = \mathbf{K}\,\mathbf{E_m}(k), \tag{20}$$

where

$$\mathbf{K} = [K_H\; K_{M_1}\; \cdots\; K_i\; \cdots\; K_{M_{N-1}}\; K_L], \tag{21}$$

$$\mathbf{E_m}(k) = [e_H(k)\; e_{M_1}(k)\; \cdots\; e_i(k)\; \cdots\; e_{M_{N-1}}(k)\; e_L(k)]^T, \tag{22}$$

where N is the level of the MRPID controller.

While a classical PID control has three parameters to be tuned k_P, k_I and k_D, the MRPID control has two or more parameters and the number of parameters depends on the level of decomposition is applied to the signal error e. The schematic diagram of a plant using a MRPID control is shown in Fig. 4.

As shown in Table 2, there are a number of different wavelets, the wavelet selection affects the operation of the controller. Therefore, there are characteristics that should be taken into account, such as:

- The type of system representation (continuous or discrete).

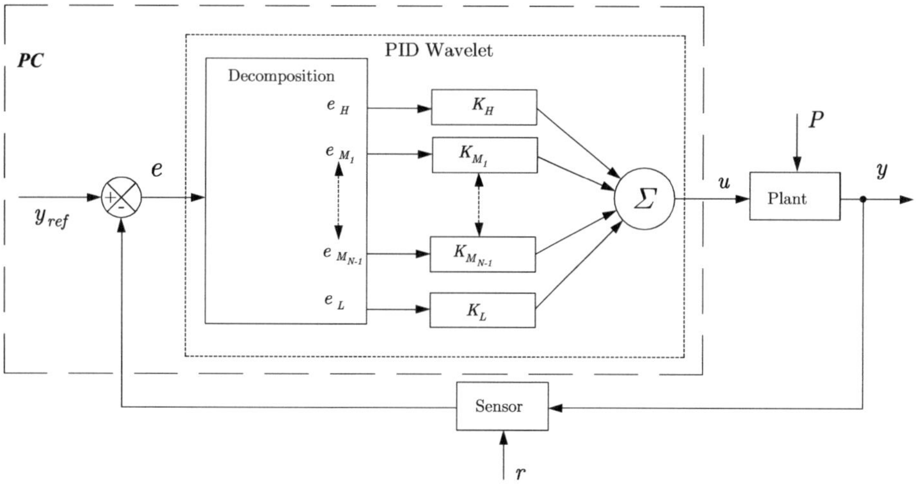

Fig. 4. Close loop block diagram of a SISO system with the MRPID controller.

- The properties of the wavelet to be used.
- The dynamics of the system.

For more details on the selection of the wavelet, see (Parvez, 2003). All physical systems are subject to any external signals or noise during the test. Therefore, when we design a control system must consider whether the system will provide greater sensitivity to noise or disturbance. In practice, disturbances and references are sometimes low frequency signals and noise is a high frequency signal, with a MRPID controller we can manipulated these signals, this means we tuning the gains directly. For example, adjusting the gain of the low scale to zero, i.e. $K_L = 0$, it produces a control signal that reduces the effects of noise on the output of the plant y, and therefore a smooth control signal which help in minimal effort to improve the life of the actuators and the whole plant performance.

3. MRPID applications

3.1 Control position of a DC motor

Here, we present an application of the MRPID controlller for a DC motor for it, we are using the Daubechies wavelet of order 2 for multiresolution signal decomposition of the error e and a level decomposition $N = 3$. The Daubechies filter coefficients of order 2 used in the multiresolution decomposition of the control are given in Table 3 and the structure of the filters h and g in the plane z are given by (23) and(24), respectively.

$$h(z) = 0.4839 + 0.8365z^{-1} + 0.2241z^{-2} - 0.1294z^{-3}, \tag{23}$$

$$g(z) = -0.1294 + 0.2241z^{-1} + 0.8365z^{-2} + 0.4839z^{-3}. \tag{24}$$

The MRPID control applied to the position control of a DC motor with the following transfer

h	0.4830	0.8365	0.2241	-0.1294
g	-0.1294	0.2241	0.8365	0.4830

Table 3. Coefficients of the Daubechies filters of order two.

function (Tang, 2001):

$$G(s) = \frac{b}{s(Js + c)} \tag{25}$$

where b is the torque constant, c is the friction coefficient and J is the total inertia of the motor and load, whose values are given in Table 4. This system is considered to implement a classical PID controller and MRPID controller for level $N = 3$, the values of the gains are shown in Table 5, which are obtained by trial and error. To analyze the behavior of the system in

Parameter	Value	Units
b	22	$N \cdot m/\text{volts}$
c	4	$kg \cdot m^2/seg \cdot rad$
J	1	$kg \cdot m^2/rad$

Table 4. Parameters of DC motor

Gains values of the PID	K_P	K_D	K_I	
	7	1.5	0	
Gains values of the MRPID	k_H	K_{M_1}	K_{M_2}	K_L
	0.15	4	10	0

Table 5. Gains values of the PID and MRPID

the presence of noise we are injected white noise signal with maximum amplitude of ± 0.16 radians, as shown in Fig. 5 in the measurement of output for both the classic controllers PID and MRPID. The results are shown in Figures 6 and 7.

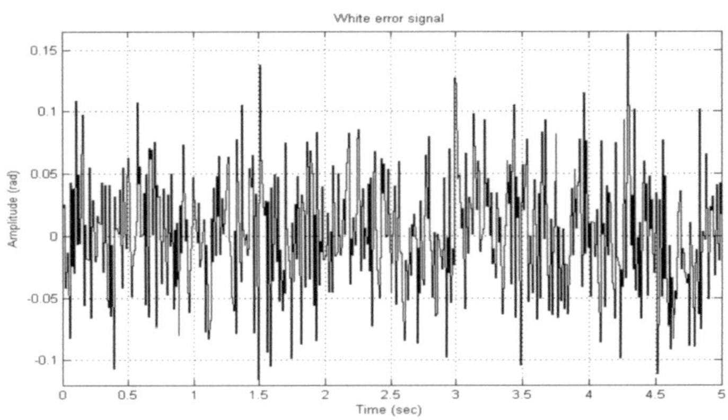

Fig. 5. White noise signal

From the Fig. 6a we observe that the output signal of the system with both controller is similar in behavior, with some variations generated by the noise in measuring the output signal. Fig. 6b, is observed as the classic PID control signal varies about ± 15 volts with abrupt changes, which may generate stress and wear gradually engine life. While the control signal of a MRPID control is a smooth signal compared to the classic PID control signal.

This is because although the error signals are similar as shown in Fig. 7a, showing that both contain the noise, the MRPID control signal as shown in Fig. 7b and discriminates the noise contained in the error signal, by scaling the component e_L with $K_L = 0$ which is the component

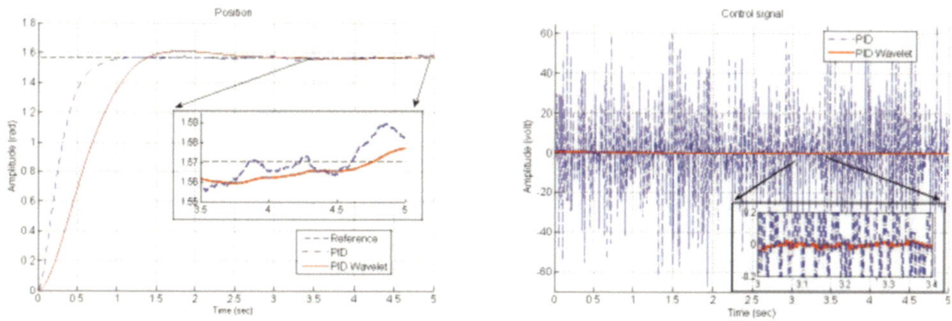

(a) Joint position (b) Control signal

Fig. 6. Results of the position and control signal on the system with PID and MRPID controller even in the presence of noise

signal contains much noise, so we can not do the same with e_{M_2} component, that is required at the start to give the necessary power to the system to overcome the inertia of the system.

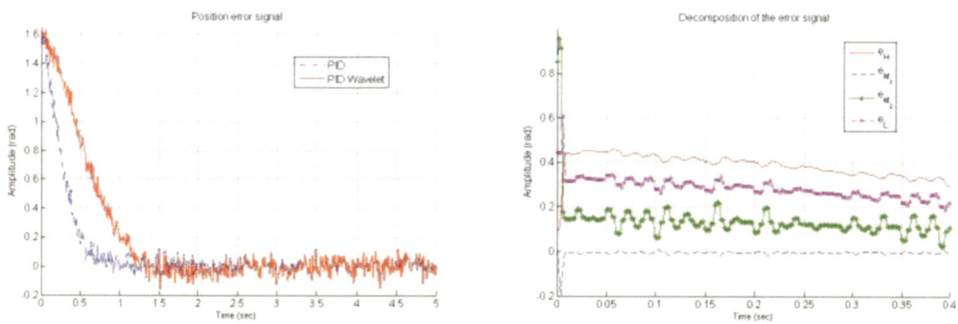

(a) Joint position error (b) Decomposition of the joint error

Fig. 7. Results and decomposition of the error signal on the system with PID and MRPID even in the presence noise

As we can see in previous simulations wavelet PID controller has the feature of being immune to the presence of noise.

The following results are obtained from the experimental implementation of the MRPID controller in a joint position system control, shown in Fig. 8a, the voltage transfer function of v for a position x can be modeled as:

$$G(s) = \frac{X(s)}{V(s)} = \frac{b}{s(Js + c)}. \tag{26}$$

where b, J, c is the torque constant, the total inertia of motor and load, and the viscous friction coefficient, respectively. The control goal is to rotate the motor from point A to point B, as shown in Fig. 8b. It is important to note that the parameters of the plant are not required for the proposed tuning controllers.

(a) Plant: DC motor

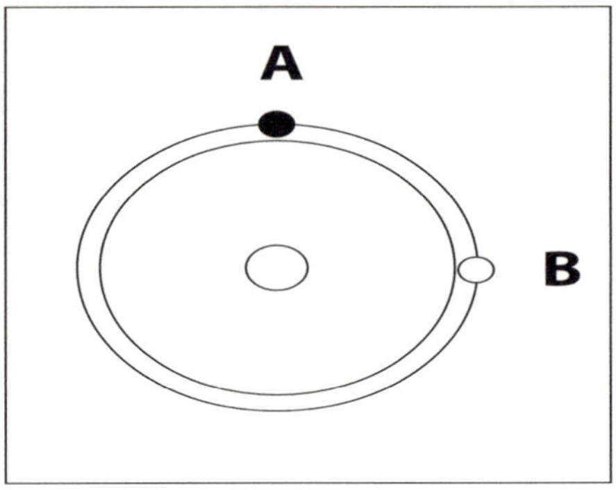

(b) Joint tracking trajectory

Fig. 8. SISO System to be controled and joint tracking

The test was performed using an experimental platform and a PC with a data acquisition card from National Instrument PCI-6024E and a servomotor. The control law is programming in Simulink environment, using an encoding sub-band scheme with Daubechies coefficients of order 2 and a level decomposition $N = 3$, where the gains of the classic PID are tuned heuristic and the MRPID tuned based on experience.

Gains values of the PID	K_P	K_D	K_I	
	10	1.6	6	
Gains values of the MRPID	k_H	K_{m_1}	K_{m_2}	K_L
	1.6	18	1.6	0

Table 6. Gains values of the PID and MRPID.

The graphs obtained are given in Figs. 9 and 10, which show the particular feature of the MRPID controller, i.e. generating a smooth control signal preventing stress and wear the engine. Besides having a better performance in the output of the plant when it is controlled with a classical PID, such as by requiring the control signal response would generate high frequency contain damage to the actuator, and some cases could not let this happen such is the case of this plant, the engine would vibrate only without generating any movement.

3.2 Control for global regulatory on a robot manipulator

3.2.1 Platform

For experimental purposes we use the system which is shown in Fig. 11, it is a planar robot with two degrees of freedom, which has two servo motors to move the links, the position is measured with resistive type sensors.

The position control law for the planar robot is given by (20) with

$$\mathbf{u} = \begin{bmatrix} u_1 \\ u_2 \end{bmatrix} \tag{27}$$

The classic PID control gains and MRPID control are given in Table 7 and Table 8, respectively.

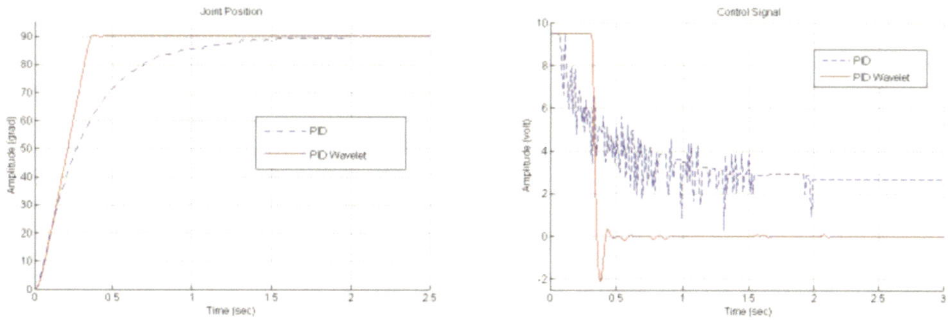

(a) Joint position signal

(b) Control signal law

Fig. 9. Result of the joint position and control signal on the system with PID and MRPID controllers

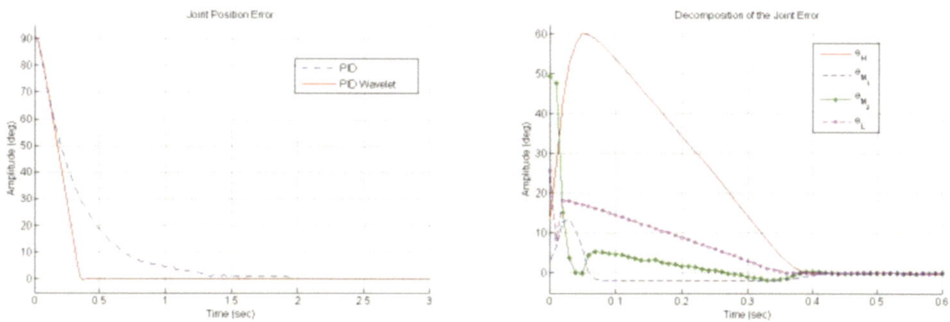

(a) Joint position error signal

(b) Decomposition of the joint error

Fig. 10. Results and decomposition of the error signal on the system with PID and MRPID controllers

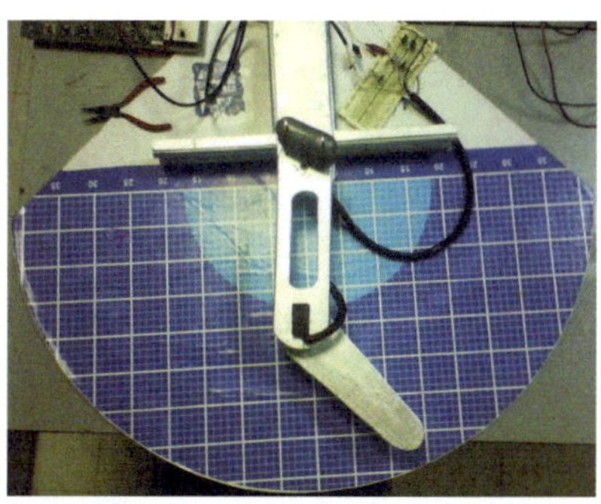

Fig. 11. Two-link planar robot arm

Gains values of the PID	k_P	k_D	k_I
u_1	3.5	2.3	0.5
u_2	9.0	2.3	0.5

Table 7. Gains values of the PID.

Gains values of the MRPID	k_H	k_{M_1}	k_{M_2}	k_L
u_1	4.7	4.8	0.8	0
u_2	12.0	12.0	0.8	0

Table 8. Gains values of the MRPID.

The results of the experimental part are shown in Fig. 12a, for the behavior of each link position with both controllers, where we can see a similarity in behavior of the system both with a classical PID controller and with MRPID controller. The most notable improvement is observed in the control signal generated by the MRPID controller, as shown in Fig. 12b, since it is a very smooth signal with respect to the signal generated by the classic PID controller.

The error signal to both controllers the classical PID as the wavelet MRPID are shown in Fig. 13a, the components of the error signal that are generated with the wavelet decomposition are shown in Fig. 13b.

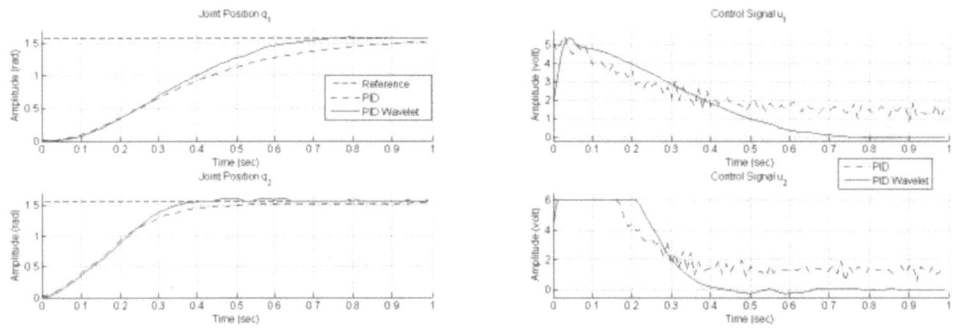

(a) Joint position signal (b) Control signal law

Fig. 12. Result of the joint position and control signal on the system with PID and MRPID controllers

In the experimental platform, we have introduced noise in the process of sensing the output signal, this is because the position sensors are resistive type and also the effects of friction of mechanical parts contributed. The noise is not very evident in the error signal which is shown in Fig. 13a, but this itself is observed in the control signal generated by the classical PID control, as it is amplified by its derivative part ($k_D * \dot{e}$), while the control signal generated by the MRPID controller the noise is filtered by the same control and therefore the signal generated is smooth, as shown in Fig. 12b.

It is worth mentioning that at present a way to tune this type of controller is through an experimental or heuristically, one alternative is employing wavelet neural networks as in (Cruz et al., 2010; O. Islas Gómez, 2010; 2011a;b). So, it is important to show the components, not only to observe how they behave, but also because we are used to tune the control of a MRPID heuristically. The components of the error signal are shown in Fig. 13b. First, we put the value of $k_l = 0$, since this gain is to scale the highest frequency component, and which

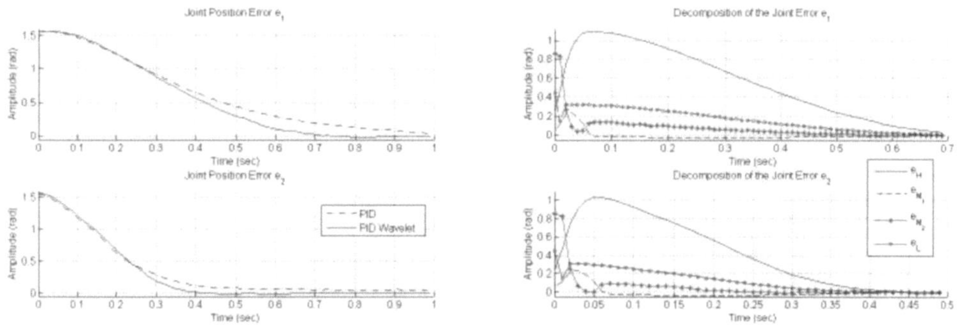

(a) Joint position error signal (b) Decomposition of the joint error

Fig. 13. Experimental results of the regulation on a robot manipulator

contains most of the noise signal frequencies. While the gains of the medium-scale signals k_{M_1} and k_{M_2}, their values are calibrated so as to scale by their respective signals, the signal generated serve to overcome the inertia of the system. The gain k_H, must be calibrated so as to be scaled to the low frequency signal handler allows the robot to reach the reference signal.

3.3 Control for regulatory tracking on a haptic interface

This section we present the description of the experimental platform and the control strategy used for path tracking.

3.3.1 Platform

To evaluate control techniques in haptic interface PHANToM premium 1.0 (see Fig. 14), to improve performance on tasks of exploration, training and telepresence, is exceeded aspects considered open system architecture, such as:

- Application programming interface (GHOST SDK 3.1).
- Kinds of input and output handlers for system control and data acquisition.
- Code kinematic and dynamic model of PHANToM.
- Code in Visual C++ for protection PHANToM.

3.3.2 Hardware

For experimental validation equipment is used with the following specifications:

- Pentium computer 4 to 1.4 *GHz* and 1 *GB* of RAM, with two processors.
- Video Card *GForce*3.
- Equipment *PHANToM* 1.0 (Sensable Technologies)

3.3.3 Software

Software features, which were developed the experiments are

- Windows XP.
- Visual C++ 6.0.
- MatLab 7.1.
- API of GHOST 3.1.

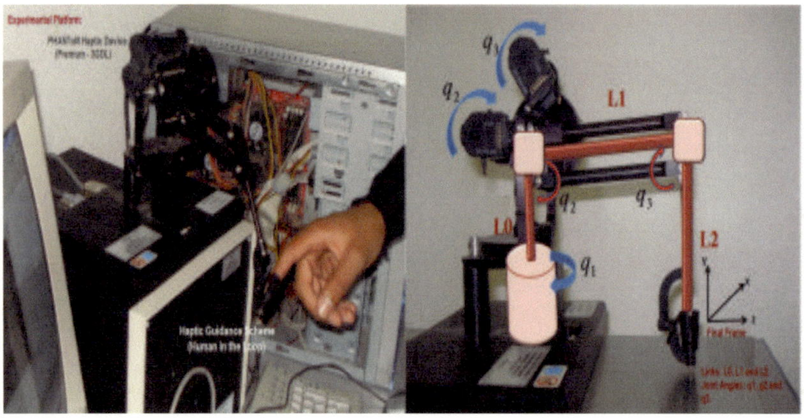

Fig. 14. Haptic interface PHANToM Premium 1.0.

3.3.4 Tracking based regulation

In this section, the use of polynomial which adjust for optimum performance in the task of regulation solves the problem of overcoming the inertia effect due to the state of rest and motion, limit the maximum stresses inherent in the robotic system during the execution of the task and allows the convergence in finite time, this idea is adopted as a regulation based on follow-up.

Tracking-based regulation is of great importance and is performed by a function $\xi(t)$, which is designed in such a way that has a smooth performance from 0 to 1 in a arbitrary finite time $t = t_b > 0$ with t_b as the time convergence arbitrarily chosen by the user and $\xi(t)$ is such that $\dot{\xi}(t_0) = \dot{\xi}(t_b) \equiv 0$. The proposed trajectory $\xi(t)$ is given by:

$$\xi(t) = a_3 \frac{(t-t_0)^3}{(t_b-t_0)^3} - a_4 \frac{(t-t_0)^4}{(t_b-t_0)^4} + a_5 \frac{(t-t_0)^5}{(t_b-t_0)^5}. \tag{28}$$

If we derive (28) yields the velocity

$$\dot{\xi}(t) = 3a_3 \frac{(t-t_0)^2}{(t_b-t_0)^3} - 4a_4 \frac{(t-t_0)^3}{(t_b-t_0)^4} + 5a_5 \frac{(t-t_0)^4}{(t_b-t_0)^5}, \tag{29}$$

and the second derivative of (28) is given by

$$\ddot{\xi}(t) = 6a_3 \frac{(t-t_0)}{(t_b-t_0)^3} - 12a_4 \frac{(t-t_0)^2}{(t_b-t_0)^4} + 20a_5 \frac{(t-t_0)^3}{(t_b-t_0)^5}, \tag{30}$$

taking as conditions $\xi(t_0) = 0$, $\xi(t_b) = 1$, $\dot{\xi}(t_0) = 0$, $\dot{\xi}(t_b) = 0$ and $\ddot{\xi}(\frac{1}{2}t_b) = 0$.

The coefficients are defined by the following equations:

$$\begin{aligned} a_3 - a_4 + a_6 &= 1, \\ 3a_3 - 4a_4 + 5a_6 &= 0, \\ 6a_3 - 12a_4 + 20a_6 &= 0, \end{aligned}$$

where $a_3 = 10$, $a_4 = 15$ y $a_6 = 6$.

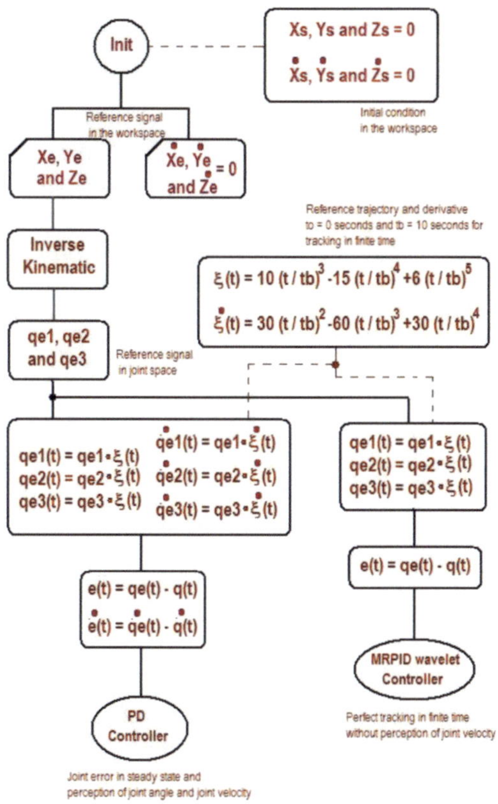

Fig. 15. Block diagram which illustrates the flow of information between the two controllers employed.

As can be seen in Fig. 15, the benefits of the MRPID wavelet control are: not required measurement of the velocity and perfect tracking in finite time, while the PD control has a steady-state error. In (B. A. Itzá Ortiz & Tolentino, 2011) they present sufficient condition for closed-loop stability for stable linear plants.

3.3.5 MRPID controller

The control law for the haptic device is given by (20) for each servo motor, as

$$\mathbf{u} = \begin{bmatrix} u_1 \\ u_2 \\ u_3 \end{bmatrix} \tag{31}$$

3.3.6 Experimental results

In this section, we will present the experiment results using the system shown in Fig. 14, which has three servo motors to move the links, the position signal is measured with encoder sensors and an optical speed.

The values of the constants of PD and the MRPID controllers are given in Table 9 and Table 10, respectively.

Gains values for the PD	k_P	k_D
u_1	0.9	0.05
u_2	0.9	0.05
u_3	0.9	0.05

Table 9. Gains values for the PD

Gains for the MRPID	k_H	k_{M_1}	k_{M_2}	k_L
u_1	7	6	5	3
u_2	7	6	5	3
u_3	7	6	5	3

Table 10. Gains values for the MRPID

The results of the experiment are shown in Fig. 16a, for the behavior of the trajectory in space with both controls, where we can see an improvement in behavior controller with the MRPID difference with PD controller. Also notable is the improvement in the performance of the joint velocity, as shown in Figs. 18a, 18b and 19. In Figure 16b shows the control signals generated by MRPID for the three actuators.

The error signal with the wavelet MRPID controller and the three actuators signals are shown in Fig. 17a, the components of the error signal is generated with the wavelet decomposition 1 actuator are shown in Fig. 17b.

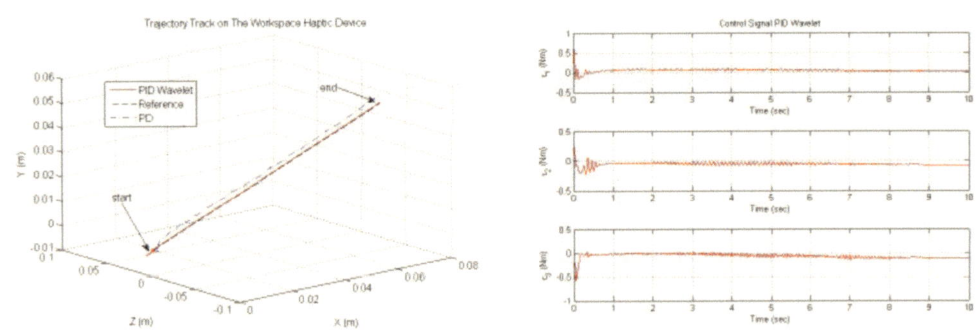

(a) Joint position signal

(b) Control signal law

Fig. 16. Result of the joint position and control signal on the system with PD and MRPID controllers

In the experimental platform, we have inserted noise into the sensing output signal, this is due to the sensors (position and speed) and the effects of friction due to the mechanical parts. The noise is very evident in the error signal which is shown in Fig. 17a, to be used to calculate the control signal generated by the PD control is amplified by the derivative part ($k_D * \dot{e}$), while the control signal generated by the MRPID wavelet controller, noise is filtered wavelet for the same control and therefore the signal generated is very smooth, as shown in Fig. 16b.

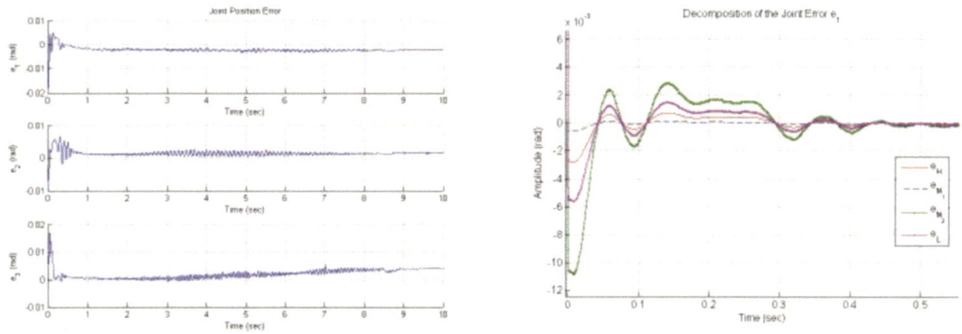

(a) Joint position error signal (b) Decomposition of the joint error

Fig. 17. Results and decomposition of the error signal on the system with MRPID

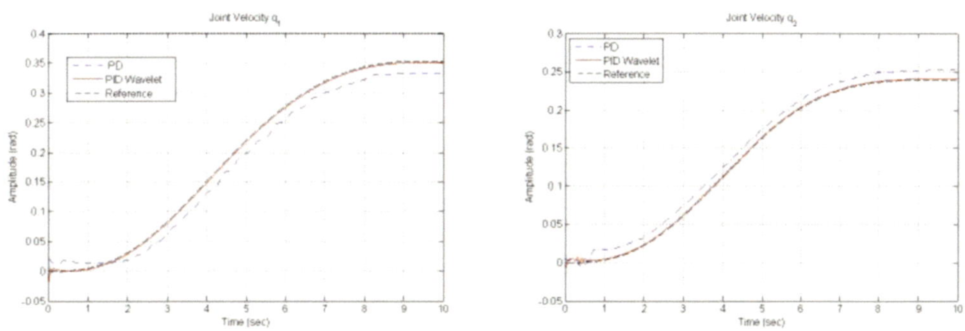

(a) Joint velocity 1 signal (b) Joint velocity 2 signal

Fig. 18. Results of the speed signal to the actuator joint 1 y 2

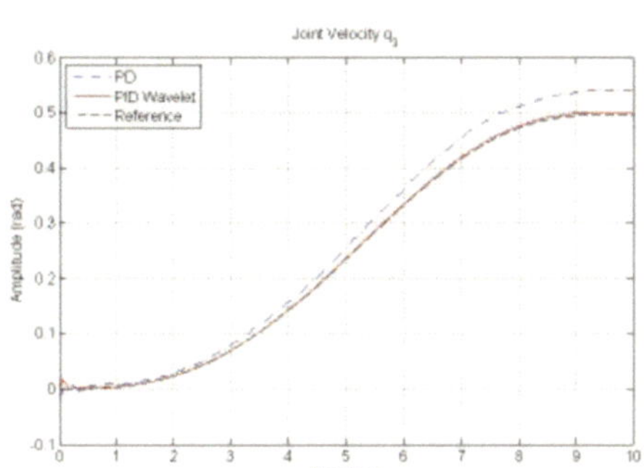

Fig. 19. Results of the speed signal to the actuator joint 3.

4. PID control based on wavelet neural network

4.1 Wavelet neural network theory

Combining the theory of wavelet transform with the basic concept of neural networks, it has a new network called adaptive wavelet neural network or *wavenet* as an alternative to the neural networks of feedforward to approximate arbitrary nonlinear functions.

A function $f(t) \in L^2(\mathbb{R})$ can be approximated by a linear combination using (12), which is similar to a radial basis neural network. In Figure 20 shows the architecture of the adaptive wavelet neural network, which approximates any desired signal $u(t)$ by generalizing a linear combination of a set of daughters wavelets $\psi(\tau)$, where they are generated by a dilation a and b a translation of the mother wavelet $\psi(t)$:

$$\psi(\tau) = \psi\left(\frac{t-b}{a}\right), \quad a, b \in \mathbb{R}, \quad \tau = \frac{t-b}{a}, \tag{32}$$

the dilation factor $a > 0$.

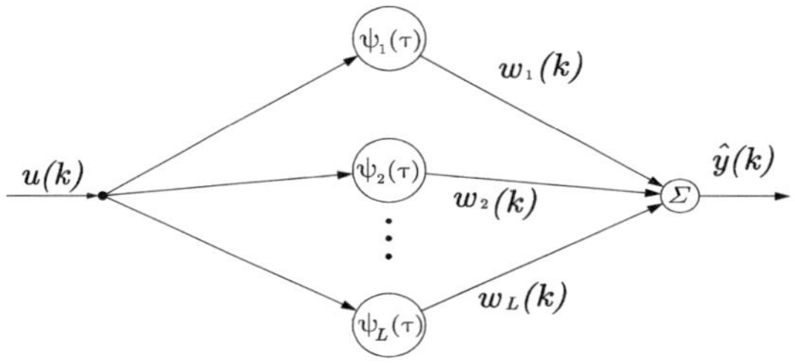

Fig. 20. Structure of wavelet network of three layers

To achieve the approximation, we assume that the output function of the network meets the admissibility condition and sufficiently close to the reference, i.e. the time-frequency region is actually covered by their L windows. The signal from the network about $\hat{y}(t)$ can be represented by:

$$\hat{y}(t) = u(t) \sum_{l=1}^{L} w_l \psi_l(\tau), \quad y, u, w \in \mathbb{R}, \tag{33}$$

in which $\psi_l(\tau) = \psi\left(\frac{t-b_l}{a_l}\right)$ for $l = 1, 2, \cdots, L$, where $L \in \mathbb{Z}$ is the number of neurons in the layer of the wavenet.

And, as a wavenet is a local network in which the output function is well localized in both time and frequency. In addition, a two local network can be achieved by a combination of neural network in cascaded with a infinite impulse response filter (IIR), which provides an efficient computational method for learning the system. In Figure 21 shows the structure of the IIR filter and Fig. 22 shows the final structure of the IIR filter wavenet.

Defining

$$\mathbf{W}(k) \triangleq [w_1(k)\, w_2(k) \;\cdots\; w_l(k) \;\cdots\; w_{L-1}(k)\, w_L(k)]^T, \tag{34}$$

$$\mathbf{A}(k) \triangleq [a_1(k)\, a_2(k) \;\cdots\; a_l(k) \;\cdots\; a_{L-1}(k)\, a_L(k)]^T, \tag{35}$$

$$\mathbf{B}(k) \triangleq [b_1(k)\, b_2(k) \;\cdots\; b_l(k) \;\cdots\; b_{L-1}(k)\, b_L(k)]^T, \tag{36}$$

$$\mathbf{C}(k) \triangleq [c_0(k)\, c_1(k) \;\cdots\; c_m(k) \;\cdots\; c_{M-1}(k)\, c_M(k)]^T, \tag{37}$$

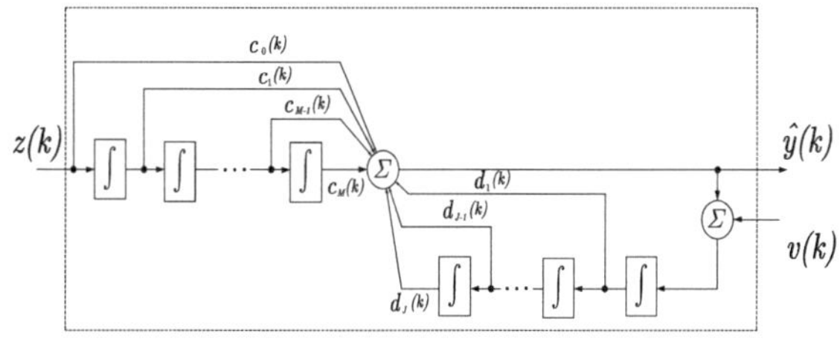

Fig. 21. Scheme block diagram of IIR filter model

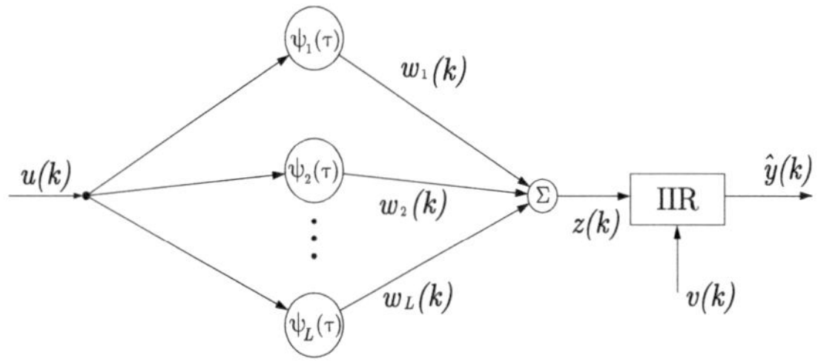

Fig. 22. Struture of wavelet network with IIR filter

$$\mathbf{D}(k) \triangleq [d_1(k)\; d_2(k)\; \cdots\; d_j(k)\; \cdots\; d_{J-1}(k)\; d_J(k)]^T, \tag{38}$$

$$\mathbf{\Psi}(k) \triangleq [\psi_1(\tau)\; \psi_2(\tau)\; \cdots\; \psi_l(\tau)\; \cdots\; \psi_{L-1}(\tau)\; \psi_L(\tau)]^T, \tag{39}$$

$$\mathbf{Z}(k) \triangleq [z(k)\; z(k-1)\; \cdots\; z(k-m)\; \cdots\; z(k-M+1)\; z(k-M)]^T, \tag{40}$$

$$\hat{\mathbf{Y}}(k) \triangleq [\hat{y}(k-1)\; \hat{y}(k-2)\; \cdots\; \hat{y}(k-j)\; \cdots\; \hat{y}(k-J+1)\; \hat{y}(k-J)]^T, \tag{41}$$

in which $\psi_l(\tau) = \psi\left(\frac{k-b_l(k)}{a_l(k)}\right)$ for $l = 1, 2, \ldots, L$, where $L \in \mathbb{Z}$ is the number of neurons in the layer of the neural network. Now the approximate signal $\hat{y}(t)$ with the cascade IIR filter can be expressed in vector form as

$$\hat{y}(k) \triangleq \mathbf{D}^T(k)\hat{\mathbf{Y}}(k)v(k) + \mathbf{C}^T(k)\mathbf{Z}(k)u(k), \tag{42}$$

$$u(k) \triangleq \mathbf{K}(k)\mathbf{E_m}(k), \tag{43}$$

$$z(k) \triangleq \mathbf{\Psi}^T(k)\mathbf{W}(k), \tag{44}$$

The wavenet parameters are optimized with minimum mean square algorithm LMS by minimizing a cost function or energy function E, along all the time k. If we define the error e_n between the plant output y and the output of wavenet $\hat{y}$, as:

$$e_n(k) = y(k) - \hat{y}(k), \tag{45}$$

which is a function of error in time k, the energy function is defined by:

$$E = \frac{1}{2}E_n^T E_n, \tag{46}$$

with

$$E_n = [e_n(1)\; e_n(2)\; \cdots\; e_n(k)\; \cdots\; e_n(\mathcal{T})]^T. \tag{47}$$

So, to minimize E using the gradient method steps down, which requires the gradients $\frac{\partial E}{\partial \mathbf{A}(k)}$, $\frac{\partial E}{\partial \mathbf{B}(k)}$, $\frac{\partial E}{\partial \mathbf{W}(k)}$, $\frac{\partial E}{\partial \mathbf{C}(k)}$ y $\frac{\partial E}{\partial \mathbf{D}(k)}$ to update the incremental changes of each parameter particular. Which are expressed as:

$$\frac{\partial E}{\partial \mathbf{W}(k)} = \left[\frac{\partial E}{\partial w_1(k)} \; \frac{\partial E}{\partial w_2(k)} \cdots \frac{\partial E}{\partial w_l(k)} \cdots \frac{\partial E}{\partial w_{L-1}(k)} \; \frac{\partial E}{\partial w_L(k)} \right]^T, \tag{48}$$

$$\frac{\partial E}{\partial \mathbf{A}(k)} = \left[\frac{\partial E}{\partial a_1(k)} \; \frac{\partial E}{\partial a_2(k)} \cdots \frac{\partial E}{\partial a_l(k)} \cdots \frac{\partial E}{\partial a_{L-1}(k)} \; \frac{\partial E}{\partial a_L(k)} \right]^T, \tag{49}$$

$$\frac{\partial E}{\partial \mathbf{B}(k)} = \left[\frac{\partial E}{\partial b_1(k)} \; \frac{\partial E}{\partial b_2(k)} \cdots \frac{\partial E}{\partial b_l(k)} \cdots \frac{\partial E}{\partial b_{L-1}(k)} \; \frac{\partial E}{\partial b_L(k)} \right]^T, \tag{50}$$

$$\frac{\partial E}{\partial \mathbf{C}(k)} = \left[\frac{\partial E}{\partial c_1(k)} \; \frac{\partial E}{\partial c_2(k)} \cdots \frac{\partial E}{\partial c_m(k)} \cdots \frac{\partial E}{\partial c_{M-1}(k)} \; \frac{\partial E}{\partial c_M(k)} \right]^T, \tag{51}$$

$$\frac{\partial E}{\partial \mathbf{D}(k)} = \left[\frac{\partial E}{\partial c_1(k)} \; \frac{\partial E}{\partial c_2(k)} \cdots \frac{\partial E}{\partial c_j(k)} \cdots \frac{\partial E}{\partial c_{J-1}(k)} \; \frac{\partial E}{\partial c_J(k)} \right]^T. \tag{52}$$

Incremental changes of each coefficient is simply the negative of their gradients,

$$\Delta \mathbf{\Theta}(k) = -\frac{\partial E}{\partial \mathbf{\Theta}(k)}, \tag{53}$$

where Θ can be **W**, **A**, **B**, **C** or **D**.

Thus wavenet coefficients are updated in accordance with the following rule:

$$\mathbf{\Theta}(k+1) = \mathbf{\Theta}(k) + \mu_{\Theta} \Delta \mathbf{\Theta}, \tag{54}$$

where $\mu \in \mathbb{R}$ is the learning rate coefficient for each parameter.

Then calculated the gradients required by (48) - (52):

The equation of the gradient for each w_l is

$$\frac{\partial E}{\partial w_l(k)} = -e_n(k)\mathbf{C}^T(k)\mathbf{\Psi}_l(\tau)u(k). \tag{55}$$

where

$$\mathbf{\Psi}_l(\tau) = [\psi_l(\tau) \; \psi_l(\tau-1) \cdots \psi_l(\tau-m) \cdots \psi_l(\tau-M)]^T, \tag{56}$$

The equation of the gradient for each b_l is

$$\frac{\partial E}{\partial b_l(k)} = -e_n(k)\mathbf{C}^T(k)\mathbf{\Psi}_{b_l}(\tau)w_l(k)u(k). \tag{57}$$

where

$$\mathbf{\Psi}_{b_l}(\tau) = \left[\frac{\partial \psi_l(\tau)}{\partial b_l(k)} \; \frac{\partial \psi_l(\tau-1)}{\partial b_l(k)} \cdots \frac{\partial \psi_l(\tau-m)}{\partial b_l(k)} \cdots \frac{\partial \psi_l(\tau-M)}{\partial b_l(k)} \right]^T, \tag{58}$$

The equation of the gradient for each a_l is

$$\frac{\partial E}{\partial a_l(k)} = -\tau_l e_n(k)\mathbf{C}^T(k)\mathbf{\Psi}_{b_l}(\tau)w_l(k)u(k), \tag{59}$$

$$= \tau_l \frac{\partial E}{\partial b_l(k)}, \tag{60}$$

The equation of the gradient for each d_j is

$$\frac{\partial E}{\partial d_j(k)} = -e_n(k)\hat{y}(k-j)v(k), \tag{61}$$

The equation of the gradient for each c_m is

$$\frac{\partial E}{\partial c_m(k)} = -e_n(k)z(k-m)u(k), \tag{62}$$

Table 11 shows some mother wavelets with their derivatives with respect to b, used to estimate (58).

Name	$\psi(\tau)$	$\frac{\partial\psi(\tau)}{\partial b}$
Morlet	$\cos(\omega_0\tau)exp(-0.5\tau^2)$	$\frac{1}{a}[\omega_0\sin(\omega_0\tau)exp(-0.5\tau^2)+\tau\psi(\tau)]$
RASP1	$\frac{\tau}{(\tau^2+1)^2}$	$\frac{3\tau^2-1}{a(\tau^2+1)^3}$

Table 11. Some mother wavelets and their derivative with respect b.

4.2 Wavenet PID controller design

Given a general SISO dynamical system represented in the following discrete time state equations (Levin & Narendra, 1993; 1996):

$$x(k+1) = f[x(k), u(k), k], \tag{63}$$

$$y(k) = g[x(k), k], \tag{64}$$

where $x \in \mathbb{R}^n$ and u, $y \in \mathbb{R}$, besides the functions f, $g \in \mathbb{R}$ are unknown and the only accessible data are the input u and the output, y. In (Levin & Narendra, 1993) showed that if the linearized system of (63) and (64) around the equilibrium state is observable, then, there is an input-output representation which has the following form:

$$y(k+1) = \Omega[\mathbf{Y}(k), \mathbf{U}(k)], \tag{65}$$

$$\mathbf{Y}(k) = [y(k)\ y(k-1)\ \cdots\ y(k-n+1)], \tag{66}$$

$$\mathbf{U}(k) = [u(k)\ u(k-1)\ \cdots\ u(k-n+1)], \tag{67}$$

i.e., there exists a function Ω that maps to the pair $(y(k), u(k))$ and $n-1$ past values within $y(k+1)$.

Then a wavelet neural network model $\hat{\Omega}$, can be trained to approximate Ω on the domain of interest. An alternative model of an unknown plant that can simplify the algorithm of the control input is described by the following equation:

$$y(k+1) = \Phi[\mathbf{Y}(k), \mathbf{U}(k)] + \Gamma[\mathbf{Y}(k), \mathbf{U}(k)] \cdot u(k), \tag{68}$$

where $y(k)$ and $u(k)$ denote the input and output at time k. If the terms $\Phi(\cdot)$ and $\Gamma(\cdot)$ are unknown, then it is here that uses a adaptive wavelets neural network model to approximate the dynamic system as follows:

$$\hat{y}(k+1) = \hat{\Phi}[\mathbf{Y}(k), \mathbf{U}(k), \Theta_{\Phi}] + \hat{\Gamma}[\mathbf{Y}(k), \mathbf{U}(k), \Theta_{\Gamma}] \cdot u(k), \tag{69}$$

comparing the model (69) with (42) we obtained that

$$\hat{y}(t) = \mathbf{C}^T \mathbf{Z}(t) u(t) + \mathbf{D}^T \hat{\mathbf{Y}}(t) v(t), \tag{70}$$

we conclude that

$$\hat{\Phi}[\mathbf{Y}(k), \mathbf{U}(k), \Theta_{\Phi}] = \mathbf{D}^T \hat{\mathbf{Y}}(k) v(k), \tag{71}$$

$$\hat{\Gamma}[\mathbf{Y}(k), \mathbf{U}(k), \Theta_{\Gamma}] = \mathbf{C}^T \mathbf{Z}(k), \tag{72}$$

$$z(k) = \mathbf{\Psi}^T \mathbf{W}. \tag{73}$$

After the nonlinearity $\Phi(\cdot)$ and $\Gamma(\cdot)$ are approximated with two distinct neural network functions $\hat{\Phi}(\cdot)$ and $\hat{\Gamma}(\cdot)$ with adjustables parameters (such as: weights w, the expansions a, the translations, b, the IIR filter coefficients c and d), represented by Θ_{Φ} and Θ_{Γ} respectively, control signal $u(k)$ to obtain an output signal $y_{ref}(k+1)$ can be obtained from:

$$u(k) = \frac{y_{ref}(k+1) - \hat{\Phi}[\mathbf{Y}(k), \mathbf{U}(k), \Theta_{\Phi}]}{\hat{\Gamma}[\mathbf{Y}(k), \mathbf{U}(k), \Theta_{\Gamma}]}. \tag{74}$$

The wavelet control described by (20), has the difficulty to be tuned, and is a linear control and which can not deal with complex dynamic processes such as those with larger dead time, inverse response and characteristics of high nonlinearities. To improve the performance of control, uses an algorithm to auto tune the values of each gain in $\mathbf{K}$ of the wavelet control, in the same way that autotuning PID parameters as in (Sedighizadeh & Rezazadeh, 2008b) with the difference that here we can have more than three parameters in the control level wavelet decomposition. The wavenet control scheme is shown in Fig. 23

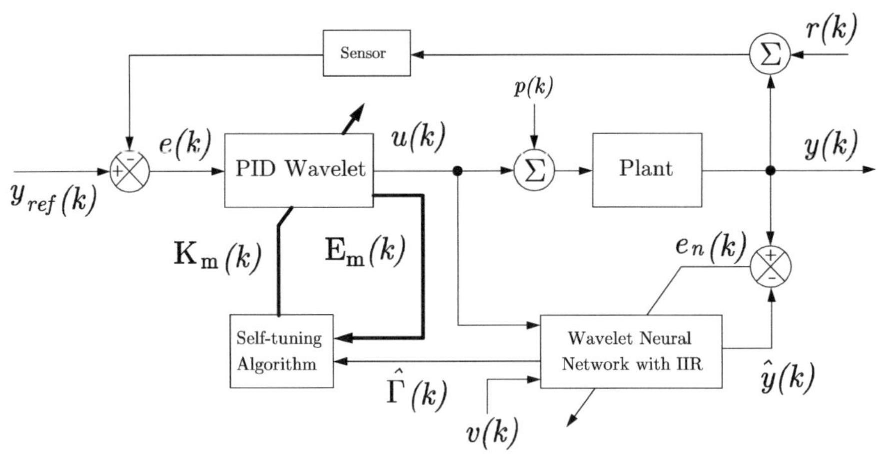

Fig. 23. Close loop block diagram of a SISO system with wavenet PID

For the wavelet control tuning gains K_i, we use the rule (54), so that the gradient $\frac{\partial E}{\partial K_i}$ is calculated as

$$\frac{\partial E}{\partial K_i(k)} = -e_n(k)\hat{\Gamma}[Y(k), U(k), \Theta_\Gamma]e_i(k), \tag{75}$$

5. Wavenet PID application

5.1 Control position on a CD motor

To validate the algorithms, we are applied to the stable linear system described by equation (25) with the parameters of the neural network shown in Table 12 and initial values are given in Table 13.

Neurons	5
Wavelet Mother	Wavelet Morlet
Feedfoward IIR coefficients c	2
Feedback IIR coefficients d	2

Table 12. Parameter of the adaptive wavelet neural network.

W	[-0.5 -0.5 -0.5 -0.5 -0.5]	$\triangle$ **W**	0.2
A	[10 10 10 10 10]	$\triangle$ **A**	0.2
B	[0 30 60 90 120]	$\triangle$ **B**	0.2
C	[0.1 0.1]	$\triangle$ **C**	0.2
D	[0.1 0.1]	$\triangle$ **D**	0.2
K	[0.1 0.5 0.5 0.1]	$\triangle$ **K**	0.05

Table 13. Initial values of the parameters of the adaptive wavelet neural network and gains of the wavenet PID.

The experimental results with a wavelet neural network without prior training to be used with the scheme of Fig. 23 to tune the gains of MRPID, but if you have a learning period $0 \leq t \leq 120$ seconds to identify the system.

From the Fig. 24a, it is observed that after the learning period we get an acceptable response with the wavenet PID and the overshoot is no more than 0.2. In Figure 24b shows as the control signal is much smaller than that generated in MRPID controller and still has a rapid response as seen in Fig. 24a, and this is due to tuning of the gains made by the wavelet neural network algorithm.

In the decomposition of the error signal shown in Fig. 25, is very similar to the decomposition of the error shown in Fig. 10a generated by the MRPID without autotuning.

The Figures 26a, 26b and 27 show the behavior of the wavenet and the IIR filter parameters, and also the behavior of MRPID gains that are tuned during the training period. We observed that with little training period and without prior training of the wavelet neural network, have achieved a good response in the system output as shown in Fig. 24a.

The final values are shown in Table 14:

There exist more applications in nonlinear systems as the AC motor which are given in (O. Islas Gómez, 2011a;b).

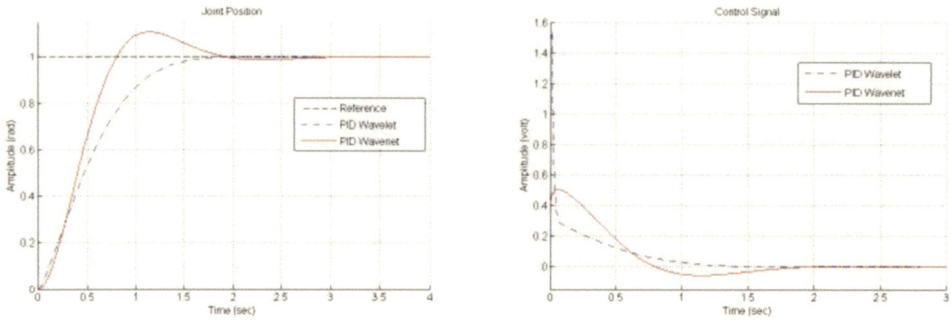

(a) Joint Position (b) Control Signal

Fig. 24. Result of the joint position and control signal on the system with MRPID and wavenet PID controller

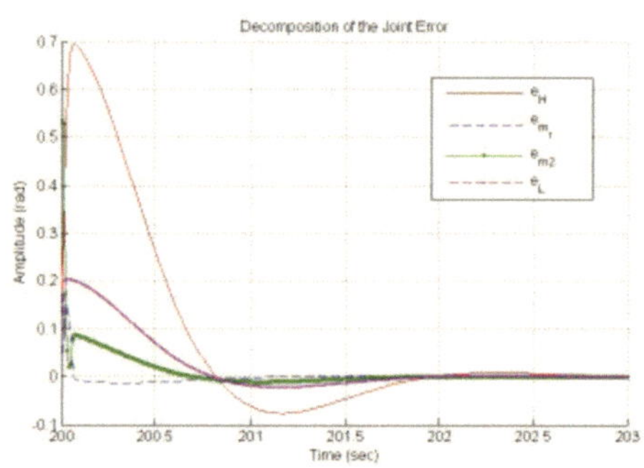

Fig. 25. Decomposition of the error signal

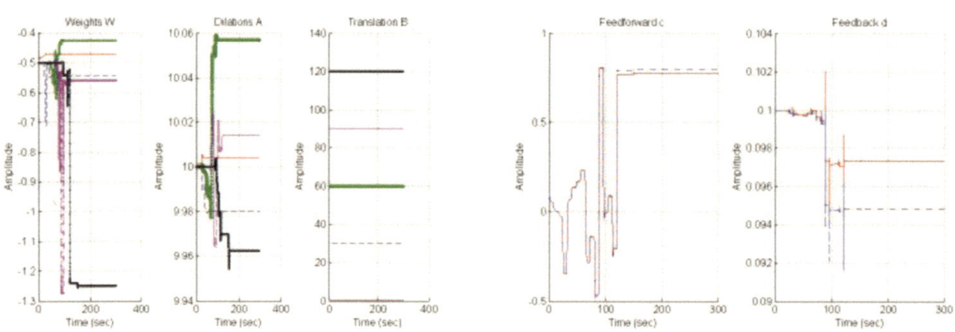

(a) Wavenet parameters (b) Parameters of the IIR filter

Fig. 26. Results of parameters of the wavelet neural network

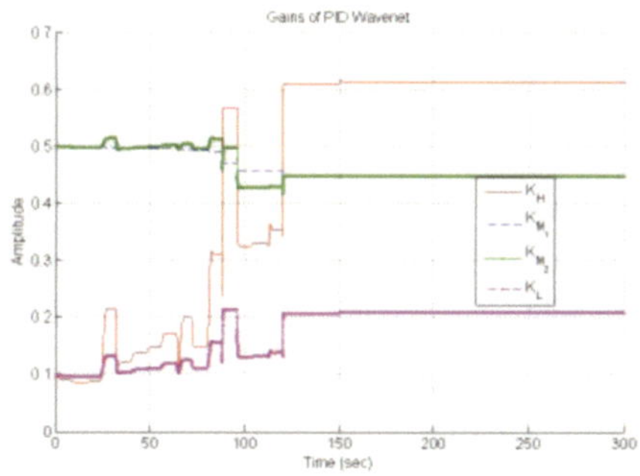

Fig. 27. Results of the wavenet PID gains

W	[-0.426 -0.47 -0.54 -0.56 -1.25]
A	[10.057 10.014 10.004 9.98 9.962]
B	[0.003 30.019 60.065 90.15 120.018]
C	[0.775 0.8]
D	[0.0973 0.0948]
K	[0.615 0.45 0.448 0.2]

Table 14. Final values parameters of the wavelet neural network and the wavenet PID gains.

6. Conclusion

The properties have the wavelets, makes a mathematical tool very useful, not only for image filtering, image compression, seismic signal analysis, denoising of audio signals, nonlinear function approximation with neural networks (Hans, 2005; Hojjat & Karim, 2005; Li et al., 2005; Mallat, 2008; Mertins, 1999), etc., but also in new areas such as automatic control, where from the results obtained shows a great first step in the experimental implementation on on Euler-Lagrange systems. The experimental implementation of the MRPID wavelet control on haptics interfaces results in an opening in the study and analysis of experimental platforms in the motor rehabilitation area, due at stability and robustness than shows the MRPID wavelet the during the tracking planned trajectories. The most notable advantage that we can see is that the MRPID wavelet controller not requires the joint velocity vector for following trajectory, only to measure the position, makes a very good following. Another advantage is that the characteristics of the control signal generates a soft even in the presence of noise, so it does not require any additional filters for signal control, avoiding a filtering stage after sensing or after generating the control signal and before the power amp. So with this type of control, you have two in one, since you can filter the signal and generates a control signal. Also, we summarize the theory of the adaptive wavelet neural network to get a wavenet PID controller which it was testing in the DC motor control.

7. References

Aström, K. & Hägglund, T. (2007). *PID Controllers: Theory, Design and Tuning*, Instrumentation Systems and Automation Society.

B. A. Itzá Ortiz, L. E. Ramos Velasco, H. R. R. & Tolentino, J. A. C. (2011). Stability analysis for a pid wavelet controller, *IEEE Transaction on Automatic Control* p. submitted.

Cruz, J. A., Ramos, L. E. & Espejel, M. A. (2010). A selft-tuning of a wavelet pid controller, *20th International Conference on Electronics, Communications and Computer CONIELECOMP'2010* pp. 73–78.

Daubechies, I. (1992). *Ten Lectures on Wavelets*, Society for Industrial and Applied Mathematics, Philadelphia, Pennsylvania.

Hans, G. (2005). *Wavelets and Signal Processing: An Application-Based Introduction*, Springer-Verlag, Netherlands.

Hojjat, A. & Karim, A. (2005). *Wavelet in Intellent Transportation System*, Wiley.

Levin, A. U. & Narendra, K. S. (1993). Control of nonlinear dynamical systems using neural networks: Controllability and stabilization, *IEEE Transactions on Neural Networks* 4(2): 192–207.

Levin, A. U. & Narendra, K. S. (1996). Control of nonlinear dynamical systems using neural networks–part ii: Observability, identification and control, *IEEE Transactions on Neural Networks* 7(1): 30–42.

Li, H., Jin, H. & Guo, C. (2005). Title pid control based on wavelet neural network identification and tuning and its application to fin stabilizer, *Proceedings of the IEEE International Conference Mechatronics and Automation*, Vol. 4, IEEE International Conference on Mechatronics and Automation, Canada location, pp. 1907–1911.

Mallat, S. (1989a). Multiresolution aproximations and wavelet orthonormal bases of $L^2(R)$, *Transactions American Mathetical Society* 315(1): 69–87.

Mallat, S. (1989b). A theory multiresolution signal decomposition: The wavelet representation, *IEEE Transactions Pattern Analysis and Machine Intelligence* 11(7): 674–693.

Mallat, S. (2008). *A Wavelets Tour of Signal Processing*, Academic Press.

Mertins, A. (1999). *Signal Analysis: Wavelets, Filter Banks, Time-Frequency Transforms and Application*, Wiley.

O. Islas Gómez, L. E. R. V. y. J. G. L. (2010). Identificación y control wavenet de un motor de cd, *Congreso Anual de la Asociación de México de Control Automático, AMCA 2010*, 1.

O. Islas Gómez, L. E. R. V. y. J. G. L. (2011a). Implementation of different wavelets in an auto-tuning wavenet pid controller and its application to a dc motor, *Electronics, Robotics and Automotive Mechanics Conference, CERMA 2011*, 1.

O. Islas Gómez, L. E. Ramos Velasco, J. G. L. J. R. F. y. M. A. E. R. (2011b). Identificación y control wavenet de un motor de ac, *Congreso Anual de la Asociación de México de Control Automático, AMCA 2011*, 1.

Parvez, S. (2003). *Advanced Control Techniques for Motion Control Problem*, PhD thesis, Cleveland State University.

Parvez, S. & Gao, Z. (2005). A wavelet-based multiresolution pid controller, *IEEE Transactions on Industry Applications* 41(2): 537–543.

Sedighizadeh, M. & Rezazadeh, A. (2008a). Adaptive PID control of wind energy conversion systems using RASP1 mother wavelet basis function network, *Proceeding of World Academy of Science, Engineering and Technology* pp. 269–273.

Sedighizadeh, M. & Rezazadeh, A. (2008b). Title adaptive pid control of wind energy conversion systems using raspl mother wavelet basis function networks, *Proceedings of World Academy of Science, Engineering and Technology*, Vol. 27, World Academy of Science, Engineering and Technology, pp. 269–273.

Tang, J. (2001). Pid controller using the tms320c31 dsk with on-line parameter adjustment for real-time dc motor speed and position control, *IEEE International Symposium on Industrial Electronics ISIE'2001* 2(1): 786–791.

Vetterli, M. & Kovačević, J. (1995). *Wavelets and Subband Coding*, Prentice Hall PTR.

Visioli, A. (2006). *Practical PID Control*, Springer.

2

Tuning Three-Term Controllers for Integrating Processes with both Inverse Response and Dead Time

K.G. Arvanitis[1], N.K. Bekiaris-Liberis[2], G.D. Pasgianos[1] and A. Pantelous[3]
[1]Department of Agricultural Engineering, Agricultural Univeristy of Athens,
[2]Department of Mechanical & Aerospace Engineering, University of California-San Diego,
[3]Department of Mathematical Sciences, University of Liverpool,
[1]Greece
[2]USA
[3]UK

1. Introduction

Most industrial processes respond to the actions of a feedback controller by moving the process variable in the same direction as the control effort. However, there are some interesting exceptions where the process variable first drops, then rises after an increase in the control effort. This peculiar behaviour that is well known as "inverse response", is due to the non-minimum phase zeros appearing in the process transfer function and representing part of the process dynamics. Second order dead-time inverse response process models (SODT-IR) are used to represent the dynamics of several chemical processes (such as level control loops in distillation columns and temperature control loops in chemical reactors), as well as the dynamics of PWM based DC-DC boost converters in industrial electronics. In the extant literature, there is a number of studies regarding the design and tuning of three-term controllers for SOPD-IR processes (Chen et al, 2005, 2006; Chien et al, 2003; Luyben, 2000; Padma Sree & Chidambaram, 2004; Scali & Rachid, 1998; Waller & Nygardas, 1975; Zhang et al, 2000). On the other hand, integrating models with both inverse response and dead-time (IPDT-IR models) was found to be suitable for a variety of engineering processes, encountered in the process industry. The common examples of such processes are chemical reactors, distillation columns and, especially, level control of the boiler steam drum. In recent years, identification and tuning of controllers for such process models has not received the appropriate attention as compared to other types of inverse response processes, although some very interesting results have been reported in the literature (Gu et al, 2006; Luyben, 2003; Shamsuzzoha & Lee, 2006; Srivastava & Verma, 2007). In particular, the method reported by (Luyben, 2003) determines the integral time of a series form PID controller as a fraction of the minimum PI integral time, while the controller proportional gain is obtained by satisfying the specification of +2 dB maximum closed-loop log modulus, and the derivative time is given as the one maximizing the controller gain. In the work proposed by (Gu et al, 2006), PID controller tuning for IPDT-IR processes is performed based on $H\infty$ optimization and Internal Model Control (IMC) theory. In the work proposed by (Shamsuzzoha & Lee, 2006), set-point weighted PID controllers are designed on the basis of the IMC theory. Finally, in the work proposed by (Srivastava & Verma, 2007), a method involving numerical integration has been proposed in order to identify process parameters of IPDT-IR process models.

From the preceding literature review, it becomes clear that results on controller tuning for IPDT-IR processes are limited, and that there is a need for new efficient tuning methods for such processes. The aim of this work is to present innovative methods of tuning three-term controllers for integrating processes incorporating both time-delay and a non-minimum phase zero. The three-term controller configuration applied in this work is the well known I-PD (or Pseudo-Derivative Feedback, PDF) controller configuration (Phelan, 1978) due its advantages over the conventional PID controller configuration (Paraskevopoulos et al, 2004; Arvanitis et al, 2005; Arvanitis et al, 2009a). A series of innovative controller tuning methods is presented in the present work. These methods can be classified in two main categories: (a) methods based on the analysis of the phase margin of the closed-loop system, and (b) methods based on a direct synthesis approach. According to the first class of proposed tuning methods, the controller parameters are selected in order to meet the desired specifications in the time domain, in terms of the damping ratio or by minimizing various integral criteria (Wilton, 1999). In addition, the proportional gain of the controller is chosen in such a way, that the resulting closed-loop system achieves the maximum phase margin for the given specification in the time domain, thus resulting in robust closed-loop performance. Controller parameters are involved in nonlinear equations that are hard to solve analytically. For that reason, iterative algorithms are proposed in order to obtain the optimal controller settings. However, in order to apply the proposed methods in the case of on-line tuning, simple approximations of the exact controller settings obtained by the aforementioned iterative algorithms are proposed, as functions of the process parameters. The second class of proposed tuning methods is based on the manipulation of the closed-loop transfer function through appropriate approximations and cancellations, in order to obtain a second order dead-time closed-loop system. On the basis of this method, the parameters of the I-PD controller is obtained in terms of an adjustable parameter that can be further appropriately selected in order either to achieve a desired damping ratio for the closed-loop system or to ensure the minimization of conventional integral criteria. Finally, In order to assess the effectiveness of the proposed control scheme and associated tuning methods and to provide a comparison with existing tuning methods for PID controllers, a simulation study on the problem of controlling a boiler steam drum modelled by an IPDT-IR process model is presented. Simulation results reveal that the proposed controller and tuning methods provide considerably smoother response than known design methods for standard PID controllers, in case of set point tracking, as well as lower maximum error in case of regulatory control. This is particular true for the proposed direct synthesis method, which outperforms existing tuning methods for PID controllers.

2. IPDT-IR process models and the I-PD controller configuration

This work elaborates on IPDT-IR process models of the form

$$G_P(s) = \frac{\overline{K}\left(-s\overline{\tau}_Z + 1\right)\exp(-\overline{d}s)}{s\left(s\overline{\tau}_p + 1\right)} \tag{1}$$

where $\bar{K}$, $\bar{d}$, $\bar{\tau}_P$ and $\bar{\tau}_Z$ are the process gain, the time delay, the time constant and the zero's time constant, respectively, controlled using the configuration of Fig. 1, i.e. the so-called I-PD or PDF controller. In this controller configuration, the three controller actions are separated. Integral action, which is dedicated to steady state error elimination, is located in the forward path of the loop, whereas proportional and derivative actions, which are mainly dedicated in assigning the desired closed-loop performance in terms of stability, responsiveness, disturbance attenuation, etc, are located in the feedback path. This separation leads to a better understanding of the role of each particular controller action. Moreover, the I-PD controller has some distinct advantages over the conventional PID controller, as reported in the works by (Paraskevopoulos et al, 2004; Arvanitis et al, 2005; Arvanitis et al, 2009a).

Observe now that applying the I-PD control strategy to the process model of the form (1), the following closed-loop transfer function is obtained

$$G_{CL}(s)=\frac{\bar{K}\bar{K}_I\left(-s\bar{\tau}_Z+1\right)\exp(-\bar{d}s)}{s^2\left(\bar{\tau}_P s+1\right)+\bar{K}\left(\bar{K}_D s^2+\bar{K}_P s+\bar{K}_I\right)\left(-s\bar{\tau}_Z+1\right)\exp(-\bar{d}s)} \tag{2}$$

It is not difficult to see that the action of the I-PD controller is equivalent to that of a PID controller in series form having the transfer function

$$G_{C,PID}(s)=\bar{K}_C\left(1+\frac{1}{\bar{\tau}_I s}\right)\left(1+\bar{\tau}_D s\right) \tag{3}$$

with a second order set-point pre-filter of the form $G_{C,SPF}(s)=1/\left(\bar{\tau}_I s+1\right)\left(1+\bar{\tau}_D s\right)$, provided that the following relations hold

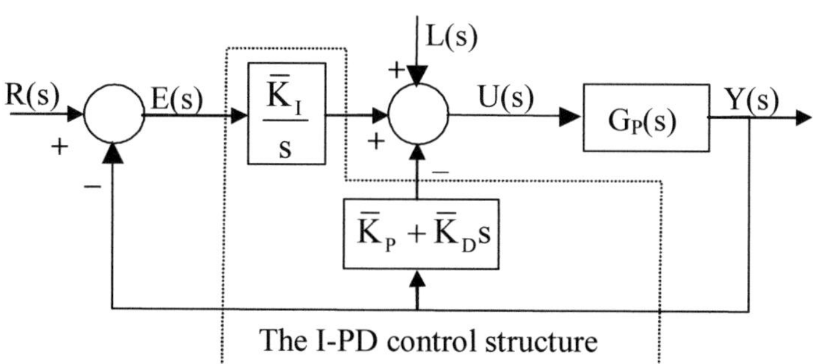

Fig. 1. The I-PD or PDF control strategy.

$$\bar{K}_P=\bar{K}_C\left(\bar{\tau}_D+\bar{\tau}_I\right)/\bar{\tau}_I\ ,\ \bar{K}_I=\bar{K}_P/\left(\bar{\tau}_D+\bar{\tau}_I\right)=\bar{K}_C/\bar{\tau}_I\ ,\ \bar{K}_D=\bar{K}_P\bar{\tau}_D\bar{\tau}_I/\left(\bar{\tau}_D+\bar{\tau}_I\right)=\bar{K}_C\bar{\tau}_D \tag{4}$$

Taking into account the above equivalence, the loop transfer function of the proposed feedback structure is given by

$$G_L(s)=\frac{\bar{K}\bar{K}_C\left(-s\bar{\tau}_Z+1\right)\left(\bar{\tau}_I s+1\right)\left(\bar{\tau}_D s+1\right)\exp(-\bar{d}s)}{\bar{\tau}_I s^2\left(\bar{\tau}_P s+1\right)} \tag{5}$$

Relations (2) and (5) are used in the next Sections for the derivation of the tuning methods proposed in this work.

3. Frequency domain analysis for closed-loop IPDT-IR processes

The equivalence between the PD-1F controller and the set-point pre-filtered PID controller provides us the possibility, to work with $\overline{K}_C$, $\overline{\tau}_I$ and $\overline{\tau}_D$ and not directly with $\overline{K}_P$, $\overline{K}_I$ and $\overline{K}_D$. Furthermore, in order to facilitate comparisons, let all system and controller parameters be normalized with respect to $\overline{\tau}_P$ and $\overline{K}$. Thus, the original process and controller parameters are replaced with the dimensionless parameters shown in Table 1. Then, relations (2) and (5) yield

$$G_{CL}(\hat{s}) = \frac{K_C(-\hat{s}\tau_Z+1)\exp(-d\hat{s})}{\tau_I\hat{s}^2(\hat{s}+1)+K_C(\tau_I\hat{s}+1)(\tau_D\hat{s}+1)(-\hat{s}\tau_Z+1)\exp(-d\hat{s})} \tag{6}$$

$$G_L(\hat{s}) = \frac{K_C(-\hat{s}\tau_Z+1)(\tau_I\hat{s}+1)(\tau_D\hat{s}+1)\exp(-d\hat{s})}{\tau_I\hat{s}^2(\hat{s}+1)} \tag{7}$$

From relation (7), the argument and the magnitude of the loop transfer function are given by

$$\varphi_L(w) = -\pi - dw - atan(w) - atan(\tau_Z w) + atan(\tau_I w) + atan(\tau_D w) \tag{8}$$

Original Parameters	Normalized Parameters	Original Parameters	Normalized Parameters
$\overline{\tau}_p$	τ_p=1	S	$\hat{s} = s\overline{\tau}_P$
$\overline{\tau}_Z$	$\tau_Z = \overline{\tau}_Z / \overline{\tau}_P$	$\overline{K}$	K=1
$\overline{d}$	$d = \overline{d} / \overline{\tau}_P$	$\overline{K}_C$	$K_C = \overline{K}\overline{K}_C$
$\overline{\tau}_I$	$\tau_I = \overline{\tau}_I / \overline{\tau}_P$	$\overline{K}_D$	$K_D = \overline{K}\overline{K}_D$
$\overline{\tau}_D$	$\tau_D = \overline{\tau}_D / \overline{\tau}_P$	$\overline{K}_P$	$K_P = \overline{\tau}_p\overline{K}\overline{K}_P$
ω	$w = \omega\overline{\tau}_P$	$\overline{K}_I$	$K_I = \overline{\tau}_p^2\overline{K}\overline{K}_I$

Table 1. Normalized vs. original system parameters.

$$A_L(w) = |G_L(jw)| = K_C\frac{\sqrt{1+(\tau_I w)^2}\sqrt{1+(\tau_D w)^2}\sqrt{1+(\tau_Z w)^2}}{\tau_I w^2\sqrt{1+w^2}} \tag{9}$$

In Fig. 2, the Nyquist plots of $G_L(\hat{s})$ for typical IPDT-IR processes are depicted for several values of the parameter τ_I. From this figure, it becomes clear that, for specific d and τ_Z, and for τ_I greater than a critical value, say $\tau_{I,min}$, there exists a crossover point of the Nyquist plot with the negative real axis. In this case, the system can be stabilized, with an appropriate choice of K_C. Moreover, from these Nyquist plots, one can observe that the stability region is reduced when τ_I is decreased, starting from the maximum region of stability when $\tau_I \to\infty$, (which corresponds to a PD-controller). The Nyquist plot does not have any crossover point with the negative real axis, in the case where $\tau_I=\tau_{I,min}$; that is, for $\tau_I \leq \tau_{I,min}$, the process cannot be stabilized. Moreover, in Fig. 3, the Nyquist plots of $G_L(\hat{s})$ are depicted for several values of the parameter τ_D. From these plots, it becomes clear that, for small values of τ_D, the stability region is increased with τ_D, whereas for larger values of τ_D the stability region decreases when τ_D increases.

Let $PM=\varphi(w_G)+\pi$ be the phase margin of the closed-loop system, where w_G is the frequency, at which the magnitude of the loop transfer function $G_L(\hat{s})$ equals unity. Taking into account (8), we obtain $PM = -dw_G - atan(w_G) - atan(\tau_Z w_G) + atan(\tau_I w_G) + atan(\tau_D w_G)$. From Fig. 2, it can be readily observed that for given d, τ_I, τ_D, τ_Z, there exists one point of the Nyquist plot corresponding to the maximum argument $\varphi_{L,max}(d, \tau_I, \tau_D, \tau_Z)$. The frequency w_P at which the argument (8) is maximized is given by the maximum real root of the equation $\left[d\varphi_L(w)/dw\right]\Big|_{w=w_P}=0$, which, after some easy algebraic manipulations, yields

$$-d-\frac{1}{1+w_P^2}-\frac{\tau_Z}{1+\tau_Z^2 w_P^2}+\frac{\tau_I}{1+\tau_I^2 w_P^2}+\frac{\tau_D}{1+\tau_D^2 w_P^2}=0 \tag{10}$$

Hence, substituting w_P, as obtained by (10), in (8), the respective argument $\varphi_L(w_P)$ is computed. Consequently, the maximum phase margin PM_{max} for given d, τ_I, τ_D, τ_Z, can be obtained if we put $w_P=w_G$, i.e. choosing the controller proportional gain K_C according to

$$K_C=\frac{\tau_I w_P^2\sqrt{1+w_P^2}}{\sqrt{1+(\tau_D w_P)^2}\sqrt{1+(\tau_I w_P)^2}\sqrt{1+(\tau_Z w_P)^2}} \tag{11}$$

With this choice for K_C, the phase margin is given by

$$\begin{aligned} PM(d,\tau_I,\tau_D,\tau_Z) &= -dw_P - a\tan(w_P) - a\tan(\tau_Z w_P) \\ &+ a\tan(\tau_I w_P) + a\tan(\tau_D w_P) = PM_{\max}(d,\tau_I,\tau_D,\tau_Z) \end{aligned} \tag{12}$$

Obviously, in the case where $\tau_I=\tau_{I,min}$ and K_C is obtained by (11), then the closed-loop system is marginally stable, that is $PM_{max}=0$. Note that, from (12), $PM_{max}=0$, when $w_P=0$. Therefore, from (10), for $w_P=0$, we obtain $\tau_{I,min}=d+\tau_Z+1-\tau_D$. Since, for all values of τ_I, larger than $\tau_{I,min}=d+\tau_Z+1-\tau_D$, it holds $\left(d\left[PM(d,\tau_I,\tau_D,\tau_Z)\right]/dw\right)\Big|_{w=0}=\tau_I+\tau_D-d-\tau_Z-1>0$, one can readily conclude that $PM_{max}>0$. Moreover, PM_{max} is an increasing function of both τ_I and τ_D. This is illustrated in Fig. 4, where PM_{max} is given as a function of τ_I and τ_D, for a typical IPDT-IR process.

As it was previously mentioned, the stability region of the closed-loop system increases with τ_D. This is due to the fact that the closed-loop system gain margin increases with τ_D, as one can verify from the variation of the crossover point of the Nyquist plot with the negative real axis as τ_D varies. Furthermore, there is one value of τ_D, say $\tau_{D,GM_{\max}}(\tau_I,\tau_Z,d)$, for which the closed loop gain margin starts to decrease. In addition, from Fig. 2, it can be easily verified that the closed-loop gain margin increases arbitrarily as τ_I increases. These observations can also become evident from Fig. 5 that illustrates the maximum closed-loop gain margin GM_{max} as a function of the controller parameters τ_D and τ_I.

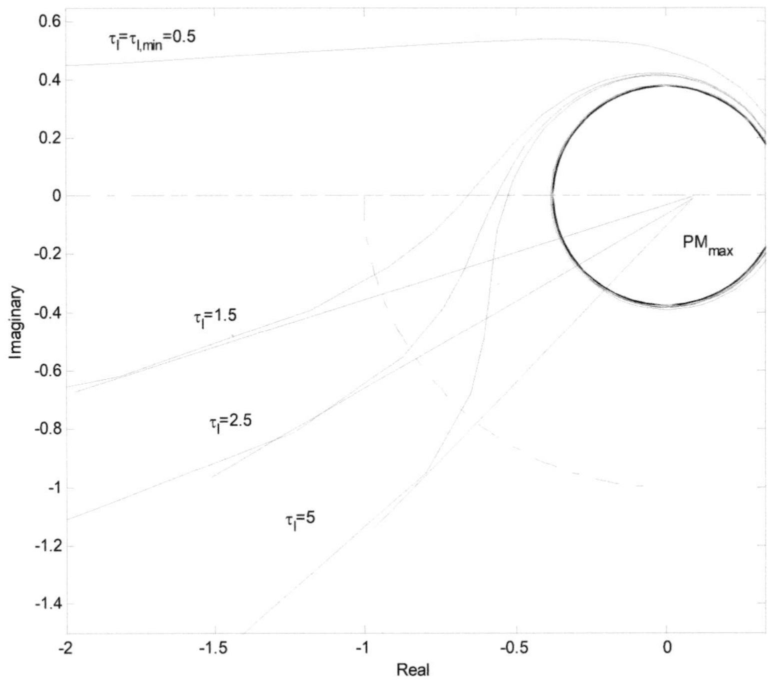

Fig. 2. Nyquist plots of a typical IPDT-IR process controlled by a PD1-F controller with K_C=0.5, d=0.5, τ_Z=0.5, τ_D=1.5, for various values of τ_I.

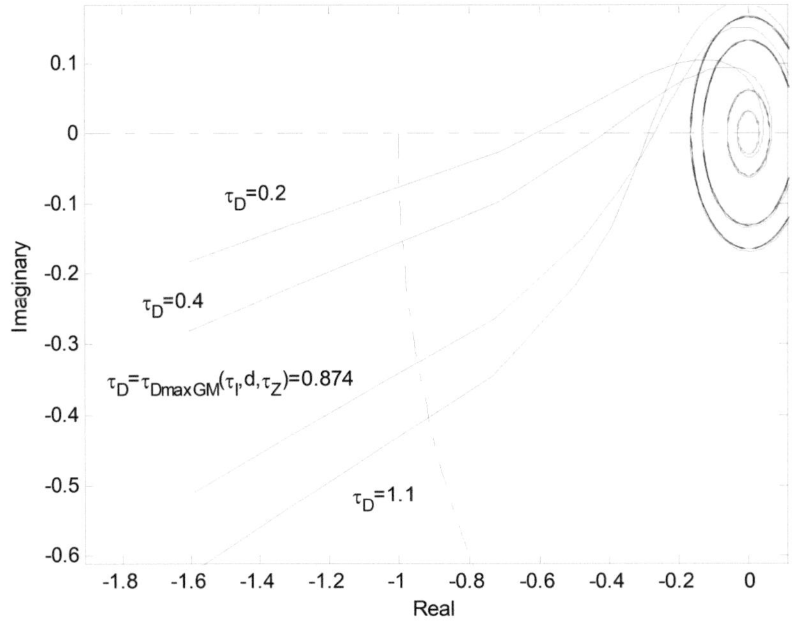

Fig. 3. Nyquist plots of a typical IPDT-IR process controlled by a PD-1F controller with K_C=0.3, d=0.5, τ_Z=0.5, τ_I=2.5, for various values of τ_D.

It can also be observed, from Fig. 5, that the value of $\tau_{D,GM_{max}}(\tau_I,\tau_Z,d)$ decreases as τ_I increases and it takes its minimum value, denoted by $\min\tau_{D,GM_{max}}(\tau_Z,d)$, when $\tau_I\to\infty$. This is also evident from Fig. 6, where $\tau_{D,GM_{max}}(\tau_I,\tau_Z,d)$ is depicted, as a function of τ_I, for several values of τ_Z and d. Applying optimization techniques using MATLAB®, it is plausible to provide accurate approximations of that limit as a function of the normalized parameters d and τ_Z. These approximations are summarized in Table 2. Note that, the maximum normalized error (M.N.E.), defined by

$$\max\left\{\frac{\left|\min\hat{\tau}_{D,GM_{max}}(\tau_Z,d)-\min\tau_{D,GM_{max}}(\tau_Z,d)\right|}{\min\tau_{D,GM_{max}}(\tau_Z,d)}\right\},\ \text{where}\ \min\hat{\tau}_{D,GM_{max}}(\tau_Z,d)$$

denotes the approximate value, never exceeds 3%, for a wide range of d and τ_Z.

Finally, another interesting value relative to τ_D, is $\tau_{D,maxGM,I\text{-}P}(\tau_I,\tau_D,d)$ that denotes the maximum value of τ_D, for which the gain margin obtained by an I-PD controller is larger than the gain margin obtained by an I-P controller. It is worth noticing that the value of $\tau_{D,maxGM,I\text{-}P}(\tau_I,\tau_D,d)$ decreases as τ_I increases and it takes its minimum value, which is denoted by $\min\tau_{D,maxGM,I\text{-}P}(\tau_I,\tau_D,d)$, when $\tau_I\to\infty$. This is evident from Fig. 7, which illustrates $\tau_{D,maxGM,I\text{-}P}(\tau_I,\tau_D,d)$ as a function of τ_I, for several pairs (τ_Z,d). Application of optimization techniques yields some useful approximations of this parameter, which are summarized in Table 3. These approximations are quite accurate, since the respective maximum normalized errors never exceed 5%, for a wide range of d and τ_Z.

4. Controller tuning based on the maximum phase margin specification

The above analysis provides us the means to propose an efficient method for tuning I-PD controllers for IPDT-IR processes. The main characteristic of the proposed tuning method, which is designated as Method I in the sequel, is the selection of the controller gain K_C using (11). The remaining two parameters τ_I and τ_D can be selected such that a specific closed-loop performance is achieved. In particular, one can select the parameter τ_I in order to achieve a specific closed-loop response while selecting the parameter τ_D in order to improve this response in terms of the achievable gain margin or in terms of the minimization of several integral criteria, such as the well-known ISE criterion, the integral of squared error plus the normalized squared controller output deviation from its final value u_∞ (ISENSCOD criterion) of the form

$$J_{ISENSCOD}=\int_0^\infty\left\{\left[y(t)-r(t)\right]^2+K^2\left[u(t)-u_\infty\right]^2\right\}dt \tag{13}$$

or the integral of squared error plus the normalized squared derivative of the controller output (ISENSDCO criterion), having the form

$$J_{ISENSDCO}=\int_0^\infty\left\{\left[y(t)-r(t)\right]^2+K^2\bar{\tau}_p^2\dot{u}(t)^2\right\}dt \tag{14}$$

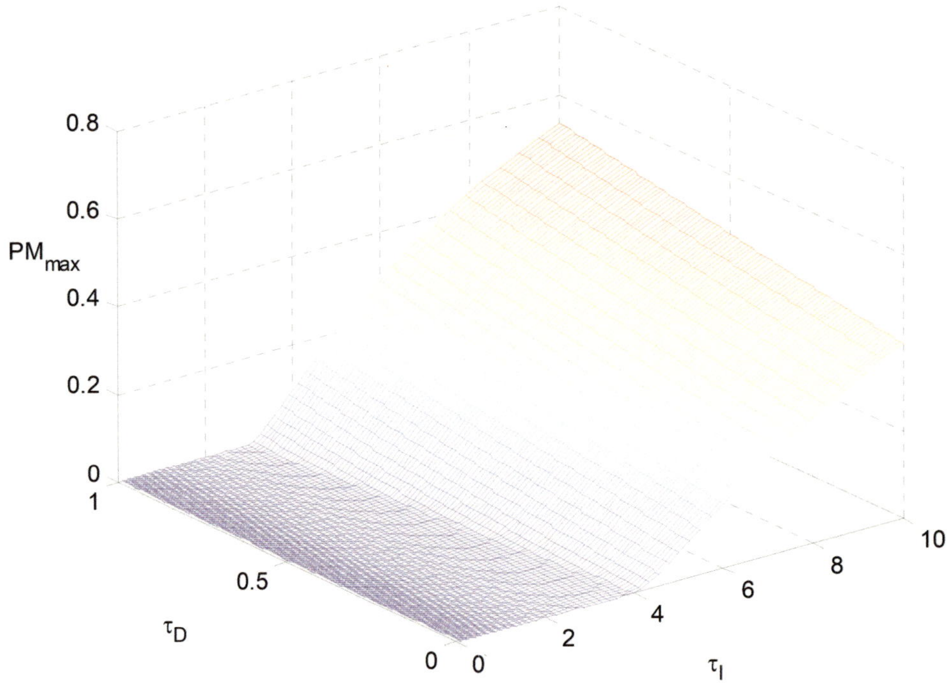

Fig. 4. PM_{max} as a function of τ_I and τ_D for a typical IPDT-IR with d=τ_Z=1.5.

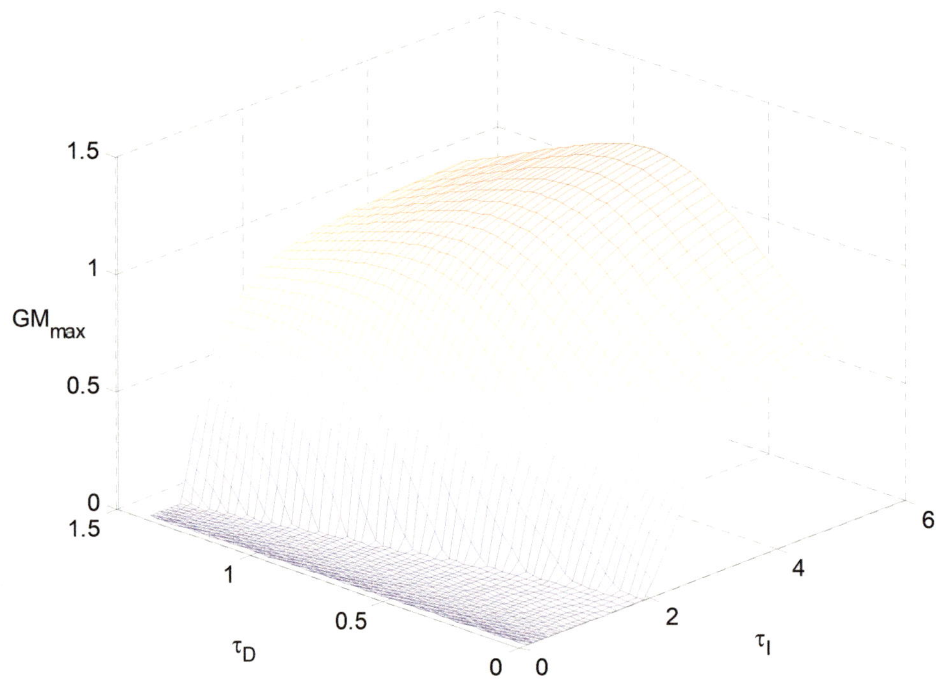

Fig. 5. GM_{max} as a function of τ_I and τ_D for a typical IPDT-IR process with d=τ_Z=0.5.

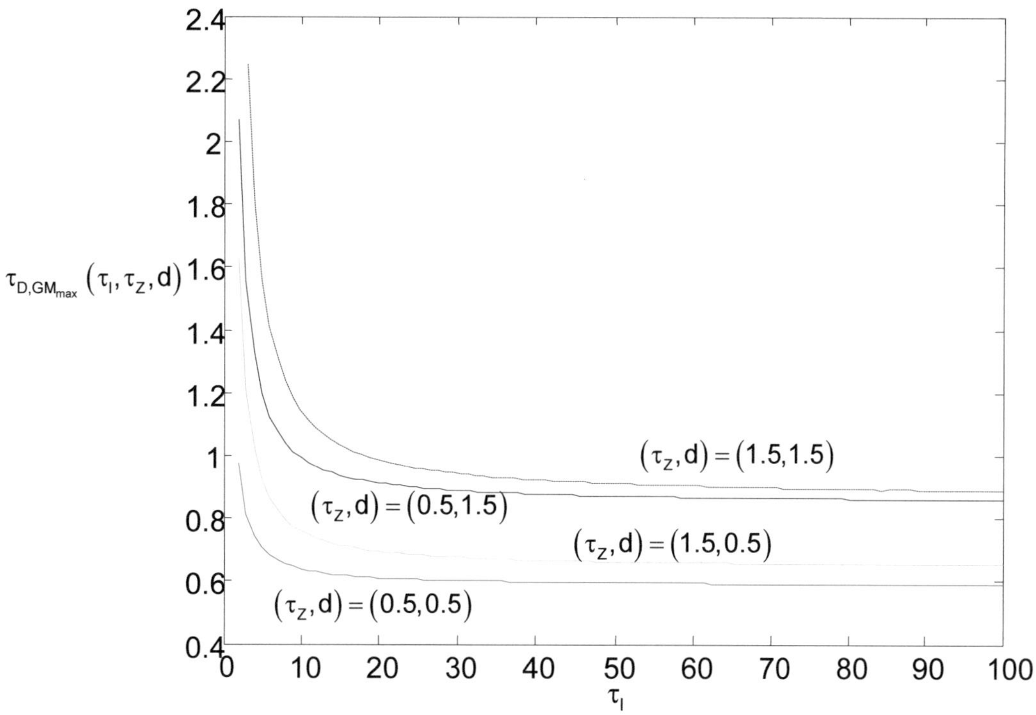

Fig. 6. Parameter $\tau_{D,GM_{max}}(\tau_I,\ \tau_Z, d)$ as a function of τ_I, for several values of τ_Z and d.

$\min \tau_{D,GMmax}(d,\tau_Z)$	M.N.E.
$-0.0561+0.555d-0.0629\tau_Z^2+0.1308\tau_Z-0.372d^2-0.1171(d+\tau_Z)^3+0.5144(d+\tau_Z)^{0.5}$ for $0<\tau_Z<0.5$ & $0<d<0.5$	2.8%
$0.4722+0.2213d-0.0045\tau_Z^2+0.0375\tau_Z+0.0026d^2+0.00004093(d+\tau_Z)^3$ for $0.5<\tau_Z<5$ & $0.5<d<5$	2.5%

Table 2. Proposed approximations for $\min \tau_{D,GMmax}(d,\tau_Z)$ for several values of τ_Z and d.

$\min \tau_{D,maxGM,I\text{-}P}(d,\tau_Z)$	M.N.E.
$1.3335-1.311d-0.755\tau_Z^2+1.7764\tau_Z+0.634d^2+1.484d^{0.5}-1.7531\tau_Z^{0.5}$ for $0<\tau_Z<0.5$ & $0<d<0.5$	2.8%
$1.1393+0.4528d-0.107\tau_Z+0.0114\tau_Z^2+0.0044d^2$ for $0.5<\tau_Z<5$ & $0.5<d<5$	2.5%

Table 3. Proposed approximations for $\min \tau_{D,maxGM,I\text{-}P}(d,\tau_Z)$ for several values of τ_Z and d.

It is worth to notice, at this point, that tuning methods based on the minimization of ISE guarantee small error and very fast response, particularly useful in the case of regulatory control. However, the closed-loop step response is very oscillatory, and the tuning can lead to excessive controller output swings that cause process disturbances in other control loops. In contrast, minimization of criteria (13) or (14) leads to smoother closed-loop responses that are less demanding for the process actuators.

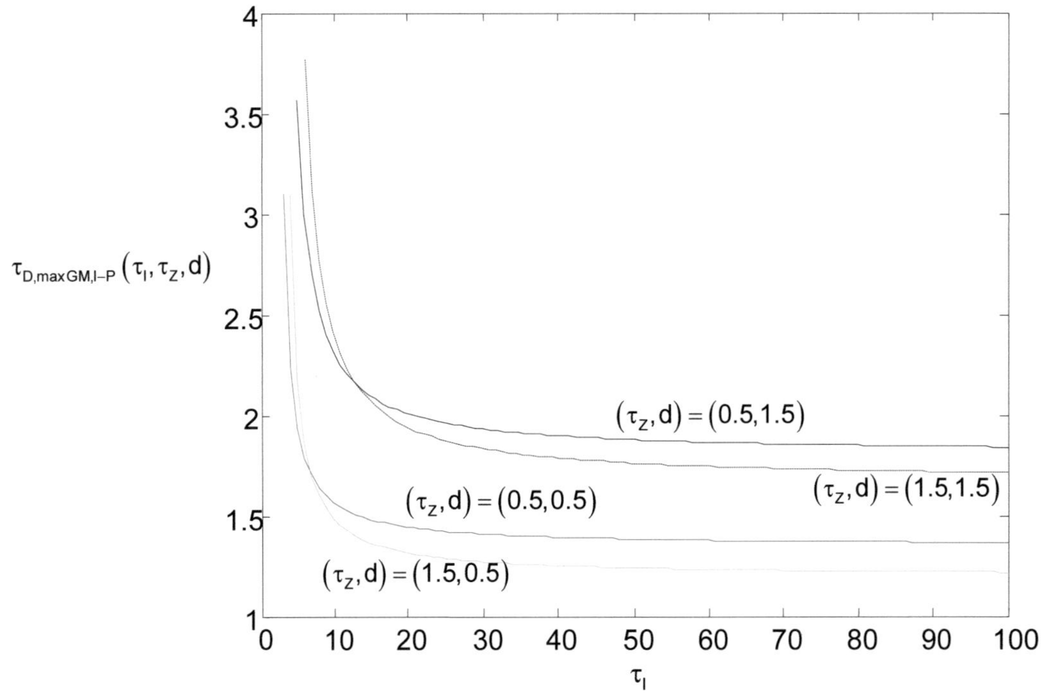

Fig. 7. Parameter $\tau_{D,maxGM,I-P}(\tau_I,\tau_D,d)$ as a function of τ_I, for several values of τ_Z and d.

In order to present systematically Method I, observe first that (6) can also be written as

$$G_{CL}(\hat{s}) = \frac{(-\hat{s}\tau_Z+1)(\hat{s}+1)^{-1}\exp(-d\hat{s})}{K_C^{-1}\tau_I\hat{s}^2+(-\hat{s}\tau_Z+1)(\tau_I\hat{s}+1)(\tau_D\hat{s}+1)(\hat{s}+1)^{-1}\exp(-d\hat{s})} \tag{15}$$

Setting $(-\hat{s}\tau_Z+1)(\hat{s}+1)^{-1}\approx 1-(1+\tau_Z)\hat{s}$ in the numerator of (15), we obtain

$$G_{CL}(\hat{s}) \simeq \frac{\left[1-(1+\tau_Z)\hat{s}\right]\exp(-d\hat{s})}{K_C^{-1}\tau_I\hat{s}^2+P(\hat{s})} \tag{16}$$

where $P(\hat{s})=(-\hat{s}\tau_Z+1)(\tau_I\hat{s}+1)(\tau_D\hat{s}+1)(\hat{s}+1)^{-1}\exp(-d\hat{s})$. Let us now perform a second order MacLaurin approximation of $P(\hat{s})$. That is $P(\hat{s}) \approx P_0+P_1\hat{s}+(P_2/2)\hat{s}^2$. Simple algebra yields $P_0=1$, $P_1=\tau_I+\tau_D-\tau_Z-d-1$, $P_2=2\tau_I(\tau_D-\tau_Z-d-1)+2(d+\tau_Z+1)(1-\tau_D)+d(2\tau_Z+d)$. Substituting the above relations in (16), we obtain

$$G_{CL}(\hat{s}) \simeq \frac{\left[1-(1+\tau_Z)\hat{s}\right]}{\tau_e^2\hat{s}^2+2\zeta\tau_e\hat{s}+1}\exp(-d\hat{s}) \tag{17}$$

where

$$\tau_e = \sqrt{\tau_I(K_C^{-1}+\tau_D-\tau_Z-d-1)+(d+\tau_Z+1)(1-\tau_D)+d(\tau_Z+0.5d)} \tag{18}$$

$$\zeta = \frac{(\tau_I + \tau_D - \tau_Z - d - 1)}{2\sqrt{\tau_I(K_C^{-1} + \tau_D - \tau_Z - d - 1) + (d + \tau_Z + 1)(1 - \tau_D) + d(\tau_Z + 0.5d)}} \tag{19}$$

It is now clear that, the parameter τ_I can be selected in such a way that a desired damping ratio ζ_{des} is obtained for the second order approximation (17), of the closed-loop transfer function. With this design specification, the parameter τ_I must be selected as the maximum real root of the quadratic equation

$$\begin{aligned}&\tau_I^2 + \tau_I\left[2(\tau_D - \tau_Z - d - 1) - 4\zeta_{des}^2(K_C^{-1} + \tau_D - \tau_Z - d - 1)\right] \\ &+(\tau_D - \tau_Z - d - 1)^2 - 4\zeta_{des}^2\left[(d + \tau_Z + 1)(1 - \tau_D) + d(\tau_Z + 0.5d)\right] = 0\end{aligned} \tag{20}$$

Relation (20) suggests that τ_I depends on τ_D and K_C. Since, here, it is proposed to select K_C according to (10) and (11), it remains to consider how to select τ_D. With regard to the parameter τ_D, there are several possible alternative ways to select it:

A first way, is to select $\tau_D = \min\tau_{D,maxGM,I\text{-}P}(\tau_Z,d)$. In this case, in order to calculate the parameters K_C and τ_I, it is necessary to solve equations (10), (11) and (20). To this end, the following iterative algorithm is applied:

4.1 Algorithm I

Step 1. Set $\tau_D = \min\tau_{D,maxGM,I\text{-}P}(\tau_Z,d)$ and $\zeta = \zeta_{des}$.
Step 2. Start with an initial values of τ_I. An appropriate choice is $\tau_{I,init} = 1.2\tau_{I,min} = 1.2(d+\tau_Z+1-\tau_D) = 1.2[d+\tau_Z+1-\min\tau_{D,maxGM,I\text{-}P}(\tau_Z,d)]$
Step 3. For these values of τ_D and τ_I, calculate K_C from relation (11), using relation (10).
Step 4. Calculate the integral term τ_I as the maximum real root of (20).
Step 5. Repeat Steps 3 and 4 until the algorithm converges to a certain value for τ_I and K_C.

Note that the above algorithm always converges to values of K_C and τ_I that satisfy equations (10), (11) and (20). The parameters K_C and τ_I, obtained by the above algorithm, for several values of the desired damping ratio ζ_{des} can be approximated by the functions summarized in Table 4. These functions have been obtained by applying optimization techniques, which intent to minimize the respective M.N.E.s. These errors never exceed 5%. For intermediate values of ζ_{des}, a simple linear interpolation provides sufficiently accurate estimates of $\bar{K}_C$ and $\bar{\tau}_I$. The above approximations are very useful, since the need of iterations is avoided, when the proposed method is applied.

It is worth to notice, at this point, that the values of the parameter $\tau_D = \min\tau_{D,maxGM,I\text{-}P}(\tau_Z,d)$ are, in general, a bit small. Additionally, in the case of regulatory control, the I-PD controller settings obtained by Algorithm I, give somehow large maximum errors, although provide rather satisfactory settling times,. Simulation results show that larger values of τ_D give a better performance in terms of maximum error, whereas the settling time is larger. In this case, a good trade-off between settling time and maximum error, is obtained when $\tau_D = 1.25\min\tau_{D,maxGM,I\text{-}P}(\tau_Z,d)$. Table 5 summarizes the estimations of K_C and τ_I as functions of τ_Z

ζ_{des}	$K_C(\tau_Z,d)$ (for $0.2<\tau_Z<2$ and $0.1<d<2$)	M.N.E.
0.6	$0.1994+0.0832d-0.0143\tau_Z-0.0179d^2-0.1038\tau_Z^{0.5}+0.3591(d+\tau_Z+0.1)^{-1}$ $-0.197d(d+\tau_Z)^{-1}$	3.8%
0.707	$0.1711+0.0648d+0.0041\tau_Z-0.0157d^2-0.1135\tau_Z^{0.5}+0.3412(d+\tau_Z+0.1)^{-1}$ $-0.1353d(d+\tau_Z)^{-1}$	3.82%
0.85	$0.0885+0.0372d-0.0128\tau_Z-0.0094d^2-0.0449\tau_Z^{0.5}+0.3131(d+\tau_Z+0.1)^{-1}$ $-0.0484d(d+\tau_Z)^{-1}$	4%
1	$0.0124-0.0048d-0.008\tau_Z-0.0002d^2-0.005\tau_Z^{0.5}+0.2823(d+\tau_Z+0.1)^{-1}$ $-0.0509d(d+\tau_Z)^{-1}$	3.86%
1.2	$-0.0887-0.0339d+0.0072\tau_Z+0.0039d^2+0.0282\tau_Z^{0.5}+0.256(d+\tau_Z+0.1)^{-1}$ $+0.1593d(d+\tau_Z)^{-1}$	3.1%
ζ_{des}	$\tau_I(\tau_Z,d)$ (for $0.2<\tau_Z<2$ and $0.1<d<2$)	M.N.E.
0.6	$1.4954+3.1027d-3.8488\tau_Z-0.1729d^2-0.5138\tau_Z^{0.5}-1.4686(d+\tau_Z+1)^{-1}$	2.6%
0.707	$2.5628+4.329d+6.2521\tau_Z-0.4398d^2-2.5004\tau_Z^{0.5}-1.585(d+\tau_Z+1)^{-1}$	2.86%
0.85	$2.2014+7.5718d+10.1797\tau_Z-0.9624d^2-4.3353\tau_Z^{0.5}+0.3528(d+\tau_Z+1)^{-1}$	3.32%
1	$1.0227+4.2832d+9.2825\tau_Z+0.6779d^2-0.1484\tau_Z^{0.5}+2.0726\tau_Z^2+5.7469d^{0.5}$	3.14%
1.2	$3.0653+11.6507d+22.9691\tau_Z+0.9966d^2-3.453\tau_Z^{0.5}+3.0497\tau_Z^2+4.7546d^{0.5}$	3.15%

Table 4. Proposed approximations for $K_C(\tau_Z,d)$ and $\tau_I(\tau_Z,d)$ obtained by Algorithm I, for $0.2<\tau_Z<2$ and $0.1<d<2$, in the case where $\tau_D=\min\tau_{D,maxGM,I\text{-}P}(\tau_Z,d)$.

and d, in the case where, in Step 1 of Algorithm I, parameter τ_D is selected as $\tau_D=1.25\min\tau_{D,maxGM,I\text{-}P}(\tau_Z,d)$ instead of $\tau_D=\min\tau_{D,maxGM,I\text{-}P}(\tau_Z,d)$.

A second obvious way to select τ_D is through the relation $\tau_D=\min\tau_{D,GM_{\max}}(\tau_Z,d)$. However, in this case, the obtained values of τ_D, are quite small. It is worth to remember that the value $\min\tau_{D,GM_{\max}}(\tau_Z,d)$ is obtained when $\tau_I\to\infty$. So, in order to obtain a larger and more efficient value for the parameter τ_D, we propose here to select $\tau_D=2\tau_{D,GM_{\max}}(\tau_I,\tau_Z,d)$. This value of τ_D, which is twice the one that maximizes the gain margin for a given τ_I, is obtained without the assumption $\tau_I\to\infty$, and of course it is always greater than the value $\min\tau_{D,GM_{\max}}(\tau_Z,d)$. However, that choice causes additional difficulties on our attempts to obtain the parameters K_C, τ_I and τ_D, since τ_D, now, depends implicitly on τ_I, whereas the choice of $\tau_D=\min\tau_{D,GM_{\max}}(\tau_Z,d)$ or $\tau_D=\min\tau_{D,\max GM,I-P}(\tau_Z,d)$ are independent of τ_I. Therefore, additional iterations are necessary, in the case where the selection $\tau_D=2\tau_{D,GM_{\max}}(\tau_I,\tau_Z,d)$ is made. More precisely, in this case, the following algorithm is proposed to obtain admissible controller settings:

ζ_{des}	$K_C(\tau_Z,d)$	M.N.E.
0.6	$0.2574+0.0349d-0.0356\tau_Z-0.0085d^2-0.0559\tau_Z^{0.5}+0.2538(d+\tau_Z+0.1)^{-1}$ $-0.1702d(d+\tau_Z)^{-1}$ for $0.3<\tau_Z<2$ & $0.3<d<2$	3%
0.6	$0.0439+0.1017d+0.0268\tau_Z-0.0155d^2-0.1347\tau_Z^{0.5}+0.9583(d+\tau_Z+1)^{-1}$ $-0.2309d(d+\tau_Z)^{-1}$ for $0.4<\tau_Z<2$ & $0.2<d<2$	1.3%
0.6	$0.0225+0.0943d-0.0025\tau_Z-0.0152d^2-0.0707\tau_Z^{0.5}+0.9161(d+\tau_Z+1)^{-1}$ $-0.2117d(d+\tau_Z)^{-1}$ for $0.2<\tau_Z<2$ & $0.4<d<2$	1.1%
0.707	$0.2814+0.0362d-0.046\tau_Z-0.0087d^2-0.0552\tau_Z^{0.5}+0.2225(d+\tau_Z+0.1)^{-1}$ $-0.1841d(d+\tau_Z)^{-1}$ for $0.2<\tau_Z<2$ & $0.06<d<2$	4.8%
0.85	$0.1698+0.0457d-0.0349\tau_Z-0.0099d^2-0.0413\tau_Z^{0.5}+0.3256(d+\tau_Z+0.3)^{-1}$ $-0.1381d(d+\tau_Z)^{-1}$ for $0.2<\tau_Z<2$ & $0.1<d<2$	4%
1	$0.0939+0.0305d-0.0366\tau_Z-0.0065d^2-0.0076\tau_Z^{0.5}+0.3198(d+\tau_Z+0.3)^{-1}$ $-0.0709d(d+\tau_Z)^{-1}$ for $0.2<\tau_Z<2$ & $0.1<d<2$	4.72%
1.2	$0.0195+0.0205d-0.0153\tau_Z-0.0047d^2-0.0159\tau_Z^{0.5}+0.3219(d+\tau_Z+0.3)^{-1}$ $+0.0024d(d+\tau_Z)^{-1}$ for $0.2<\tau_Z<2$ & $0.12<d<2$	4.96%
ζ_{des}	$\tau_I(\tau_Z,d)$	M.N.E.
0.6	$-0.4713+1.8218d+1.8329\tau_Z+0.0067d^2+1.4335\tau_Z^{0.5}+0.3006\tau_Z^2+1.5613d^{0.5}$ for $0.3<\tau_Z<2$ & $0.3<d<2$	1.13%
0.707	$0.2867+0.2494d-3.2074\tau_Z+0.4602d^2+6.44\tau_Z^{0.5}+1.5289\tau_Z^2+3.6189d^{0.5}$ $-2.2731(d+\tau_Z+1)^{-1}$ for $0.2<\tau_Z<2$ & $0.06<d<2$	4.17%
0.85	$1.9692+1.1533d+5.4948\tau_Z+0.4359d^2-2.9027\tau_Z^{0.5}+0.998\tau_Z^2+5.6043d^{0.5}$ $+0.8839(d+\tau_Z+1)^{-1}$ for $0.2<\tau_Z<2$ & $0.1<d<2$	1.7%
1	$2.0741+3.515d+3.3445\tau_Z+0.3835d^2+0.2227\tau_Z^{0.5}+3.2956\tau_Z^2+6.4788d^{0.5}$ $+2.326(d+\tau_Z+1)^{-1}$ for $0.2<\tau_Z<2$ & $0.1<d<2$	2.4%
1.2	$-1.3108+8.9972d+19.7612\tau_Z+0.2125d^2-10.1307\tau_Z^{0.5}+4.4869\tau_Z^2$ $+9.6879d^{0.5}+12.725(d+\tau_Z+1)^{-1}$ for $0.2<\tau_Z<2$ & $0.12<d<2$	2.5%

Table 5. Proposed approximations for $K_C(\tau_Z,d)$ and $\tau_I(\tau_Z,d)$ obtained by Algorithm I, in the case where $\tau_D=1.25\min\tau_{D,maxGM,I\text{-}P}(\tau_Z,d)$.

4.2 Algorithm II

Step 1. Set $\zeta=\zeta_{des}$ and start with some initial values of τ_D and τ_I. Appropriate choices are

$$\tau_{D,init} = \min \tau_{D,GM_{\max}}(\tau_Z, d) \tag{21}$$

$$\tau_{I,init} = 1.2\tau_{I,\min} = 1.2\left[d + \tau_Z + 1 - \min \tau_{D,GM_{\max}}(\tau_Z, d)\right] \tag{22}$$

Step 2. For these values of τ_D and τ_I, calculate K_C from relation (11) using relation (10).
Step 3. Calculate the integral term τ_I as the maximum real root of (20).
Step 4. Repeat Steps 2 and 3 until the algorithm converges to a certain value for τ_I and K_C.
Step 5. For the obtained value of τ_I, select $\tau_D = 2\tau_{D,GM_{\max}}(\tau_I, \tau_Z, d)$.
Step 6. Repeat Steps 2 to 5 until convergence.

Similarly to Algorithm I, Algorithm II always converges to values of K_C, τ_I and τ_D that satisfy equations (10), (11) and (20). The parameters K_C, τ_I and τ_D, obtained by the above algorithm, for several values of the desired damping ratio ζ_{des} can be approximated by the functions summarized in Table 6, using optimization techniques. Once again, the maximum normalized errors never exceed 5%. For intermediate values of ζ_{des}, a simple linear interpolation provides sufficiently accurate estimates of K_C, τ_I and τ_D.

Finally, a third way to choose τ_D, is in order to minimize some integral criteria, such as the criteria (13) or (14). In this case, the following iterative algorithm can be applied to achieve admissible controller settings:

4.3 Algorithm III

Step 1. Set ζ=ζ_{des} and start with some initial values of τ_D and τ_I. Appropriate initial values are given by relations (21) and (22), respectively. Alternatively, on can initialize the algorithm by using $\tau_{D,init}$=min$\tau_{D,maxGM,I\text{-}P}(\tau_Z,d)$ and $\tau_{I,init}$=1.2$\tau_{I,min}$=1.2[d+τ_Z+1-min$\tau_{D,maxGM,I\text{-}P}(\tau_Z,d)$].
Step 2. For these values of τ_D and τ_I, calculate K_C from relation (11) using relation (10).
Step 3. Calculate the integral term τ_I as the maximum real root of (20).
Step 4. Repeat Steps 2 and 3 until the algorithm converges to a certain value for τ_I and K_C.
Step 5. For the obtained value of τ_I, select τ_D in order to minimize (13) or (14).
Step 6. Repeat Steps 2 to 5 until convergence.

Note that, Step 5 of the above iterative algorithm is an iterative algorithm by itself. This is due to the fact that optimization algorithms as well as extensive simulations (since there are no close form solution for such integrals in the case of time-delay systems), are used to obtain the optimal values of τ_D that minimize them. Iterative algorithms that minimize the aforementioned integral criteria are usually based on the golden section method.

Algorithm III always converges to values of K_C, τ_I, τ_D that satisfy equations (10), (11) and (20). Note also that convergence of the algorithm is independent of the initial pair ($\tau_{D,init}$, $\tau_{I,init}$). The parameters K_C, τ_I and τ_D, obtained by the above algorithm, for several values of the desired damping ratio ζ_{des} can be approximated by the functions summarized in Table 7, using optimization techniques. Once again, the maximum normalized errors never exceed 5%. For intermediate values of ζ_{des}, a simple linear interpolation provides sufficiently accurate estimates of K_C, τ_I and τ_D.

Finally, once K_C, τ_I and τ_D are obtained according to the Algorithms I-III, the original I-PD controller parameters $\bar{K}_P$, $\bar{K}_I$ and $\bar{K}_D$, can be easily obtained by using the relations summarized in Table 1, as well as relations (4), interrelating controller gains for the I-PD controller and the series PID controller with set-point pre-filter.

ζ_{des}	$K_C(\tau_Z,d)$ (for $0.2<\tau_Z<2$ & $0.1<d<2$)	M.N.E.
0.6	$0.1432+0.0072d-0.0363\tau_Z-0.0035d^2-0.0207\tau_Z^{0.5}+0.2477(d+\tau_Z-0.1)^{-1}-0.064d(d+\tau_Z)^{-1}$	4.57%
0.707	$0.2163+0.036d-0.0025\tau_Z-0.008d^2-0.0914\tau_Z^{0.5}+0.2353(d+\tau_Z-0.1)^{-1}-0.1212d(d+\tau_Z)^{-1}$	4.3%
0.85	$0.1454-0.0477d-0.0313\tau_Z+0.0082d^2+0.0092\tau_Z^{0.5}+0.1569(d+\tau_Z-0.15)^{-1}+0.0439d(d+\tau_Z)^{-1}$	4.7%
1	$-0.1131-0.0476d-0.0211\tau_Z+0.0113d^2+0.089\tau_Z^{0.5}+0.3852(d+\tau_Z+0.3)^{-1}+0.1452d(d+\tau_Z)^{-1}$	2.6%
1.2	$-0.1858-0.0375d+0.0029\tau_Z+0.0091d^2+0.0607\tau_Z^{0.5}+0.5999(d+\tau_Z+1)^{-1}+0.1459d(d+\tau_Z)^{-1}$	4.65%
ζ_{des}	$\tau_I(\tau_Z,d)$	M.N.E.
0.6	$-0.5878+1.7259d+1.5127\tau_Z+0.0187d^2+2.0613\tau_Z^{0.5}+0.11\tau_Z^2+1.6319d^{0.5}+0.0603(d+\tau_Z+1)^{-1}$ for $0.2<\tau_Z<2$ & $0.1<d<2$	1.8%
0.707	$-4.8923+3.223d+2.7225\tau_Z-0.4963d^2-0.1359\tau_Z^2+4.8271(d+\tau_Z)^{0.5}+3.418(d+\tau_Z+1)^{-0.5}$ for $0.2<\tau_Z<2$ & $0.02<d<2$	5%
0.85	$-0.7157+1.6486d+4.1163\tau_Z+0.3192d^2+0.8445\tau_Z^{0.5}+2.5935d^{0.5}+4.3918(d+\tau_Z)^{0.5}-1.5426(d+\tau_Z+1)^{-0.5}$ for $0.2<\tau_Z<1.9$ & $0.14<d<2$	3.6%
1	$-0.3735+5.9535d+11.0894\tau_Z+0.3347d^2+0.9352\tau_Z^{0.5}+3.7375d^{0.5}+0.1432(d+\tau_Z)^{0.5}+1.438(d+\tau_Z+1)^{-0.5}$ for $0.2<\tau_Z<1.9$ & $0.14<d<2$	2.38%
1.2	$-10.2543+26.9196d+21.2015\tau_Z-2.3514d^2+7.6005\tau_Z^{0.5}-6.1009d^{0.5}+9.7159(d+\tau_Z)^{-0.5}$ for $0.2<\tau_Z<2$ & $0.02<d<2$	4.37%
ζ_{des}	$\tau_D(\tau_Z,d)$ (for $0.2<\tau_Z<2$ & $0.1<d<2$)	M.N.E.
0.6	$1.664-0.1961d+0.0583\tau_Z+0.1392d^2+0.0145\tau_Z^{0.5}+0.7743d^{0.5}-1.3746(d+\tau_Z+1)^{-1}$	2.9%
0.707	$1.9179+0.524d+1.8374\tau_Z-0.0061d^2-0.2274\tau_Z^2-2.1731\tau_Z^{0.5}+0.045d^{0.5}-0.178(d+\tau_Z+1)^{-1}$	4.37%
0.85	$2.2066+0.2973d-0.0407\tau_Z+0.0079d^2+0.0105\tau_Z^2-0.1954\tau_Z^{0.5}+0.0203d^{0.5}-1.8737(d+\tau_Z+1)^{-1}$	4.4%
1	$2.2783-0.2417d-0.1019\tau_Z+0.1367d^2-0.1256\tau_Z^{0.5}+0.362d^{0.5}-2.1512(d+\tau_Z+1)^{-1}$	4.9%
1.2	$1.8375+0.3488d+0.2776\tau_Z+0.022d^2-0.5294\tau_Z^{0.5}-0.0578d^{0.5}-0.5849(d+\tau_Z)^{-0.5}$	4.8%

Table 6. Proposed approximations for $K_C(\tau_Z,d)$, $\tau_I(\tau_Z,d)$ and $\tau_D(\tau_Z,d)$ obtained by Algorithm II

ζ_{des}	$K_C(\tau_Z,d)$ $\tau_I(\tau_Z,d)$	M.N.E.
0.6	$0.1786-0.0462d+0.0034\tau_Z+0.0085d^2-0.0403\tau_Z^{0.5}+0.2234(d+\tau_Z)^{-1}$ $+0.0271(d+\tau_Z)^{-2}$ $0.0717+1.5545d+2.5829\tau_Z+0.0439d^2-0.0783\tau_Z^2$ $+0.2783\tau_Z^{0.5}+1.6007d^{0.5}+0.1138(d+\tau_Z+1)^{-1}$ for $0.2<\tau_Z<2$ & $0.2<d<2$	2.35% 3.3%
0.707	$0.1471-0.0324d+0.0012\tau_Z+0.0059d^2-0.0352\tau_Z^{0.5}+0.2616(d+\tau_Z)^{-1}$ $-0.0005(d+\tau_Z)^{-2}$ $-0.457+2.5379d+2.6464\tau_Z+0.0252d^2-0.4502\tau_Z^2$ $+1.7881d^{0.5}+2.5791\tau_Z^{0.5}$ for $0.3<\tau_Z<1.65$ & $0.3<d<1.65$	0.5% 2.12%
0.85	$0.1269-0.0244d-0.0017\tau_Z+0.0037d^2-0.0261\tau_Z^{0.5}+0.2567(d+\tau_Z)^{-1}$ $-0.0136(d+\tau_Z)^{-2}$ $-0.3892+3.6961d+3.4709\tau_Z-0.0198d^2-0.2082\tau_Z^2$ $+3.1789d^{0.5}+3.5233\tau_Z^{0.5}$ for $0.3<\tau_Z<1.69$ & $0.3<d<1.69$	0.5% 1.36%
1	$0.1101-0.0187d-0.0015\tau_Z+0.0023d^2-0.0224\tau_Z^{0.5}+0.2471(d+\tau_Z)^{-1}$ $-0.0255(d+\tau_Z)^{-2}$ $0.2357+5.5238d+5.3907\tau_Z+0.0616d^2-0.2819\tau_Z^2$ $+4.5751d^{0.5}+4.998\tau_Z^{0.5}$ for $0.3<\tau_Z<1.69$ & $0.3<d<1.69$	0.5% 1.29%
1.2	$0.1042-0.0162d-0.0068\tau_Z+0.0014d^2-0.0112\tau_Z^{0.5}+0.2081(d+\tau_Z)^{-1}$ $-0.0212(d+\tau_Z)^{-2}$ $0.3879+14.3387d+1.3617\tau_Z-1.0061d^2+0.9472\tau_Z^2$ $+1.9009d^{0.5}+16.4489\tau_Z^{0.5}$ for $0.2<\tau_Z<2$ & $0.2<d<2$	1.0% 2.5%
ζ_{des}	$\tau_D(\tau_Z,d)$	M.N.E.
0.6	$0.7388+1.1169d+1.5105\tau_Z-0.0632d^2-0.3234\tau_Z^2-0.4559\tau_Z^{0.5}$ for $0.2<\tau_Z<1.7$ & $0.2<d<2$	2.3%
0.707	$0.3239+0.758d+0.5251\tau_Z-0.0065d^2-0.0819\tau_Z^2+0.393d^{0.5}+0.5963\tau_Z^{0.5}$ for $0.3<\tau_Z<1.65$ & $0.3<d<1.65$	1.42%
0.85	$0.333+0.7693d+0.4648\tau_Z+0.007d^2-0.043\tau_Z^2+0.3902d^{0.5}+0.7295\tau_Z^{0.5}$ for $0.3<\tau_Z<1.69$ & $0.3<d<1.69$	1.35%
1	$0.3745+0.7965d+0.4051\tau_Z+0.0164d^2+0.0029\tau_Z^2+0.4242d^{0.5}+0.0029\tau_Z^{0.5}$ for $0.3<\tau_Z<1.69$ & $0.3<d<1.69$	1.2%
1.2	$0.6193+0.9377d+1.0697\tau_Z+0.006d^2-0.1332\tau_Z^2+0.4388d^{0.5}-0.1332\tau_Z^{0.5}$ for $0.2<\tau_Z<2$ & $0.2<d<2$	2.1%

Table 7. Proposed approximations for $K_C(\tau_Z,d)$, $\tau_I(\tau_Z,d)$, $\tau_D(\tau_Z,d)$ obtained by Algorithm III.

5. A direct synthesis tuning method

The tuning methods presented in the previous Section are somehow complicated, since they are based on iterative algorithms. In what follows, our aim is to present a rather simpler tuning method, which is called here Method II, and it is based on the direct synthesis approach. To this end, observe that after parameter normalization according to Table 1 relation (2) may further be written as

$$G_{CL}(\hat{s}) = \frac{(-\hat{s}\tau_Z + 1)(\hat{s}+1)^{-1}\exp(-d\hat{s})}{K_I^{-1}\hat{s}^2 + \left(K_D K_I^{-1}\hat{s}^2 + K_P K_I^{-1}\hat{s} + 1\right)(-\hat{s}\tau_Z + 1)(\hat{s}+1)^{-1}\exp(-d\hat{s})} \tag{23}$$

Relation (23) may be approximated as

$$G_{CL}(\hat{s}) = \frac{\left[1-(\tau_Z+1)\hat{s}\right]\exp(-d\hat{s})}{K_I^{-1}\hat{s}^2 + \left(K_D K_I^{-1}\hat{s}^2 + K_P K_I^{-1}\hat{s} + 1\right)(-\hat{s}\tau_Z + 1)(\hat{s}+1)^{-1}(1+d\hat{s})^{-1}} \tag{24}$$

where the approximations $(-\hat{s}\tau_Z+1)(\hat{s}+1)^{-1} \approx 1-(\tau_Z+1)\hat{s}$ and $\exp(-d\hat{s}) \approx 1/(d\hat{s}+1)$ are used to obtain relation (24). In what follows, define $d_{max}=\max\{1,d\}$ and $d_{min}=\min\{1,d\}$. Then, relation (24) yields

$$G_{CL}(\hat{s}) = \frac{\left[1-(\tau_Z+1)\hat{s}\right]\exp(-d\hat{s})}{K_I^{-1}\hat{s}^2 + \left(K_D K_I^{-1}\hat{s}^2 + K_P K_I^{-1}\hat{s} + 1\right)\left[1-(\tau_Z+d_{\min})\hat{s}\right](\hat{s}d_{\max}+1)^{-1}} \tag{25}$$

where, the approximation $(-\hat{s}\tau_Z+1)(\hat{s}d_{min}+1)^{-1} \approx 1-(\tau_Z+d_{min})\hat{s}$ is used to produce (25). By performing appropriate division, relation (25) becomes

$$G_{CL}(\hat{s}) = \frac{\left[1-(\tau_Z+1)\hat{s}\right]\exp(-d\hat{s})}{K_I^{-1}\hat{s}^2 + \left[d_{\max}^{-1}K_D K_I^{-1}\hat{s} + \left(d_{\max}^{-1}K_P - d_{\max}^{-2}K_D K_I^{-1}\right) + Q(\hat{s})\right]\left[1-(\tau_Z+d_{\min})\right]\hat{s}} \tag{26}$$

where $Q(\hat{s}) = \left(1 - d_{\max}^{-1}K_P K_I^{-1} + d_{\max}^{-2}K_D K_I^{-1}\right)\left(d_{\max}\hat{s}+1\right)^{-1}$. Now, selecting

$$K_D = d_{\max}K_P - d_{\max}^2 K_I \tag{27}$$

we obtain $Q(\hat{s})=0$ and $G_{CL}(\hat{s}) = \dfrac{\left[1-(\tau_Z+1)\hat{s}\right]\exp(-d\hat{s})}{K_I^{-1}\hat{s}^2 + \left[\left(K_p K_I^{-1} - d_{\max}\right)\hat{s} + 1\right]\left[1-(\tau_Z+d_{\min})\hat{s}\right]}$, which yields

$$G_{CL}(\hat{s}) = \frac{\left[1-(\tau_Z+1)\hat{s}\right]\exp(-d\hat{s})}{\lambda^2\hat{s}^2 + 2\xi\lambda\hat{s} + 1} \tag{28}$$

where, $\lambda = \sqrt{K_I^{-1} - \left(K_p K_I^{-1} - d_{\max}\right)(\tau_Z + d_{\min})}$ and $\xi = \left(K_p K_I^{-1} - d - \tau_Z - 1\right)/(2\lambda)$, since $d_{max}+d_{min}=d+1$.

The Routh stability criterion about relation (28) yields

$$K_P>(d+\tau_Z+1)K_I=K_{P,min} \text{ and } K_P<d_{max}K_I+(\tau_Z+d_{min})^{-1}=K_{P,max} \tag{29}$$

Hence, as for K_P, one can select the middle value of the range given by inequalities (29), i.e. $K_P=(K_{P,min}+K_{P,max})/2=[(d+\tau_Z+d_{max}+1)K_I+(\tau_Z+d_{min})^{-1}]/2$. This relation, further yields

$$\theta=K_P/K_I=[(d+\tau_Z+d_{max}+1)K_I+(\tau_Z+d_{min})^{-1}K_I^{-1}]/2 \qquad (30)$$

Solving (30) with respect to K_I and taking into account the definition of θ and (27), we obtain

$$K_I=[(\tau_Z+d_{min})(2\theta-d-\tau_Z-d_{max}-1)]^{-1}$$

$$K_P=\theta[(\tau_Z+d_{min})(2\theta-d-\tau_Z-d_{max}-1)]^{-1} \qquad (31)$$

$$K_D=(\theta d_{max}-d^2_{max})[(\tau_Z+d_{min})(2\theta-d-\tau_Z-d_{max}-1)]^{-1}$$

provided that $\theta>(d+\tau_Z+d_{max}+1)/2=\theta_{min}$.

Now, it only remains to specify θ. Note that, θ can be arbitrarily selected as an adjustable parameter in the interval $[\theta_{min}, \infty)$, thus permitting on-line tuning. However, it would be useful for the designer to follow certain rules, based on some criteria relative to the closed-loop system performance, in order to select the adjustable parameter θ.

A first criterion for the selection of θ is related to the responsiveness of the closed-loop system. In particular, parameter θ can be selected in such a way that a desired damping ratio ξ_{des} is obtained for the second order approximation (28), of the closed-loop transfer function. In this case, using the definitions of θ, λ and ξ and the first of (31), after some trivial algebraic manipulations, on can conclude that parameter θ must be selected as the maximum positive real root of the quadratic equation

$$\theta^2-2\left[2\xi^2_{des}\left(\tau_Z+d_{\min}\right)+d+\tau_Z+1\right]\theta+4\xi^2_{des}\left(d+\tau_Z+1\right)=0 \qquad (32)$$

An alternative tuning can be obtained from the minimization of the integral criteria (13) and (14). Let w_u be the ultimate frequency of the normalized open-loop system, with transfer function $G_p(\hat{s})=(-\hat{s}\tau_Z+1)\exp(-d\hat{s})/[\hat{s}(\hat{s}+1)]$, i.e. the frequency at which $\arg\left(G_P(\hat{s})\big|_{\hat{s}=jw_u}\right)=-\pi$, and set $\theta=2\pi/(w_u\beta)$, where β is a parameter to be specified. Note that w_u is the solution of

$$dw_u+a\tan(\tau_Z w_u)+a\tan(w_u)=\pi/2 \qquad (33)$$

By defining $\tau_{min}=\min\{\tau_Z,1\}$, $\tau_{max}=\max\{\tau_Z,1\}$, and by appropriately using the approximations $atan(\tau_{min}w_u)\approx\tau_{min}w_u$, $atan(x)\approx x+x^2[\pi(0.5\pi-x)]^{-1}$ and the fact that $\tau_{min}+\tau_{max}=\tau_Z+1$, we finally obtain (see Arvanitis et al, 2009b, for details)

$$w_u\approx\frac{(0.5\pi-1)}{2\left[d+\tau_z+1-\left(d+\tau_{\min}\right)\pi^{-1}\right]}\left[1+\sqrt{1+\frac{\pi\left[d+\tau_z+1-\left(d+\tau_{\min}\right)\pi^{-1}\right]}{\left(d+\tau_{\min}\right)(0.5\pi-1)^2}}\right] \qquad (34)$$

Having obtained a sufficiently accurate estimation of w_u as a function of the normalized process parameters, we now focus our attention on the determination of β that minimizes (13) or (14). Since there is no closed form solution for the minimization of the above integrals in the case of time-delay systems, simulation must be used instead. Here, optimization algorithms are used to obtain the parameter β that minimizes (13) or (14), as a function of d and τ_Z. Table 8 summarizes the estimated parameters β minimizing criteria (13) and (14) in the case of regulatory control (ISENSCOD-L and ISENDCO-L criteria), together with the respective maximum normalized errors.

Criterion	$\beta(\tau_Z,d)$	M.N.E.
ISENSCOD-L (Eq. (13))	$-0.1131-1.4126d-2.0935\tau_Z-0.2022d^2+0.0486\tau_Z^{0.5}$ $-0.5246d^{0.5}+4.2676(d+\tau_Z)^{0.5}+0.0233(d+\tau_Z)^3$ for $0<\tau_Z<2$ & $0<d<1$	4.26%
ISENSCOD-L (Eq. (13))	$2.1108+1.4284d+0.5002\tau_Z-0.6778d^2+0.9347\tau_Z^{0.5}$ $+1.5618d^{0.5}-2.651(d+\tau_Z)^{0.5}+0.0209(d+\tau_Z)^3-0.269\tau_Z^2$ for $0<\tau_Z<2$ & $1<d<2$	4.96%
ISENSDCO-L (Eq. (14))	$-0.0409+4.5635d+4.1371\tau_Z-0.4161d^2+0.8091\tau_Z^{0.5}+2.8912d^{0.5}$ $-3.8217(d+\tau_Z)^{0.5}-1.5426(d+\tau_Z)^2+0.1682(d+\tau_Z)^3+0.304\tau_Z^2$ for $0<\tau_Z<2$ & $0.13<d<1$	4.5%
ISENSDCO-L (Eq. (14))	$0.886-0.6479d-1.0903\tau_Z-0.5933d^2+1.2604\tau_Z^{0.5}$ $+2.6081d^{0.5}-0.759(d+\tau_Z)^{0.5}+0.2357(d+\tau_Z)^3-0.2545\tau_Z^2$ for $0<\tau_Z<2$ & $1<d<2$	4.7%

Table 8. Proposed approximations of the parameter $\beta(\tau_Z,d)$ for Method II.

6. Simulation application to a boiler steam drum

The most typical example of an IPDT-IR process is a boiler steam drum. The level (output) is controlled by manipulating the boiler feed water (BFW) to the drum. The drum is located near the top of the boiler and is connected to it by a large number of tubes. Liquid and vapour water circulate between the drum and the boiler as a result of the density difference between the liquid in the down-comer pipes leading from the bottom of the drum to the base of the boiler and the vapour/liquid mixture in the riser pipes going up through the boiler and back into the steam drum. In has been suggested by Luyben (2003) that the transfer function model of the process takes the form (1), with $\bar{K}=0.547$, $\bar{\tau}_P=1.06$, $\bar{\tau}_Z=0.418$, $\bar{d}=0.1$.

We next apply the tuning methods presented in Sections 4 and 5 in order to tune I-PD controllers for the above IPDT-IR model of boiler steam drum, as well to provide a comparison with existing tuning methods for PID controllers. Note that, for the above process model, the method proposed in the work (Luyben, 2003), for an integral time constant equal to 25% of the minimum PI integral time, yields the conventional PID controller settings $K_C=1.61$, $\tau_I=11.5$ and $\tau_D=1.15$. The method reported in the work (Shamsuzzoha & Lee, 2006), for $\lambda=0.798$, $\xi=1$, $\psi=25$, yields the two-degrees-of-freedom PID controller settings, $K_C=2.3892$, $\tau_I=3.5778$, $\tau_D=0.7249$, in the case where the set-point weighting parameter is selected as $b=0.3$. Finally, the method proposed in the work (Gu et al, 2006), yields the following IMC based PID controller settings $K_C=2.0883$, $\tau_I=3.8664$, $\tau_D=0.6879$, with the filter time constant having the value $\lambda=0.8$.

Applying Algorithm I of Method I, with $\tau_D=\min\tau_{D,maxGM,I\text{-}P}(\tau_Z,d)=1.1535$ (as calculated by Table 3), $\tau_{I,init}=0.4022$ and with the desired damping ratio of (17) having the value $\zeta_{des}=0.6$, we obtain the I-PD controller settings $\bar{K}_P=2.104$, $\bar{K}_I=0.6391$ and $\bar{K}_D=1.5876$. Moreover, applying the same Algorithm with $\tau_D=1.25\min\tau_{D,maxGM,I\text{-}P}(\tau_Z,d)=1.4419$ (as calculated by Table

3), $\tau_{I,init}$=0.0561 and with the same closed-loop system specification, we obtain the I-PD controller settings $\bar{K}_P$=2.0026, $\bar{K}_I$=0.5695 and $\bar{K}_D$=1.7289. Fig. 8 illustrates the set-point tracking responses as well as the regulatory control responses of the closed-loop system, in the case of a step load disturbance L=1 at time t=30. Obviously, Method I based on Algorithm I with τ_D=1.25$min\tau_{D,maxGM,I\text{-}P}(\tau_Z,d)$ provides a better performance, as compared to Method I relying on Algorithm I with τ_D=min$\tau_{D,maxGM,I\text{-}P}(\tau_Z,d)$. Both methods give a better performance, in terms of overshoot, maximum error and settling time, as compared to the method reported in the work by (Luyben, 2003). They also provide less overshoot in terms of set-point tracking as compared to the method reported in the work by (Gu et al, 2006). The method reported in the work by (Shamsuzzoha & Lee, 2006) gives the better overall performance, in the present case. However, the proposed method shows a better performance, as compared to known tuning methods, in terms of the initial jump in the closed-loop response.

We next apply Algorithm II of Method I in order to obtain admissible I-PD controller settings in the case where ζ_{des}=0.6. Algorithm II has been initialized with $\tau_{D,init}$=0.3807 and $\tau_{I,init}$=1.3296 and yields the controller parameters $\bar{K}_P$=2.1071, $\bar{K}_I$=0.6521 and $\bar{K}_D$=1.504. The set-point tracking responses as well as the regulatory control responses of the closed-loop system are illustrated in Fig. 9. The proposed method outperforms the method reported by (Luyben, 2003), and it is comparable to that reported by (Gu et al, 2006). Our method shows a better performance in terms of initial jump. The method by (Shamsuzzoha & Lee, 2006) shows, once again, a better performance in terms of overshoot and settling time in case of set-point tracking, while it performs better in terms of maximum error in the case of regulatory control.

The above conclusions hold also in the case where Algorithm III of Method I is applied in order to obtain ζ_{des}=0.6 and, simultaneously, to minimize the integral criterion of the form (13) in case of disturbance rejection. Algorithm III is initialized here $\tau_{D,init}$=0.3807, and $\tau_{I,init}$=1.3296, and provides the controller parameters $\bar{K}_P$=2.1069, $\bar{K}_I$=0.6426 and $\bar{K}_D$=1.5732. Fig. 10 illustrates the closed-loop set-point tracking and regulator control responses for the various methods applied to control the process.

Let us now apply Method II reported in Section 5 to the above IPDT-IR process model of the boiler steam drum and perform a comparison with known methods of tuning PID controllers. To this end, suppose first that the design specification is given in terms of the damping ratio of the closed-loop system, which is selected here as $\xi_{des}=\sqrt{2}/2$=0.707. In this case, application of Method II gives the I-PD controller parameters $\bar{K}_P$=3.5624, $\bar{K}_I$=1.363 and $\bar{K}_D$=2.2447. An alternative design is obtained by applying Method II with the specification of minimizing the integral criterion (13) in case of disturbance rejection (ISENCOD-L) criterion. In this case, the obtained controller parameters are $\bar{K}_P$=3.352, $\bar{K}_I$=1.2034 and $\bar{K}_D$=2.2009. Figs. 11 and 12 illustrate the closed-loop set-point tracking as well as the regulatory control responses obtained from the application of above I-PD controllers to the boiler steam drum process, together with those obtained from the application of known tuning methods for conventional and appropriately modified PID controllers. From these figures, it becomes obvious that the proposed Method II gives a very smooth response and outperforms most existing tuning methods in terms of overshoot,

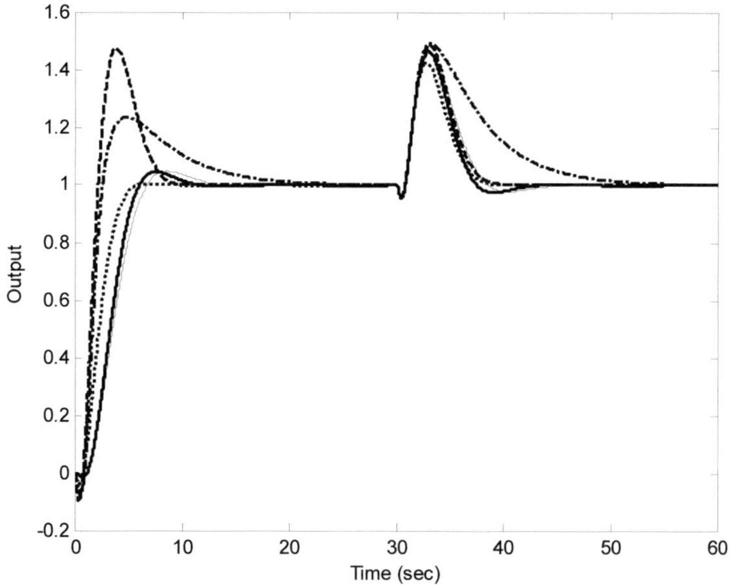

Fig. 8. Set-point tracking and regulatory control closed-loop responses of the boiler steam drum in the case where Algorithm I of Method I is applied to tune the I-PD controller. Solid-thick line: Algorithm I of Method I, with $\tau_D = \min \tau_{D,maxGM,I\text{-}P}(\tau_Z, d)$ and $\zeta_{des}=0.6$. Solid-thin line: Algorithm I of Method I, with $\tau_D = 1.25 \min \tau_{D,maxGM,I\text{-}P}(\tau_Z, d)$ and $\zeta_{des}=0.6$. Dash line: Method of Gu et al (2006). Dash-dot line: Method of Luyben (2003). Dot line: Method of Shamsuzzoha & Lee (2006).

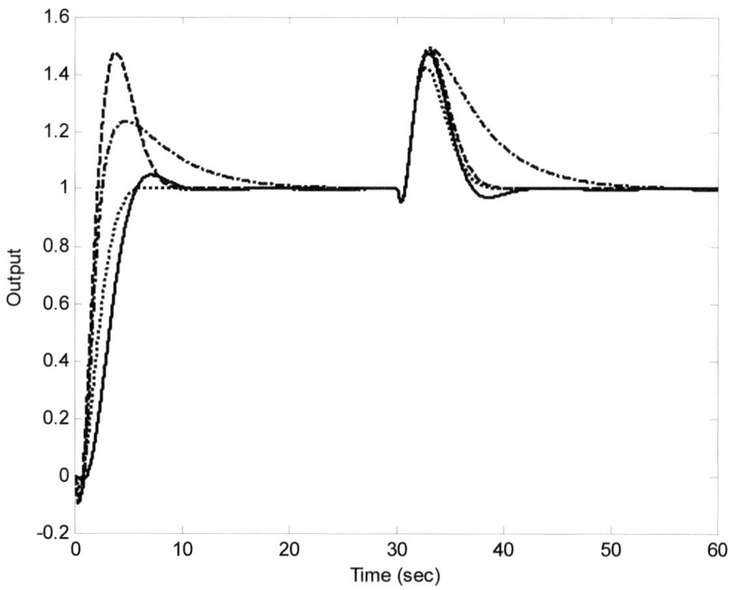

Fig. 9. Set-point tracking and regulatory control responses of the closed-loop system, when Algorithm II of Method I is applied for $\zeta_{des}=0.6$. Solid line: Proposed method. Other legend as in Fig. 8.

settling time, maximum error (in case of regulatory control) as well as initial jump. Finally, it provides the same set-point tracking capabilities as the method reported in the work (Shamsuzzoha & Lee, 2006), through the design of a simpler three-term controller structure.

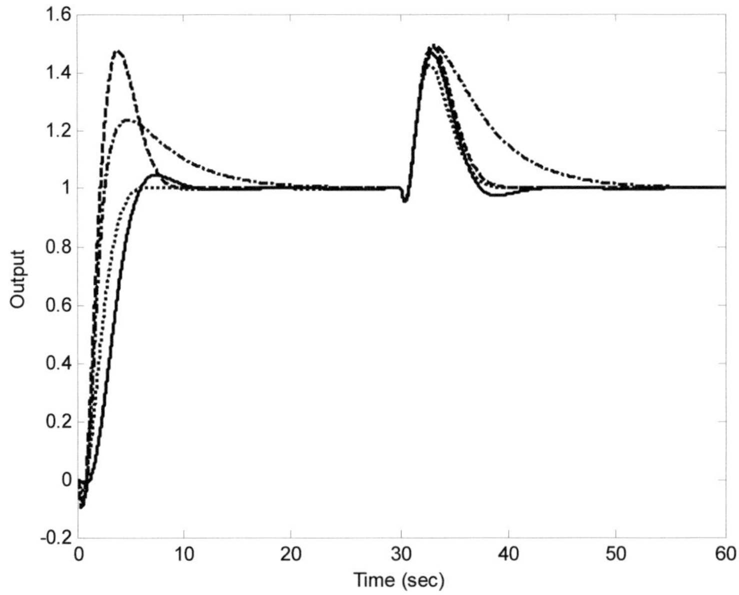

Fig. 10. Closed-loop set-point tracking and regulatory control responses, when Algorithm III is applied to minimize (13). Solid line: Proposed method. Other legend as in Fig. 8.

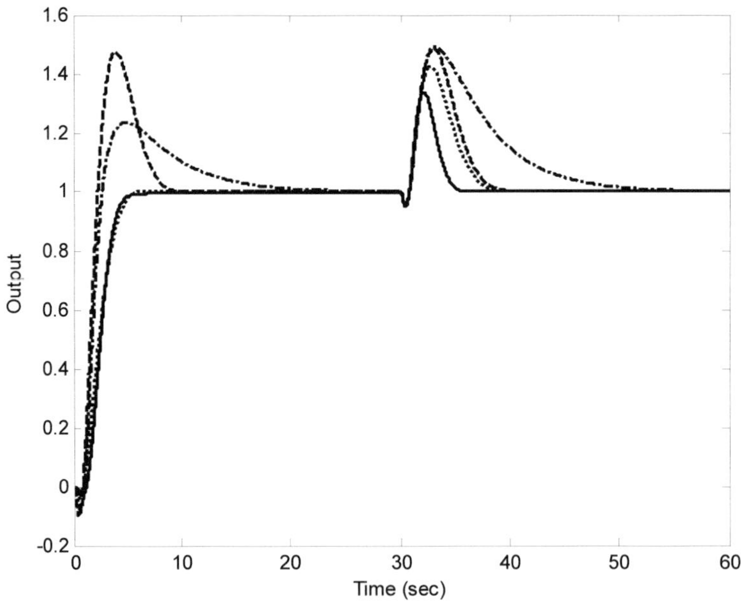

Fig. 11. Closed-loop set-point tracking and regulatory control responses, when Method II is applied with ξ_{des}=0.707. Solid line: Proposed method. Other legend as in Fig. 8.

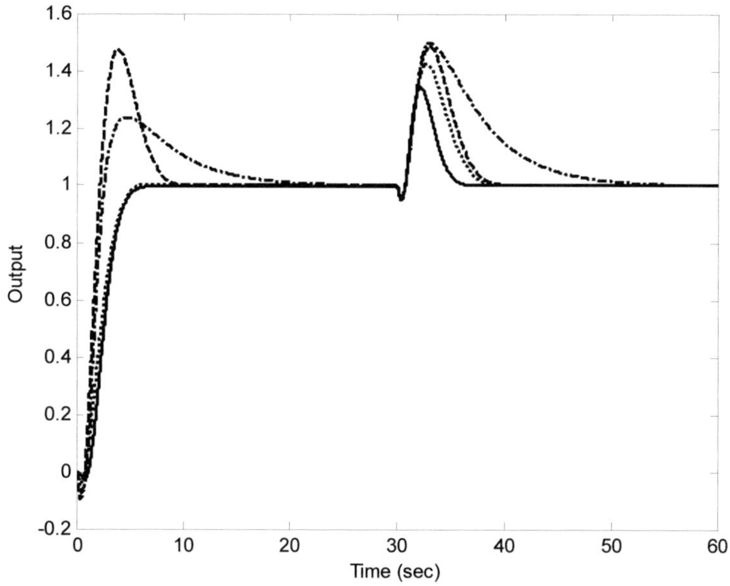

Fig. 12. Closed-loop set-point tracking and regulatory control responses, when Method II is applied in order to minimize (13). Solid line: Proposed method. Other legend as in Fig. 8.

7. Conclusions

In this work, a variety of new methods of tuning three-term controllers for integrating dead-time processes with inverse response have been presented. These methods can be classified in two main categories: (a) methods that guarantee the maximum phase margin specification in the frequency domain together with some desired specification in the time domain, and (b) methods based on direct synthesis. With regard to the first class of tuning methods, in order to obtain the optimal controller settings, iterative algorithms are used. In addition, several accurate approximations of these settings, useful for on-line tuning, are derived as functions of the process parameters. On the other hand, the proposed method based on direct synthesis provides explicit relations of the three-term controller settings in terms of the process parameters and of adjustable parameters that can be appropriately assigned to obtain the desired closed-loop performance. The performance of the above tuning methods is tested through their simulation application on a typical IPDT-IR process model of a boiler steam drum. Several successful comparisons of the proposed methods with existing tuning formulas for conventional, IMC-based, and two-degrees-of freedom PID controllers are also reported. From these comparisons, on can readily conclude that the proposed direct synthesis method outperforms existing tuning methods, while the methods based on the analysis of the phase margin of the closed-loop system provide a performance comparable to that of most PID controller tuning methods reported in the extant literature.

8. References

Arvanitis, K.G.; Pasgianos, G.D. & Kalogeropoulos, G. (2005). New Simple Methods of Tuning Three-Term Controllers for Dead-Time Processes. *WSEAS Transactions on Systems*, Vol.4, pp. 1143-1162.

Arvanitis, K.G.; Pasgianos, G.D., Boglou, A.K & Bekiaris-Liberis, N.K (2009a). A New Method of Tuning Three-Term Controllers for Dead-Time Processes with a Negative/Positive Zero. *Proceedings of the 6th International Conference on Informatics in Control, Automation and Robotics (ICINCO '09)*, pp. 74-83, Milan, Italy, July 2-5, 2009.

Arvanitis, K.G.; Boglou, A.K; Bekiaris-Liberis, N.K & Pasgianos, G.D. (2009b). A Simple Method of Tuning Three-Term Controllers for Integrating Dead-Time Processes with Inverse Response. *Proceedings of 2009 European Control Conference (ECC'09)*, pp. 4157-4162, Budapest, Hungary, August 23-26, 2009.

Chen, P.-Y.; Tang, Y.-C.; Zhang, Q.-Z. & Zhang, W.-D. (2005). A New Design Method of PID Controller for Inverse Response Processes with Dead Time. *Proceedings 2005 IEEE Conference on Industrial Technology (ICIT 2005)*, pp. 1036-1039, Hong Kong, China, December 14-17, 2005.

Chen, P.-Y.; Zhang, W.-D. & Zhu, L.-Y. (2006). Design and Tuning Method of PID Controller for a Class of Inverse Response Processes. *Proceedings of 2006 American Control Conference*, pp. 274-279, Minneapolis, Minnesota, U.S.A., June 14-16, 2006.

Chien, I.-L.; Chung, Y.-C.; Chen B.-S. & Chuang, C.-Y. (2003). Simple PID Controller Tuning Method for Processes with Inverse Response plus Dead Time or Large Overshoot Response plus Dead Time. *Industrial Engineering Chemistry Research*, Vol.42, pp. 4461-4477.

Gu, D.; Ou, L.; Wang, P. & Zhang, W. (2006). Relay Feedback Auto Tuning Method for Integrating Processes with Inverse Response and Time Delay. *Industrial Engineering Chemistry Research*, Vol.45, pp. 3119- 3132.

Luyben, W.L. (2000). Tuning Proportional-Integral Controllers for Processes with Both Inverse Response and Dead-Time. *Industrial Engineering Chemistry Research*, Vol.39, pp. 973-976.

Luyben, W.L. (2003). Identification and Tuning of Integrating Processes with Dead Time and Inverse Response. *Industrial Engineering Chemistry Research*, Vol.42, pp. 3030-3035.

Padma Sree, R. & Chidambaram, M. (2004). Simple Method of Calculating Set Point Weighting Parameter for Unstable Systems with a Zero. *Computers and Chemical Engineering*. Vol.28, pp. 2433-2437.

Paraskevopoulos, P.N.; Pasgianos, G.D. & Arvanitis, K.G. (2004). New Tuning and Identification Methods for Unstable First Order plus Dead Time Processes Based on Pseudo-Derivative Feedback Control. *IEEE Transactions on Control Systems Technology*, Vol.12, No.3, pp. 455-464.

Phelan, R.M. (1978). *Automatic Control Systems*, Cornell University Press, New York, USA.

Scali, C. & Rachid, A. (1998). Analytical Design of Proportional-Integral-Derivative Controllers for Inverse Response Processes. *Industrial Engineering Chemistry Research*, Vol.37, pp. 1372-1379.

Shamsuzzoha, M. & Lee, M. (2006). Tuning of Integrating and Integrating Processes with Dead Time and Inverse Response. *Theories and Applications of Chemical Engineering*, Vol.12, No. 2, pp. 1482-1485.

Srivastava, A. & Verma, A.K. (2007). Identification of Integrating Processes with Dead Time and Inverse Response. *Industrial Engineering Chemistry Research*, Vol. 46, pp. 8270-8272.

Waller, K.V.T. & Nygardas, C.G. (1975). On Inverse Response in Process Control. *Industrial Engineering Chemical Fundamentals*. Vol. 4, pp. 221-223.

Wilton, S.R. (1999). Controller Tuning. *ISA Transactions*, Vol.38, pp. 157-170.

Zhang, W.; Xu, X. & Sun, Y. (2000). Quantitative Performance Design for Inverse-Response Processes. *Industrial Engineering Chemistry Research*, Vol.39, pp. 2056-2061.

3

Design of a Golf Putting Pneumatic Mechanism Integer *vs* Fractional Order *PID*

Micael S. Couceiro, Carlos M. Figueiredo, Gonçalo Dias, Sara M. Machado and Nuno M. F. Ferreira
RoboCorp, Department of Electrical Engineering,
Engineering Institute of Coimbra (ISEC),
Faculty of Sport Sciences and Physical Education (CIDAF),
University of Coimbra,
Portugal

1. Introduction

According to Dave Pelz, one of the foremost short game and putting instructors in golf, the putting technique, or simply the putt, is defined as a light golf stroke made on the putting green in an effort to place the ball into the hole (Pelz, 2000). Hence, the putt is used in short distance shots on or near the green, as seen in Fig. 1. Similarly, putter may refer to a golf club used in the putting stroke.

The golf putting is an important aspect of golf because it can greatly affect a player's game performance and overall score. In the last years, an increasing number of researchers have been studying this gesture in order to understand its biomechanical characteristics (Pelz, 2000; Hume et al., 2005). However, the relative importance of the phases that describes the putt (Fig. 1) shows some inconsistencies (Pelz & Mastroni, 1989; Pelz, 2000).

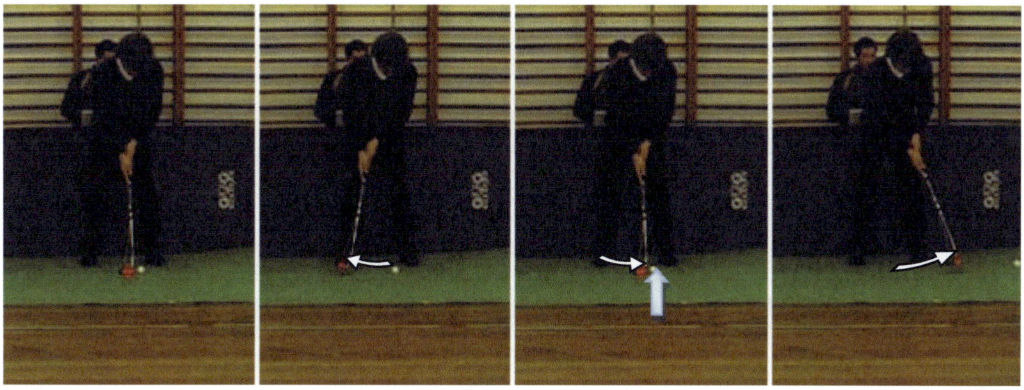

a) b) c) d)

Fig. 1. Phases of the putting: a) Initial stage; b) Backswing; c) Downswing and ball impact; d) Follow-through.

For instance, most golf experts consider that the key to a successful putt is in the power of the follow-through (Pelz, 2000). For instance, James Braid (1907), five time winner of *The Open Championship*, highlights this viewpoint by saying that "*the success of the drive is not only made by what has gone before, but it is also due largely to the course taken by the club after the ball has been hit*". However, the importance of the follow-through may be only an indication that the first part of the stroke (*i.e.*, backswing and downswing) was well played.

Also, it is not clear if the vertical trajectory of the putter is relevant for the success of the putting. Being a pendulum-like movement it is known that when the putter reaches the ball (*i.e.*, angle of inclination of the putter near 90 degrees) the vertical velocity is zero or near zero. Instead of using a regular putter, can we say that we could obtain the same performance if applying the exact same force on the golf ball using, for instance, a snooker cue?

To fill the niche area which lies between classical engineering and sports science, researchers has been exploring a recently emerged field denoted as sports engineering. The main purpose behind this new field is to apply engineering principles to understand, modify or control human biological systems directly or indirectly involved in activities related with sports, designing and producing auxiliary tools, such as monitoring, diagnosis and training of the athlete.

Thus, in this book chapter, a novel testbed for evaluation of the golf putting is proposed. The developed putting mechanism consists on a pneumatic system that emulates the golf putting based on real reference data of expert golf players previously studied in (Dias et al., 2010). All the reference data was retrieved using a detection technique to track the putter's head and an estimation technique to obtain the kinematical model of each trial which was further explained in (Couceiro et al., 2010a).

Though pneumatic actuators are often employed in industrial automation for reasons related to their good power/weight ratio, easy maintenance and assembly operations, clean operating conditions and low cost, it is not easy to control them, due to the nonlinearities. The presence of the air along with its natural compressibility introduces complexities such as friction forces, losses and time delays in the cylinder and transmission lines (Richer & Hurmuzlu, 2001).

The pneumatic servo-system is a very nonlinear, time-variant, control system because of the compressibility of air, the friction forces between the piston and the cylinder, air mass flow rate through the servo-valve and many other effects caused by the high nonlinearity of pneumatic systems. Furthermore recent improvements of digital technologies have opened new scenarios about pneumatic systems. In particular the use of the Pulse Width Modulation (*PWM*) technique is particularly attractive considering the possible use of cheap on/off valves driven by a *PWM*.

Nevertheless, the complexity of designing a controller for a system involving a complex dynamic behaviour such as a pneumatic actuator needs to be robust and efficient (Shen et al., 2006). To that end, both integer and fractional order Proportional-Integral-Derivative (*PID*) controllers will be studied and implemented on the open-source electronics prototyping platform *Arduino* which will be used to control the pneumatic device.

The controllers' gains will be obtained using the Particle Swarm Optimization (*PSO*) technique in order to achieve the minimum Integral Time Absolute Error (*ITAE*) when the pneumatic putter follows a desired trajectory. The optimization process will be accomplished using a *MatLab* script that iteratively calculates the *ITAE* between the desired putter's trajectory sent to the *Arduino* board over *USB* and the real trajectory performed by the pneumatic putter sent back from the *Arduino* board to the computer.

Bearing these ideas in mind, this book chapter is organized as follows. Section 2 presents the state-of-the-art of several experimental devices used in sport context while the proposed golf putting mechanism based on a pneumatic cylinder is described in section 3. The control architecture and optimization methodology is presented in section 4. Section 5 presents the evaluation of the putting mechanism. Section 6 outlines the main conclusions.

2. Related work

In the last few years, several devices have been developed to improve athlete's performance and to reveal particular features of a given sport. Many sport such as tennis (Salansky, 1994), table tennis (Lu, 1996) and baseball (Rizzo & Rizzo, 2001) were the first ones having their own training machine. However, more recently, many other sports had benefited from such devices.

Therefore, this section presents a selection of several mechanisms used in sport context mainly focusing on the design and controller characteristics. Furthermore, the state-of-the-art of experimental devices used to replicate the putting is thoroughly presented and discussed.

2.1 Sport devices

Similarly to tennis, badminton requires a high level of footwork and speed. One of the training machines used in badminton is called the Automatic shuttle feeder (*ASF*) (Kjeldsen, 2009). *ASF* can feed all over the court fulfilling the technical training as well as the physical and reaction training. As the presented work, *ASF*'s uses compressed air, thus requiring an external compressor to supply compressor air for the machine.

Nevertheless, just like in tennis, spring-like strategies have also been explored in badminton (Yousif & Kok, 2011). The springs are the source of the force which is controlled by an AT89S51 microcontroller. However, just like most high-speed shooting mechanisms, controller strategies are not considered and basically consist on a common launcher.

The development of a cricket bowling machine is presented in (Roy et al., 2006). The machine transfers the kinetic energy to the ball by frictional gripping between two rotating wheels whose speed is controlled by varying the analog voltages generated through a micro-controller 89C51 and associated peripherals. The authors claim that the machine is portable and low cost. However, it weights around 34Kg having 2 meters height with a cost of approximately 1200€.

The authors in (Kasaei et al., 2010) present a solenoid based multi power kicking system that enables loop and varies shooting power. The device takes use of a solenoid system to control the shooting power applying Pulse Width Modulation (*PWM*) on the pulse source in the control circuit. However, experimental results do not depict the precision or accuracy of the shooting mechanism.

Actually, devices that recreate high-speed or high-power shooting mechanism lacks on precision and accuracy since researchers have paid little or no attention to control architectures. However, slower gestures, such as the putting, have been objects of study in the fields of robotics and sports engineering, thus highlighting control techniques, sensory systems and ecological validity.

2.2 Putting devices

This section presents the state-of-the-art of experimental devices that were used to replicate the putting in field and laboratory context, *i.e.*, in real teaching and learning situations. Therefore, several papers are presented focusing the area of robotics in agreement with some assumptions underlying the systems of human movement, which studied the putting through the implementation of robotic arms and other mechanisms. In addition, as a multidisciplinary approach, this section aims to describe the advantages and disadvantages of these devices when compared to the one proposed in this paper, thus highlighting their contribution, while maintaining the scientific validity of the ecological execution of the putting.

Analysing the literature, Webster and Wei (1992) presented a robot vision golfing system *ARNIE P* (Automated Robotic Navigational unit with Intelligent Eye and Putter) that uses a *3D* tracking system to analyze the putting in the laboratory context. The presented mechanism is described by a good hand-eye coordination and intelligent sensor feedback. The robot is able to store and retain the location of the ball from two separate cameras during the time interval between the golf ball initially crossing a trigger scan line and the ball coming to a complete stop. Operationally, the robot used in this study presents a human-like gesture, taking into account that it can execute the movement while performing automatic tracking using *3D* acquisition software. However, its main limitation resides in the complexity inherent to robot programming in a cartesian coordinate motion to putt the ball effectively (swing the club). Furthermore, the use of such a complex system (*e.g.*, binocular stereo vision, robot arm motion, heuristic feedback, learning) is not fully effective to carry out the technical gesture, since this study essentially relies on artificial intelligence techniques and robotics, thus ignoring ecological validity. In addition, results are inconsistent with human performance, especially in terms of putting at shorter distances (Pelz, 1990, 2000).

Khansari-Zadeh and Billard (2011) developed an industrial robot with a mechanical arm with six *dof* denoted as Katana-T. Through the Stable Estimator of Dynamical Systems (*SEDS*) and using regression techniques, e.g., Gaussian Process Regression (*GPR*) or Gaussian Mixture Regression (*GMR*), it was possible to collect data on the robotic arm performance while executing the putting. From the kinematic point of view (*i.e.*, biomechanics of human movement), the main contributions of this work show that the putting is a complex task that is under the influence of different disorders that can be studied in various trajectories and ball positioning on the green. This aspect is reinforced by the same authors from a dynamical system perspective as an open phenomenon making an analogy with human movement system. The main limitation of this work refers to the *dof* of the robotic arm, which, despite its originality and innovation, has difficulties in representing a pendulum-like motion that characterizes the putting, far from featuring the ecological validity of this gesture. For instance, comparing with the work of Pelz (2000) which describes a robot with several *dof* that reflects almost perfectly the motor execution of a human being (*e.g.*, pendulum motion, putting amplitude, velocity and acceleration), the study of Khansari-Zadeh and Billard (2011) fails in successfully representing the task.

Another work presented by Jabson et al. (2008) developed a robot that autonomously moves using two *DC* motors controlled by a remote *PC* using fuzzy logic. The *Autonomous Golf Playing Micro Robot* is equipped with a servomotor used to execute the putting. The microcontroller is used to process the information sent by the computer through radio-frequency (*RF*) communication, thus controlling the motors. The robot is equipped with a camera allowing it to play golf while avoiding obstacles. The wheels are attached at the back and a ball caster at the center in order to achieve optimum stability. This work is particularly interesting because of the developed vision system that detects the position of the robot, golf hole, golf boundaries, and the golf ball within the playing field on real time using color object recognition algorithm. A modified golf tournament between autonomous robot golf player and man operated robot golf player was conducted. Results obtained in this study show the accuracy and robustness of the autonomous robot in performing such task. However, the micro robot has a limited putting representation, taking into account its size and the functions it performs.

More recently, Mackenzie and Evans (2010) described a robot that performs the putting using a high-speed camera (*TOMI* device). This was used to measure the putter head speed and impact spot of putts executed by a live golfer. The authors stated that the putting robot generated identical putting strokes with known stroke paths and face angles at impact. This work allows the kinematical analysis of the putting and the influence of key kinematic errors. However, the authors do not go beyond the biomechanical analysis, which limits their scientific approach when compared to similar experimental devices.

For instance, researchers such as Linda and Crick (2003) presented an autonomous robot which includes a *PID* feedback control system associated to a wireless communication mechanism. This robot autonomously performs the putting using a digital control system to establish the pose of the robot. It is noteworthy that the inclusion of a *PID* control with shaft encoder sensors is a novel feature of this work. However, it is also true that the system "heavy" and complex in terms of information processing and synchronization of the several components.

Finally, Munasinghe, Lee, Usui and Egashira (2004) described a telerobotic testbed via a mechanical arm. A user-friendly operator interface and Synchronized Orientation Control (*SOC*) with multiple commands are one of the most innovative aspects of this study. A laser pointer is used to help remote operator in perceiving self-location and navigation, whereas orientation control has been completely automated and synchronized to the position commands of the teleoperator. This device is important to putting kinematical analysis, taking into account that it is very accurate. However, its main limitation is evident since this device offers small information about the process variables of motor execution (*e.g.*, putting amplitude, speed and acceleration).

The following section presents the development of a pneumatic putting mechanism that will further be compared to real data obtained by real expert golf players.

3. Experimental setup

We emphasize that the focus of this work will not be directly related with the analysis of the phases of the putting motion presented in Fig. 1 (*e.g.*, backswing, downswing, ball impact and follow-through) (Pelz, 2000). However, the proposed mechanism will allow reproducing the phases of the horizontal component of the putting.

The putting mechanism consists on a pneumatic actuator CE1B32-200 equipped with a putter's head from a Putter Jumbo Black Beauty (Fig. 2). As an important driving element, the pneumatic cylinder is widely used in industrial applications for many automation purposes thanks to their variety of advantages.

The schematic of the putting mechanism is depicted in Fig. 3. The system consists of air supply, pneumatic cylinder (SC) with encoder (CE), pressure sensors (P_X), limit switches (FC_X), electro-valves (VC_X), Interface board (IB) and a *Arduino* board (μC) (which consists of an 8-bit microcontroller with A/D and D/A converters, external interrupts and other features) connected to a main computer, $x = \{1, 2\}$.

The scale cylinder (SC) is equipped with two electro-pneumatic proportional valves VER2000-03F (VC_1 e VC_2) which allows controlling the piston position with a Pulse Width Modulation (*PWM*) current signal used to represent an analog current value. VC_1 controls the cylinder to move forward (*i.e.*, downswing, ball impact and follow-through) while VC_2 controls the cylinder to move backward (*i.e.*, backswing).

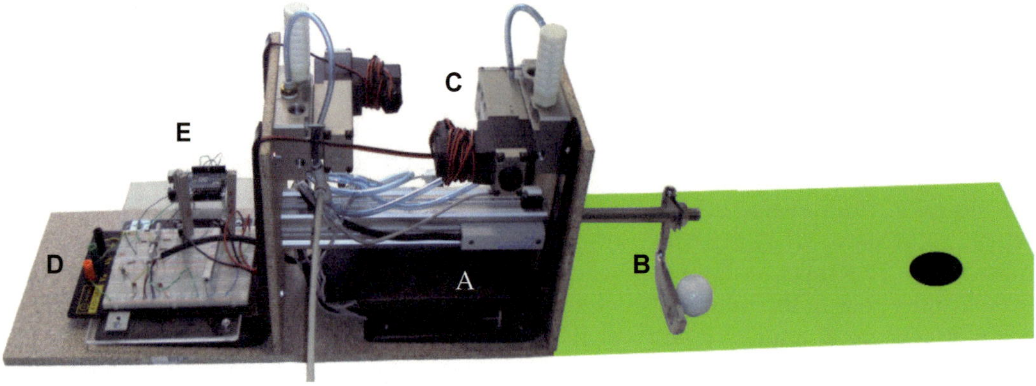

Fig. 2. Putting Mechanism: A - High Precision Scale Cylinder CE1B32-200 with encoder; B - Putter's head from Putter Jumbo Black Beauty; C - Electro-Valves VER2000-03F; D - Interface Board; E - *Arduino* Control Board.

As an important driving element, the pneumatic cylinder is widely used in industrial applications for many automation purposes thanks to their variety of advantages. In fact it is simple and clean, has low cost, high speed, high power to weight ratio, it is easy to maintain and has inherently compliance.

The schematic representation of the Interface board (IB) from Fig. 3 shows that the *PWM* current signal from the Interface board is proportional to the *PWM* voltage signal from the *Arduino* board (μC). A voltage value between 0v and 5v from the *Arduino* corresponds to a current value between 0A and 1A from the Interface board.

The duty cycle of the *PWM* voltage signal of the *Arduino* board (µC) will be the controller output $u(t)$ while the error $e(t)$ will be the difference between the reference trajectory (*i.e.*, controller input) sent by the computer and the real trajectory of the pneumatic putter provided by the encoder (CE).

The encoder *MODEL CE1 from MONOSASHI-KUN* have a resolution of 0.1mm/impulse, an accuracy of ± 0.05mm, with an open collector output of 12V of two impulses 90 degrees out of phase. This is an incremental rotary encoder with a magnetic resistance element which function is to provide the relative position of the piston rod (Fig. 4).

When the sensor passes through the magnetic section, it presents an output described by a 2-phase signal of sine and cosine by the piston rod movement. For this waveform, 1 pitch (0.8 mm) is equal to one cycle. This signal is then amplified and divided into 1/8. As a result, 90 degrees phase difference pulse signal is output. This signal is represented by two impulses (phase A and phase B) and works like an ordinary incremental rotary encoder (quadrature encoder).

Since the output of the encoder has a logical high level of 12v voltage, the Interface board (IB) is once again used to convert this signal to a standard Transistor–Transistor Logic (*TTL*) of 5v to be compatible with the external interrupts of the *Arduino* board (µC).

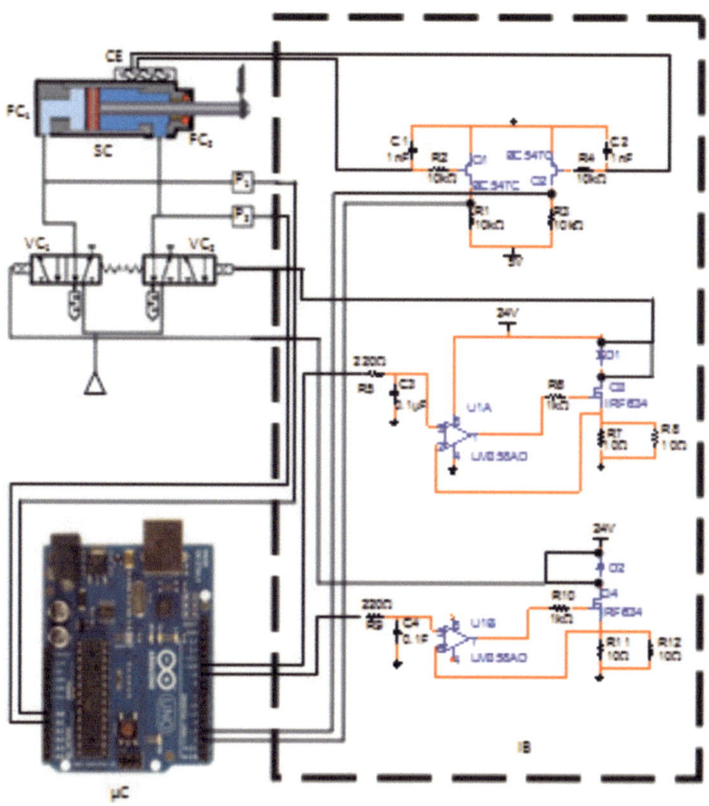

Fig. 3. Schematic of the Putting Mechanism.

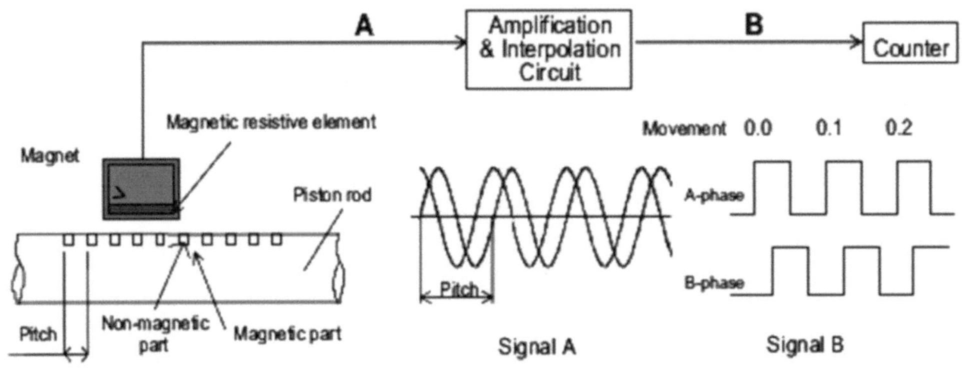

Fig. 4. Feedback System using encoders.

The hardware specifications of the developed putting mechanism are summarized in Table 1.

Action	*Double acting single rod (non-rotating piston)*
Fluid	*Air*
Operating pressure	*0.15MPa {1.5kgf/cm²} to 1.0MPa {10.2kgf/cm²}*
Maximum Putting speed	*3000 mm.s⁻¹*
Power supply	*24V DC (±10%) (Power supply ripple: 1% or less)*
Maximum Current consumption	*600 mA*
Encoder Resolution	*0.1mm/pulse ± 0.05mm*
Output signal	*A/B phase difference output*
Communication	*Serial RS-232*
Response time	*0.05 sec.*
Weight	*7 kg*
Dimensions (Length x Height x Width)	*634mm* x *325mm* x *258mm*

Table 1. Putting Mechanism Specifications.

Given the above, the proposed mechanism respects the ecological validity of putting performance with regard to the ball impact velocity which is very accurate. Furthermore, this mechanism allows executing consistent replications of the movement (Delay et al., 1997; Coello et al., 2000).

Therefore, this novel approach suggests that it will be possible to study the golf putting, thus revealing the mechanics of this gesture in field and laboratory context.

As next section shows, the proposed putting mechanism will benefit from fractional order controllers whose dynamic behavior is described thorough differential equations of non integer order.

4. Control architecture

Though pneumatic actuators are often employed in industrial automation for reasons related to their good power/weight ratio, easy maintenance and assembly operations, clean operating conditions and low cost, it is not easy to control them, due to the nonlinearities.

The presence of the air along with its natural compressibility introduces complexities such as friction forces, losses and time delays in the cylinder and transmission lines (Shearer, 1956; Richer & Hurmuzlu, 2001).

The pneumatic servo-system is a very nonlinear, time-variant, control system because of the compressibility of air, the friction forces between the piston and the cylinder, air mass flow rate through the servo-valve and many other effects caused by the high nonlinearity of pneumatic systems.

Furthermore recent improvements of digital technologies have opened new scenarios about pneumatic systems. In particular the use of the Pulse Width Modulation (*PWM*) technique is particularly attractive considering the possible use of cheap on/off valves driven by a *PWM*.

Nevertheless, the complexity of designing a controller for a system involving a complex dynamic behaviour such as a pneumatic actuator needs to be robust and efficient (Åström, 1980; Chien et al., 1993; Shen et al., 2006).

In this work, both classical (*aka*, integer order) and fractional order Proportional-Integral-Derivative (*PID*) controllers were compared while emulating the putting gesture in order to overcome the nonlinearities inherent to pneumatic systems.

4.1 Integer *PID* controller

In general, a classical *PID* controller, usually known as integer *PID* controller, takes as its inputs the error, or the difference, between the desired set point and the output. It then acts on the error such that a control output, u is generated. Gains K_p, K_i and K_d are the Proportional, Integral and Derivative gains used by the system to act on the error.

The Proportional Integral and Derivative PID control action can be expressed in time domain as:

$$u(t) = Ke(t) + \frac{K}{T_i}\int e(t)\,dt + KT_d\frac{de(t)}{dt} \tag{1}$$

Taking the Laplace transform yields:

$$G_c(s) = \frac{U(s)}{E(s)} = K\left(1 + \frac{1}{T_i s} + T_d s\right) \tag{2}$$

4.2 Fractional *PID* controller

Fractional order controllers are algorithms whose dynamic behaviour is described thorough differential equations of non integer order. Contrary to the classical *PID*, where we have three gains to adjust, the fractional *PID* (*aka*, $PI^{\lambda}D^{\mu}$) has five tuning parameters, including the derivative and the integral orders to improve de design flexibility (Couceiro et al., 2010c).

The mathematical definition of a derivative of fractional order a has been the subject of several different approaches. The Grünwald-Letnikov definition is perhaps the best suited for designing directly discrete time algorithms.

$$D^{\alpha}[x(t)] = L^{-1}\{s^{\alpha}X(s)\} \tag{3}$$

$$D^{\alpha}[x(t)] = \lim_{k \to 0}\left[\frac{1}{h^{\alpha}}\sum_{k=1}^{\infty}\frac{(-1)^{k}\Gamma(\alpha+1)}{\Gamma(k+1)\Gamma(\alpha-k+1)}x(t-kh)\right] \tag{4}$$

where Γ is the gamma function and h is the time increment. The implementation of the $PI^{\lambda}D^{\mu}$ is then given by:

$$G_c(s) = K\left(1 + \frac{1}{T_i s^{\lambda}} + T_d s^{\mu}\right) \tag{5}$$

we adopt a 4th-order discrete-time Pade approximation (a_i, b_i, c_i, $d_i \in \Re$, k = 4):

$$G_P[z] \approx K\left(\frac{a_0 z^k + a_1 z^{k-1} + \cdots + a_k}{b_0 z^k + b_1 z^{k-1} + \cdots + b_k}\right) \tag{6}$$

where K_P is the gain.

If both λ and μ are 1, the result is a classical *PID* (henceforth called integer *PID* as opposed to a fractional *PID*). If $\lambda = 0$ ($T_i = 0$) we obtain a PD^{μ} controller. All these types of controllers are particular cases of the $PI^{\lambda}D^{\mu}$ controller.

It can be expected that $PI^{\lambda}D^{\mu}$ controller may enhance the systems control performance due to more tuning knobs introduced. Actually, in theory, $PI^{\lambda}D^{\mu}$ itself is an infinite dimensional linear filter due to the fractional order in differentiator or integrator.

In order to implement this control methodology in *Arduino*'s 8-bit microcontrollers, an easy to use C library of the fractional order *PID* controller was fully developed for *Arduino* boards and can be found in (Couceiro, 2011). The library consists on a collection of functions but can be easily used as it follows:

$$u(t) = fopid\left(e(t), \frac{de(t)}{dt}, \int e(t)\,dt, K, KT_d, \frac{K}{T_i}, \mu, \lambda\right) \tag{7}$$

As previously stated, the controller output $u(t)$ will be directly related to the input signal (*i.e.*, duty cycle of the *PWM* wave) of the pneumatic cylinder, thus controlling its position. The methodology used to tune the proportional, derivative and integral gains of the controller, respectively denoted as K, KT_d and $\frac{K}{T_i}$, and the fractional derivative and integral parameters μ, λ is presented in next section.

4.3 Controller evaluation

For controller tuning techniques, we decided to use the Particle Swarm Optimization (*PSO*) since it is a very attractive technique among many other algorithms based on population, with only some few parameters to adjust (Couceiro et al., 2009; Tang et al., 2005; Pires et al., 2006, Alrashidi & El-Hawary, 2006).

The *PSO* was developed by Kennedy and Eberhart (Kennedy & Eberhard, 1995). This optimization technique, based on a population research, is inspired by the social behavior of birds. An analogy is established between a particle and an element of a swarm. These particles fly through the search space by following the current optimum particles. At each iteration of the algorithm, a movement of a particle is characterized by two vectors representing the current position x and velocity v (Fig. 5).

The velocity of a particle is changed according to the cognitive knowledge b (the best solution found so far by the particle) and the social knowledge g (the best solution found by the swarm). The weight of the knowledge acquired in the refresh rate is different according to the random values ϕ_i, $i = \{1, 2\}$. These values are a random factor that follow a uniform probability function $\phi_i \sim U[0,\phi_{i\,max}]$.

```
Initialize Swarm
repeat
    forall particles do
        Calculate fitness f
    end
    forall particles do
        v_{t+1} = I v_t + φ1(b-x) + φ2(g-x)
        x_{t+1} = x_t + v_{t+1}
    end
until stopping criteria
```

Fig. 5. *PSO* Algorithm.

where I and t are the inertia and the time of iteration, respectively.

In order to evaluate the control architecture, we can use performance criteria (fitness f) such as the Integral Time Absolute Error (*ITAE*) proposed by Graham and Lathrop (1953).

$$ITAE = \int_0^{\infty} t.|e(t)|\,dt \tag{8}$$

Minimizing the *ITAE* is commonly referred as a good performance metric in the design of *PID* controllers since it can be easily applied for different processes modelled by different process models (Seborg et al., 2004). Using the *PSO* to minimize the *ITAE* offer advantages since the search of controller parameters can be obtained for particular types of loads and set points changes faster than using different metrics and different optimization methods such as the Gradient Descent (Couceiro et al., 2009).

Next section presents the experimental results of how the gains of the $PI^{\lambda}D^{\mu}$ were tuned as well as some trials performed by our putting mechanism.

5. Experimental results

In order to analyze the controllers' performances, a real-time data acquisition program was designed in *MatLab* to capture the system output data through the communication interface between the *PC* and the *Arduino* controller.

Experimental results were divided in two stages: *i*) Optimization and comparison of the integer and fractional order *PID* controllers; and *ii*) Evaluation of the proposed Putting Mechanism while simulating, under the same conditions, a set of 30 trials previously performed by an expert subject when facing a ramp constraint.

In all experimental results the system pressure was set to 6 bar and the controllers were updated each time the external interrupts were activated, thus computing the time between pulses.

5.1 Controller optimization

In this section we compare the performances of the classical and fractional order *PID* controllers (Ferreira et al., 2002) (Couceiro et al., 2010b).

Therefore, the duty cycle of the *PWM* wave is set as the controller output of the putting device. The *PSO* was set with a population of 100 particles with $\phi_i = 1$ and $I = 0.9$. The stopping criteria considered was a maximum iteration number of 200.

In order to study the device response to velocity inputs, two separated rectangular pulses of 400 mm.s^{-1} (*i.e.*, low velocity) and 1500 mm.s^{-1} (*i.e.*, high velocity) were applied, thus considering common values of putting impact velocities (Delay et al., 1997; Coello et al., 2000).

Under the last conditions, the following *PID* and $PI^{\lambda}D^{\mu}$ controller parameters depicted in Table 2 were obtained as being the ones that minimizes the *ITAE*.

Figures 6 and 7 presents a trial obtained using both *PID* and $PI^{\lambda}D^{\mu}$ controllers for each condition. It is noteworthy that controllers performance improves at higher velocities since the pneumatic cylinder used in the proposed putting device usually works at velocities near 1000 mm.s^{-1}.

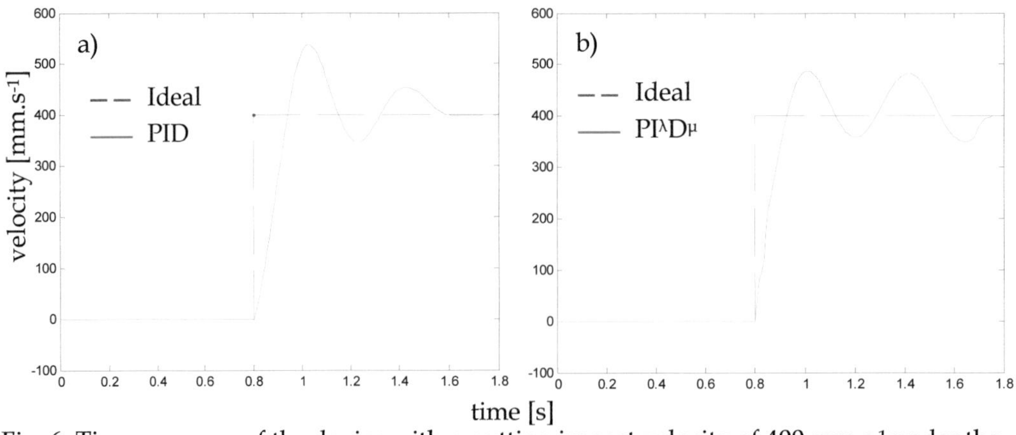

Fig. 6. Time response of the device with a putting impact velocity of 400 mm.s^{-1} under the action of the: a) *PID* controller; b) $PI^{\lambda}D^{\mu}$ controller.

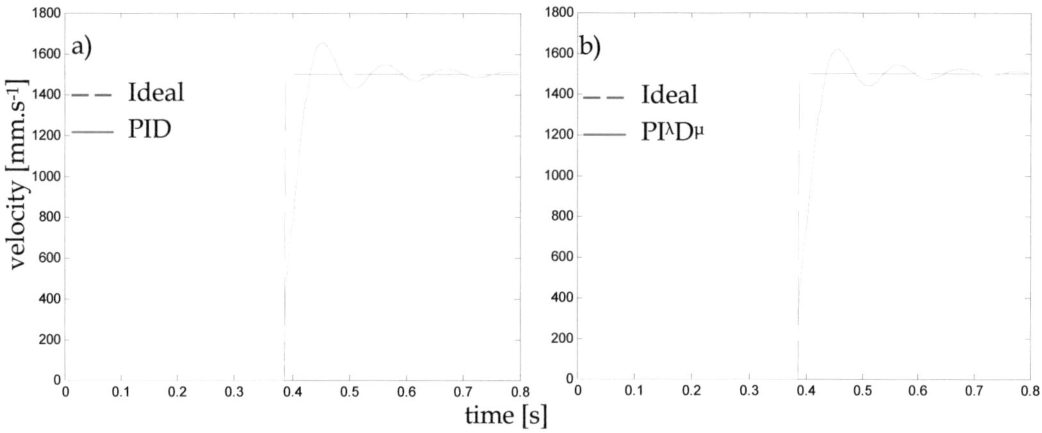

Fig. 7. Time response of the device with a putting impact velocity of 1500 mm.s^{-1} under the action of the: a) *PID* controller; b) $PI^{\lambda}D^{\mu}$ controller.

	K	K/T_i	KT_d	μ	λ
PID	0,040	0,004	0,020	-	-
$PI^{\lambda}D^{\mu}$	0,004	0,040	0,015	0,76	0,64

Table 2. *PID* and $PI^{\lambda}D^{\mu}$ controller parameters.

To analyze more clearly the dynamical response to the step perturbation, Table 3 compares the time response characteristics of the integer and the fractional *PID* controllers, namely the percent overshoot *PO*, the rise time t_r, the peak time t_p, the settling time t_s and the *ITAE*.

		PO [%]	t_r [s]	t_p [s]	t_s [s]	*ITAE*
400 mm.s^{-1}	*PID*	33,750	0,125	0,212	0,780	1,279x10^4
	$PI^{\lambda}D^{\mu}$	21,750	0,132	0,204	0,910	7,582x10^3
1500 mm.s^{-1}	*PID*	14,790	0,045	0,085	0,325	5,300x10^3
	$PI\lambda D\mu$	13,730	0,045	0,070	0,325	2,830x10^3

Table 3. Time response parameters of the device under the action of the *PID* and $PI^{\lambda}D^{\mu}$ controllers.

Table 3 shows that, generally, the fractional order controller leads to a reduction of the overshoot, the peak time and the *ITAE*. However, it should be noted that the fractional order *PID* increases the computational cost of the microcontroller. While an iteration of the *PID* can easily run at each external interruption between two pulses at the maximum putting velocity of 3000 mm.s^{-1}, the developed $PI^{\lambda}D^{\mu}$ *Arduino* library sometimes loses pulses. The rise time t_r (*i.e.*, time required by the putting mechanism to reach the specified velocity) may be a consequence of this problem. Nevertheless, and since the putting mechanism benefit from a higher encoder resolution, the computational cost imposed by the $PI^{\lambda}D^{\mu}$ *Arduino* library does not jeopardize the performance of a given putting execution.

5.2 Evaluation of the putting mechanism

This section presents the accuracy of the putting device comparing it with real data obtained from 30 trials performed by an expert golf player with a handicap of 5.

In order to allow a straightforward comparison with the golf player, the mechanism was deployed in an artificial green to hit the ball two meters away from the hole (Fig. 8). The reference trajectory performed by the golf player at each trial was sent to the microcontroller through serial communication[1].

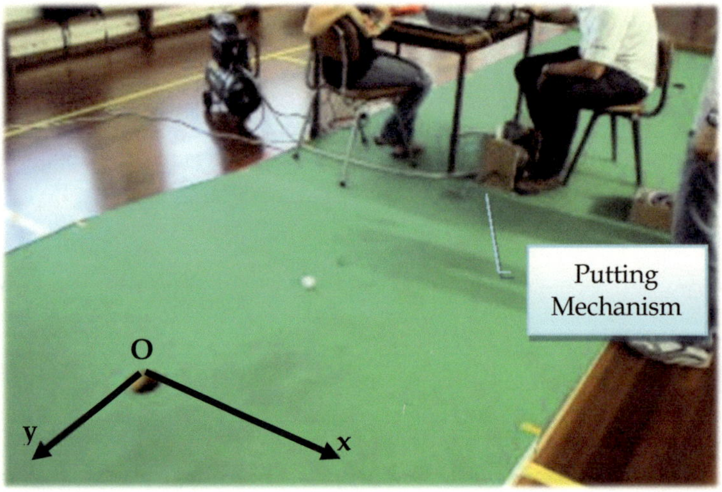

Fig. 8. Experimental setup to evaluate the putting mechanism in an artificial green with ramp.

The analysis of a set of trials is not directly accessible and need a graphical or geometrical representation. To analyze the radial error, which may be calculated using the lateral error (x-axis) and longitudinal error (y-axis) within the sport context, one of the most common representations is the error ellipse. The error ellipse allows a two-dimensional graphical analysis representing the influence of the lateral and longitudinal error (*i.e.*, accuracy) and the variability (*i.e.*, precision) of a given player (Mendes et al., 2011). By observing the shape, size and orientation of the ellipse, one can easily compare different players or, as it is presented in this book chapter, compare a man with a machine. Figure 9 depicts the error ellipse of both the golf player and the putting mechanism.

As it can be observed, there is a high similarity in the shape of both ellipses. It is noteworthy that the golf player was more accurate than the developed mechanism since it only missed 4 trials (accuracy of 86,67%) against 11 missed trials from the device (accuracy of 63,33%). However, the area of the ellipse for both the player and the putting mechanism was 2,2052 m^2 and 1,7536 m^2, respectively. This means that, despite the higher accuracy of the human player, the device was more precise thus presenting a lower variability.

[1] A video of the experiments is available at http://www2.isec.pt/~robocorp/research/putting/

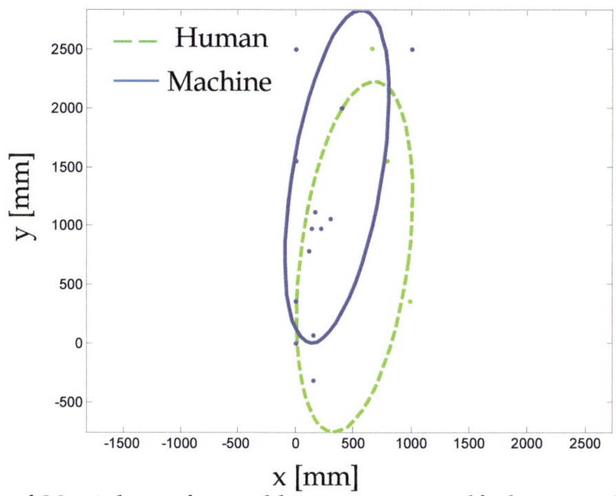

Fig. 9. Error Ellipse of 30 trials performed by an expert golf player and subsequently emulated by the putting mechanism.

6. Conclusion

This book chapter presented an experimental device to evaluate the golf putting. The proposed mechanism is similar to other mechanisms and robotic devices presented in the literature (*cf.* Related Work section), and it is the authors' opinion, that the great advantages of other mechanisms are their mobility and sensory system (*e.g.*, vision). However, since the scope of this work consists on executing the putting as an isolated movement to unveil the process and product variables, the proposed solution can be characterized by having a small size (*i.e.*, easy to transport and apply in any situation) and high reliability (*i.e.*, capable of emulating real kinematical data obtained by expert golf players). Despite these advantages, like other experimental devices, although this mechanism can simulate the putting, it can hardly represent unequivocally the motor performance of a human being, because, as expected, each individual has different characteristics and profiles that represents a "putting signature" distinct from subject to subject, which, may not be fully replicated by a robot.

Hence, man versus machine analogy is inevitable and strides for the multi and interdisciplinary research that crosses knowledge of several research fields (*e.g.*, engineering, sport science and biomechanics) to meet the challenges in science. Thus, this work, more than just presenting a mechanism or experimental device that can replicate the putting gesture, it is worth for proposing a novel creative process that can serve as support for future researchers who wish to further study this movement.

We emphasize that already in 1940, Nicolai Bernstein, a Russian physiologist that studied the "mechanics" of the human upper limb, said that it is virtually impossible to replicate two motions exactly the same way. Since then, this researcher paved the way to study the human movement in a global perspective closer to the variability that characterizes the human movement systems (Bernstein, 1967). In addition, the "body machine" designed by Descartes can become a reality in the future, winning the "body of emotion" of Benedict Spinoza (1989, 1992), which, in a society increasingly dependent on robots, may become an inevitable Matrix, where the study of the body phenomenology reported by philosophers such as Plato, Aristotle and Socrates is worth another look (Merleau-Ponty, 1964). As referred by Gaya (2005), the study of the "contemporary body" in the age of technoscience aims the hybrid body to overcome all imperfection of the biological body.

In summary, this study, which analyzes the golf putting, *i.e.*, a gesture made by the human body in sports context, may be further studied in conjunction with other scientific areas, thus benefiting from their contributions, either in laboratory context or in real teaching and learning situations.

7. Acknowledgments

This work was supported by a PhD scholarship (SFRH/BD/73382/2010) granted to the first author by the Portuguese Foundation for Science and Technology (FCT). Also, this work was made possible by the support and assistance of Rui Mendes and Miguel Luz and for their cooperation, advice and encouragement of this research in the Education School of Coimbra (ESEC) and in RoboCorp at the Engineering Institute of Coimbra (ISEC).

8. References

Alrashidi, M. R. & El-Hawary, M. E. (2006). A Survey of Particle Swarm Optimization Applications in Power System Operations, *Electric Power Components and Systems*, 34:12. 1349 - 1357 Taylor & Francis.

Åström, K.J. (1980). *A Robust Sampled Regulator for Stable Systems with Monotone Step Responses*, Automation, Vol. 16, pp 313-315, Pergamon Press Ltd.

Bernstein, N. (1940). *Biodynamics of locomotion*, In H.T.A. Whiting (Eds.), Human motor actions: Bernstein reassessed (pp. 171- 222), Amsterdam: Elsevier.

Bernstein, N. (1967). *The coordination and regulation of movements*. Oxford. UK: Pergamon.

Braid, James (1907). *How to Play Golf*. Library of Congress.

Chien, C.L.; Chien, P.C. & Chien, C.K. (1993). *A Pneumatic Model-following Control System Using a Fuzzy Adaptive Controller*, Automation, Vol. 29, No. 4, pp 1101-1105, Pergamon Press Ltd.

Coello, Y.; Delay, D.; Nougier, V. & Orliaguet, J.P. (2000). Temporal control of impact movement: The "time from departure" control hypothesis in golf putting. *International Journal of Sport Psychology*, *31*, 1, 24-46.

Couceiro, Micael S.; Mendes, Rui; Ferreira, N. M. Fonseca & Machado, J. A. Tenreiro (2009). Control Optimization of a Robotic Bird, EWOMS '09, Lisboa, Portugal.

Couceiro, M.; Luz, J.M.; Figueiredo, C.M.; Ferreira, N.M.F. & Dias, G. (2010a). Parameter Estimation for a Mathematical Model of the Golf Putting, *WACI'10 - 5th Workshop Applications of Computational Intelligence* (pp: 1-8), Portugal, ISEC/IPC.

Couceiro, Micael S.; Ferreira, N. M. Fonseca & Machado, J. A. Tenreiro (2010b). *Dynamical Analysis and Fractional Order Control of a Dragonfly-Inspired Robot*, Intelligent Systems, Control and Automation - Science and Engineering Bookseries, Springer.

Couceiro, Micael S.; Ferreira, N. M. Fonseca & Machado, J. A. Tenreiro (2010c). Aplication of Fractional Algoritms in the Control of a Robotic Bird, *Journal of Comunications in Nonlinear Science and Numerical Simulation*-Special Issue, Elsevier.

Couceiro, Micael S. (2011). *FO-PID Library for Arduino*. Retrieved in 23 June 2011 at http://www2.isec.pt/~micael/contributions/FOPID/FOPID.zip.

Delay D.; Nougier, V.; Orliaguet, J.P. & Coelho. Y. (1997). Movement control in golf putting, *Human Movement Science, 16*, 5, 597-619.

Dias, G.; Luz, M.; Couceiro, M.; Figueiredo, C.; Fernandes, O.; Iglésias, P.; Castro, M.; Ferreira, N. & Mendes, R. (2010). Visual Detection and Estimation of Golf Putting, *Proceedings of the International Conference on Mathematical Methods* (pp 137-148), ISEC.IPC: Coimbra.

Ferreira, N. M. Fonseca; Barbosa, R. & Machado, J.A. Tenreiro (2002). Fractional- Order Position/Force Control of Mechanical Manipulators. Proceedings of CIFA'02, Conférence In-ternationale Francophone d´Automatique 8-10 July, Nantes, France.

Gaya, Adroaldo (2005). *Será o corpo humano obsoleto?*, Sociologias, Porto Alegre, ano 7, n. 13, p. 324-337.

Graham, D. & Lathrop, R.C. (1953). *The Synthesis of Optimum Response: Criteria and Standard Forms, Part 2*, Transactions of the AIEE 72, pp. 273-288.

Hume, P.A.; Keogh, J. & Reid, D. (2005). *The role of biomechanics in maximising distance and accuracy of golf shots*, Sports Medicine, 35, 5, 429-49.

Jabson N. G.; Leong, K.G. B.; Licarte, S. W.; Oblepias, G. M. S.; Palomado, E. M. J. & Dadios, E. P. (2008). The autonomous golf playing micro robot: with global vision and fuzzy logic controller, *International Journal on Smart Sensing and Intelligent Systems, 1*, 4, 824-841.

Kasaei, S.H.M., Kasaei, S.M.M. & Kasaei, S.A.M. (2010). Design and Implementation Solenoid Based Kicking Mechanism for Soccer Robot Applied in Robocup-MSL, *International Journal of Advanced Robotic Systems*, Vol. 7, No. 4, pp.73-80, ISSN 1729-8806.

Kennedy, J. & Eberhart, R. (1995). *A new optimizer using particle swarm theory*. Proceedings of the IEEE Sixth International Symposium on Micro Machine and Human Science, pp. 39-43.

Khansari-Zadeh, S.M. & Billard, A. (2011). Learning to Play Mini-Golf from Human Demonstration using Autonomous Dynamical Systems, *Proceedings of the 28 th International Conference on Machine Learning*, Bellevue, WA, USA.

Kjeldsen, Michael (2009). *Sport & Teknik develop badminton in the new way*, last visited on 08 August 2011, http://www.sport-teknik.se/Technical_facts.html.

Linda, G. B., & Crick, A.P. (2003). Control Education via Autonomous Robotics, *Proceedings of the 42nd IEEE Conference on Decision and Control,* Maui, Hawaii USA.

Lu, Tzu-Hao (1996).*Table tennis training system*, United States Patent 5533722.

Mackenzie, S.J., & Evans, D.B. (2010). Validity and reliability of a new method for measuring putting stroke kinematics using the TOMI1 system, *Journal of Sports Sciences, 8*, 1-9.

Mendes P.C; Martins, F.; Facas Vicente, A.M.; Corbi, F.; Couceiro, M. S.; Mendes, R. & Trovão, J. (2011). Análise às variáveis de produto na execução do serviço de ténis sob o efeito de um escoamento aerodinâmico induzido (vento artificial), *4.º Congresso Nacional de Biomecânica*, L Roseiro, M. Augusta et al (Eds), [ISBN 978-989-97161-0-0].

Merleau-Ponty, Maurice (1964). *L'oeil et l'esprit*, Paris: Gallimard.

Munasinghe, Sudath, R; Lee, Ju-Jang; Usui, Tatsumi; Nakamura, Masatoshi & Egashira, Naruto (2004). Telerobotic Mini-Golf: System Design for Enhanced Teleoperator Performance, *Proceedings of the 2004 IEEE International Conference on Robotics & Automation*, New Orleans, LA.

Pelz, D. & Mastroni, A. (1989). *Putt like the Pro´s. Englewood Cliffs.* NJ: Penguin Books.

Pelz, D. (1990). *The long putter.* The Pelz report, 1, 3, 1-5.

Pelz, D. (2000). *Putting Bible: The complete guide to mastering the green.* New York: Publication Doubleday.

Pires, E. J. Solteiro; Oliveira, P. B. de Moura; Machado, J. A. Tenreiro & Cunha, J. Boaventura (2006). Particle Swarm Optimization versus Genetic Algorithm in Manipulator Trajectory Planning, *The 7th Portuguese Conference on Automatic Control,* pp.230. 11-13 September, Lisbon, Portugal.

Richer, E. & Hurmuzlu, Y. (2001). A High Performance Pneumatic Force Actuator System - Part 1 - Nonlinear Mathematical Model, *ASME Journal of Dynamic Systems Measurement and Control,* Vol. 122, No.3, pp. 416-425.

Rizzo, Michael J. & Rizzo, Marlene J. (2001). *Softball/baseball training machine,* United States Patent 6305366.

Roy, S. S.; Karmakar S.; Mukherjee, N. P.; Nandy, U. & Datta, U. (2006). Design and development of indigenous cricket bowling machine, *Journal of Scientific and Industrial Research,* India, vol. 65; NUMB 2, pages 148-152.

Salansky, Werner (1994). *Ball throwing device for tennis balls,* United States Patent 5347975.

Seborg, D.E.; Edgar, T. F. & Mellichamp, D. A. (2004). *Process Dynamics and Control,* Wiley, New York.

Shearer, J. L. (1956). *Study of Pneumatic Processes in the Continuous Control of Motion With Compressed Air.* Transactions of the ASME.

Shen, X.; Zhang, J.; Barth E. J. & Goldfarb, M. (2006). Nonlinear Model Based Control of Pulse Width Modulated Pneumatic Servo Systems, *ASME Journal of Dynamic Systems, Measurement, and Control,* vol. 128, no. 3, pp. 663-669.

Spinoza, Benedict (1989). *Ética,* 4 ed. Trad. de Joaquim de Carvalho, Joaquim Ferreira Gomes e Antônio Simões. São Paulo, SP: Nova Cultural.

Spinoza, Benedict (1992). *Ética,* Lisboa: Relógio d'Água.

Tang J.; Zhu J. & Sun Z. (2005). A novel path panning approach based on appart and particle swarm optimization, *Proceedings of the 2nd International Symposium on Neural Networks,* In LNCS. Editor, volume 3498.

Webster, R.W. & Wei, Y. (1992). ARNIE P. - A Robot Golfing System Using Binocular Stereo Vision and a Heuristic Feedback Mechanism, *Proceedings of the 1992 IEEE/RJS International Conference on Intelligent Robots and Systems Raleigh,* NC.

Yousif, B. F. & Kok, Soon Yeh (2011). Badminton training machine with impact mechanism, *Journal of Engineering Science and Technology,* 6 (1). pp. 61-68. ISSN 1823-4690.

4

Design and Development of PID Controller-Based Active Suspension System for Automobiles

Senthilkumar Mouleeswaran
Department of Mechanical Engineering
PSG College of Technology
Coimbatore
India

1. Introduction

Suspension systems have been widely applied to vehicles, right from the horse-drawn carriage with flexible leaf springs fixed at the four corners, to the modern automobile with complex control algorithms. Every vehicle moving on the randomly profiled road is exposed to vibration which is harmful both for the passengers in terms of comfort and for the durability of the vehicle itself. Different disturbances occur when a vehicle leans over during cornering (rolling) and dives to the front during braking (pitching). Also, unpleasant vertical vibrations (bouncing) of the vehicle body can occur while driving over road irregularities. These dynamic motions do not only have an adverse effect on comfort but can also be unsafe, because the tyres might lose their grip on the road. Therefore the main task of a vehicle suspension is to ensure ride comfort and road holding for a variety of road conditions and vehicle maneuvers. This in turn would directly contribute to the safety of the user.

A typical suspension system used in automobiles is illustrated in Figure 1. In general, a good suspension should provide a comfortable ride and good handling within a reasonable range of deflection. Moreover, these criteria subjectively depend on the purpose of the vehicle. Sports cars usually have stiff, hard suspension with poor ride quality while luxury sedans have softer suspensions but with poor road handling capabilities. From a system design point of view, there are two main categories of disturbances on a vehicle, namely road and load disturbances. Road disturbances have the characteristics of large magnitude in low frequency (such as hills) and small magnitude in high frequency (such as road roughness). Load disturbances include the variation of loads induced by accelerating, braking and cornering. Therefore, in a good suspension design, importance is given to fairly reduce the disturbance to the outputs (e.g. vehicle height etc). A suspension system with proper cushioning needs to be "soft" against road disturbances and "hard" against load disturbances.

A heavily damped suspension will yield good vehicle handling, but also transfers much of the road input to the vehicle body, whereas a lightly damped suspension will yield a more comfortable ride, but would significantly reduce the stability of the vehicle at turns, lane change maneuvers, or during negotiating an exit ramp. Therefore, a suspension design is an art of compromise between these two goals. A good design of a passive suspension can work up to some extent with respect to optimized riding comfort and road holding ability, but cannot eliminate this compromise.

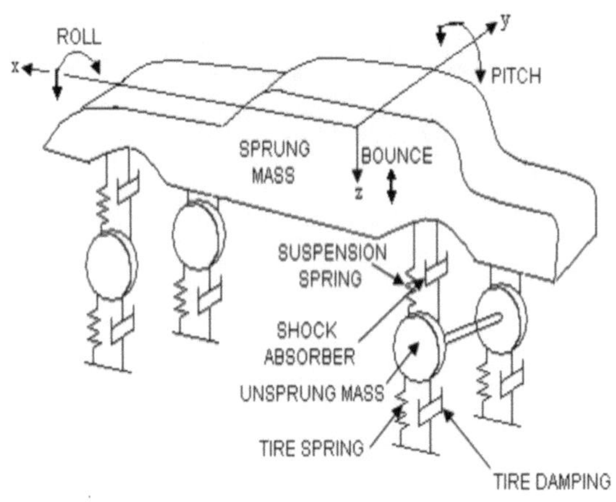

(a) Full car model

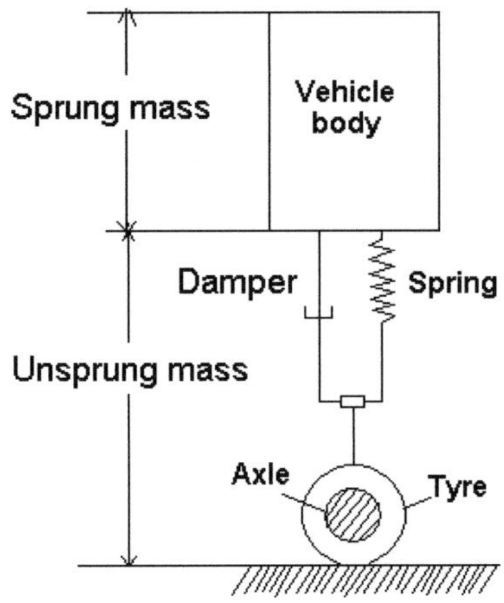

(b) Quarter car model

Fig. 1. Suspension system of a passenger car

The traditional engineering practice of designing a spring and a damper, are two separate functions that has been a compromise from its very inception in the early 1900's. Passive suspension design is a compromise between ride comfort and vehicle handling, as shown in Figure 2. In general, only a compromise between these two conflicting criteria can be obtained if the suspension is developed by using passive springs and dampers.. This also applies to modern wheel suspensions and therefore a break-through to build a safer and more comfortable car out of passive components is below expectation. The answer to this problem seems to be found only in the development of an active suspension system.

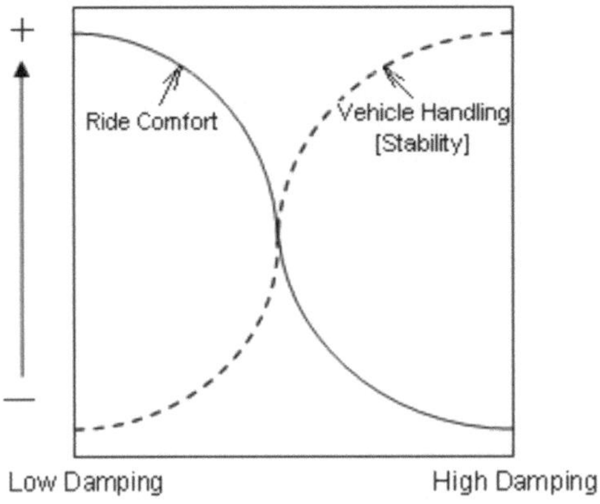

Fig. 2. Performance compromise of passive suspension system

In recent years, considerable interest has been generated in the use of active vehicle suspensions, which can overcome some of the limitations of the passive suspension systems. Demands for better ride comfort and controllability of road vehicles has motivated many automotive industries to consider the use of active suspensions. These electronically controlled active suspension systems can potentially improve the ride comfort as well as the road handling of the vehicle simultaneously. An active suspension system should be able to provide different behavioral characteristics depending upon various road conditions, and be able to do so without going beyond its travel limits.

Though the active suspension systems are superior in performance to passive suspension, their physical realization and implementation is generally complex and expensive, requiring sophisticated electronic operated sensors, actuators and controllers. Recent advances in adjustable dampers, springs, sensors and actuators have significantly contributed to the applicability of these systems. Consequently, the automobile has a better combination of ride and handling characteristics under various conditions, than cars with conventional suspension systems. Since electronic controlled suspension systems are more expensive than conventional suspension systems, they are typically found in luxury-class automobiles and high expensive sport utility vehicles. Therefore, a study has been made to develop an active suspension system for improved performance with less cost on light passenger vehicle.

Active suspension system is characterized by a built-in actuator, which can generate control forces to suppress the above-mentioned motions. In addition, the road holding has also been improved because of the dynamic behavior of the contact forces between the tyres and road.

Active vehicle suspensions have attracted a large number of researchers in the past few decades, and comprehensive surveys on related research are found in publications by (Elbeheiry et al, 1995), (Hedrick & Wormely 1975), (Sharp & Crolla 1987), (Karnopp 1995) and (Hrovat 1997). These review papers classify various suspension systems discussed in literature as passive, active (or fully active) and semi-active (SA) systems.

Some of potential benefits of active suspension were predicted decades ago by the first pioneer researchers. Indeed, the optimal control techniques that were launched with "Sputnik" and used in the aerospace industry since the 1950s and 1960s, have also been applied to the study of active suspensions, starting from about the same period by (Crossby & Karnopp 1973).

Fully active suspension system (FASS) is differentiated from semi-active suspension system(SASS) on the fact that it consists of a separate active force generator. The physical implementation of FASS is usually provided with a hydraulic actuator and power supply as shown in Figure 3. Fully active suspension systems have been designed by Wright and (Williams 1984) and (Purdy and Bulman 1993), which appear in formula one racing cars. Active suspension system has been compared with semi-active suspension system by (Karnopp 1992) and concluded that active suspensions have performance improvements, particularly in vehicle handling and control. In the process of enhancing passenger comfort and road handling, active suspensions introduce additional considerations of rattle space and power consumption, which must be factored into the overall design goals. While the ride/ handling tradeoff is prevalent in most approaches as pointed out by (Karnopp 1986), (Hrovat 1988) and (Elbeheiry 2000), the ride/ rattle space tradeoff is not explicitly addressed. Alternately, an active suspension system has been developed to improve ride comfort with rattle space limitations by (Jung-Shan Lin et al, 1995). But tire-road contact has not been studied.

However, not much literature is found on FASS as reviewed by (Pilbeam & Sharp 1996) and (Hrovat 1997). The advantage of FASS is that its bandwidth is more than that of SASS which is very much described by (Hrovat 1997). It has been shown that FASS requires considerable amount of energy to actuate and can be quite complex and bulky and therefore requires further stringent research. Recently, some research has been focused on the experimental development of the active suspension systems.The construction of an active suspension control of a one-wheel car model using fuzzy reasoning and a disturbance observer has been presented by (Yoshimura & Teramura 2005). (Senthilkumar & Vijayarangan 2007) presented the development of fully active suspension system for bumpy road input using PID controller. (Nemat & Modjtaba 2011) compared PID and fuzzy logic control of a quarter car suspension system. Non-linear active suspension systems have also been developed by (Altair & Wang 2010 & 2011). Different control strategies for developing active suspension systems have also been proposed by (Alexandru & Alexandru 2011, Lin & Lian 2011, Fatemeh Jamshidi & Afshin Shaabany 2011).

In active suspension systems, sensors are used to measure the acceleration of sprung mass and unsprung mass and the analog signals from the sensors are sent to a controller. The controller is designed to take necessary actions to improve the performance abilities already set. The controller amplifies the signals and the amplified signals are fed to the actuator to generate the required forces to form closed loop system (active suspension system), which is schematically depicted in Figure 3. The performance of this system is then compared with that of the open loop system (passive suspension system).

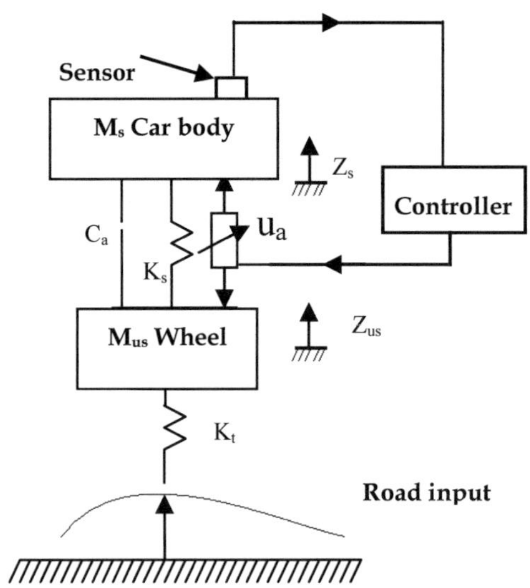

Fig. 3. Active suspension system

This chapter describes the development of a controller design for the active control of suspension system, which improves the inherent tradeoff among ride comfort, suspension travel and road-holding ability. The controller shifts its focus between the conflicting objectives of ride comfort, rattle space utilization and road-holding ability, softening the suspension when rattle space is small and stiffening it as it approaches the travel limits. The developed design allows the suspension system to behave differently in different operating conditions, without compromising on road-holding ability. The effectiveness of this control method has been explained by data from time domains. Proportional-Integral-Derivative (PID) controller including hydraulic dynamics has been developed. The displacement of hydraulic actuator and spool valve is modeled. The Ziegler – Nichols tuning rules are used to determine proportional gain, reset rate and derivative time of PID controller (Ogata 1990). Simulink diagram of active suspension system is developed and analyzed using MATLAB software. The investigations on the performance of the developed active suspension control are demonstrated through comparative simulations in this chapter.

2. Active suspension system

Active suspension systems add hydraulic actuators to the passive components of suspension system as shown in Figure 3. The advantage of such a system is that even if the active hydraulic actuator or the control system fails, the passive components come into action. The equations of motion are written as,

$$\begin{aligned} &M_s\ddot{z}_s + K_s(z_s - z_{us}) + C_a(\dot{z}_s - \dot{z}_{us}) - u_a = 0 \\ &M_{us}\ddot{z}_{us} + K_s(z_{us} - z_s) + C_a(\dot{z}_{us} - \dot{z}_s) + K_t(z_{us} - z_r) + u_a = 0 \end{aligned} \tag{1}$$

where u_a is the control force from the hydraulic actuator. It can be noted that if the control force $u_a = 0$, then Equation (1) becomes the equation of passive suspension system.

Considering u_a as the control input, the state-space representation of Equation (1) becomes,

$$\begin{aligned} &\dot{z}_1 = z_2 \\ &\dot{z}_2 = -\frac{1}{M_s}[K_s(z_1 - z_3) + C_a(z_2 - z_4)] \\ &\dot{z}_3 = z_4 \\ &\dot{z}_4 = \frac{1}{M_{us}}[K_s(z_1 - z_3) + C_a(z_2 - z_4) + K_t(z_3 - z_r)] \end{aligned} \tag{2}$$

where $z_1 = z_s, z_2 = \dot{z}_s, z_3 = z_{us}$ and $z_4 = \dot{z}_{us}$

3. Proportional - Integral - Derivative (PID) controller

PID stands for proportional, integral and derivative. These controllers are designed to eliminate the need for continuous operator attention. In order to avoid the small variation of the output at the steady state, the PID controller is so designed that it reduces the errors by the derivative nature of the controller. A PID controller is depicted in Figure 4. The set-point is where the measurement to be. Error is defined as the difference between set-point and measurement.

(Error) = (set-point) – (measurement), the variable being adjusted is called the manipulated variable which usually is equal to the output of the controller. The output of PID controllers will change in response to a change in measurement or set-point. Manufacturers of PID controllers use different names to identify the three modes. With a proportional controller, offset (deviation from set-point) is present. Increasing the controller gain will make the loop go unstable. Integral action was included in controllers to eliminate this offset. With integral action, the controller output is proportional to the amount of time the error is present. Integral action eliminates offset. Controller Output = (1/Integral) (Integral of) e(t) d(t). With derivative action, the controller output is proportional to the rate of change of the measurement or error. The controller output is calculated by the rate of change of the measurement with time. Derivative action can compensate for a change in measurement. Thus derivative takes action to inhibit more rapid changes of the measurement than proportional action. When a load or set-point change occurs, the derivative action causes the controller gain to move the "wrong" way when the measurement gets near the set-point. Derivative is often used to avoid overshoot. The different between the actual acceleration and desired acceleration is taken as error in this study.

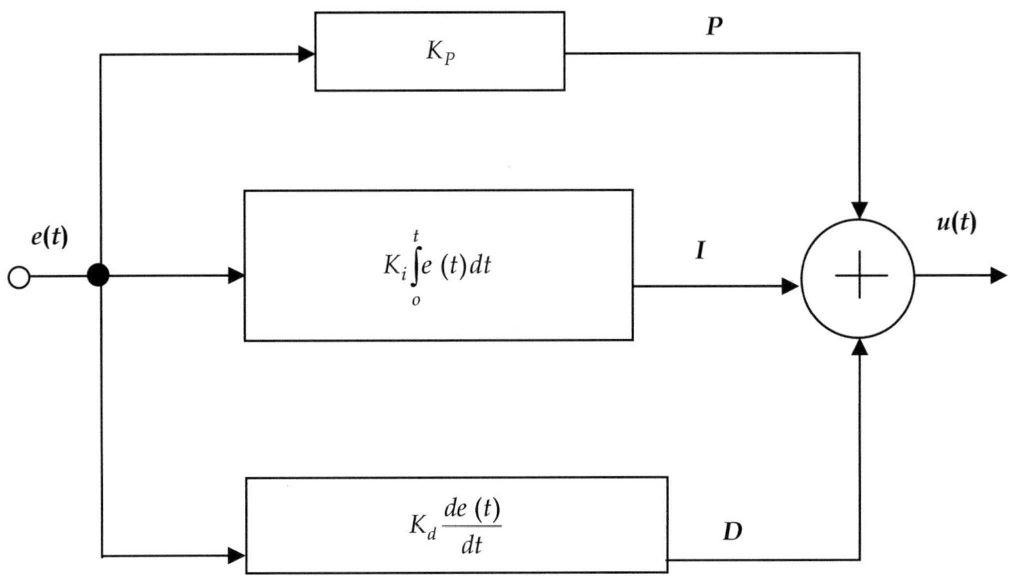

Fig. 4. PID controller

4. Hydraulic active suspension system

Block diagram of control system used to develop active suspension system is shown in Figure 5. In order to develop an active suspension system, the following hydraulic components are used.

- Pressurized hydraulic fluid source

- Pressure relief valve to control the pressure of hydraulic fluid
- Direction control valve
- Hydraulic cylinder (active actuator) to convert the hydraulic pressure into force to be transmitted between the sprung and the unsprung mass

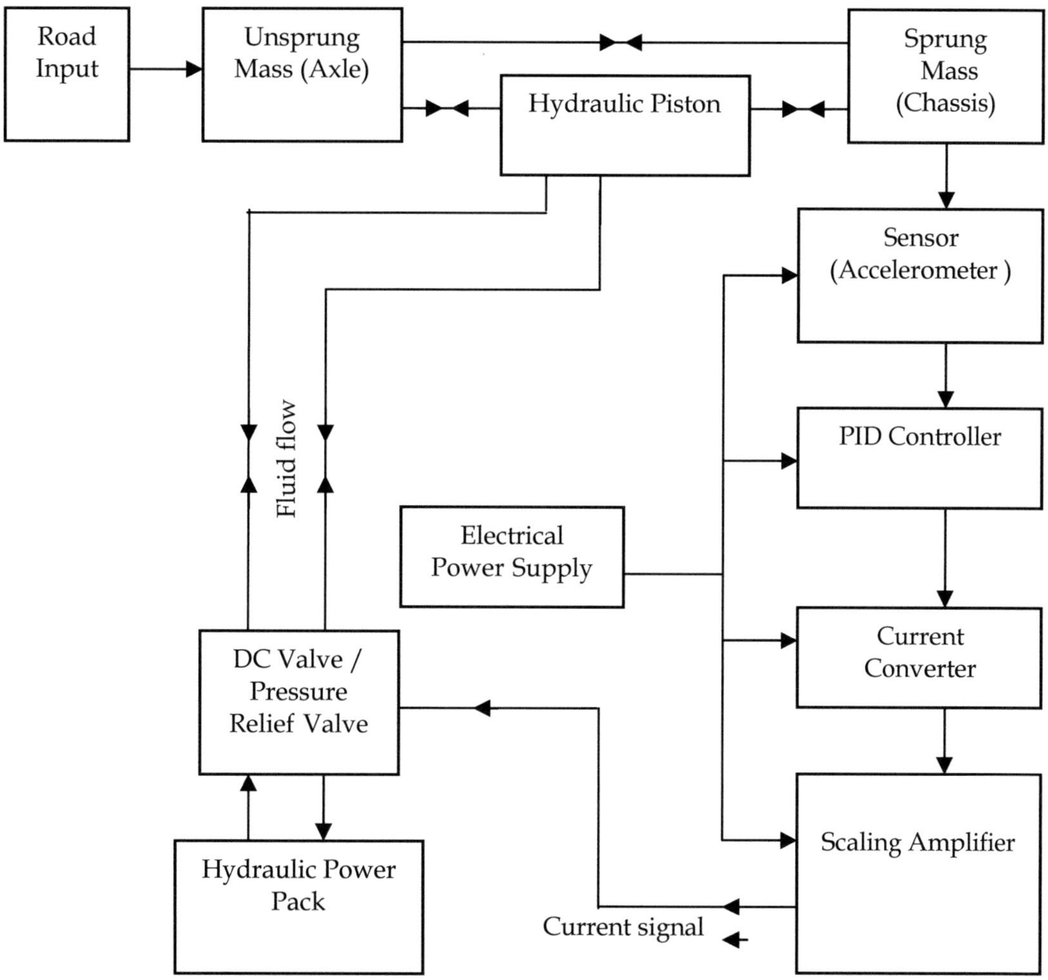

Fig. 5. Block diagram of control system

Figure 6 shows the hydraulic actuator installed in between sprung mass and unsprung mass, including a valve and a cylinder, where U_h is the actuator force generated by the hydraulic piston and x_{act} (= x_1-x_3) is the actuator displacement. U_h (equal to Ua) is applied dynamically in order to improve ride comfort as and when the road and load input vary.

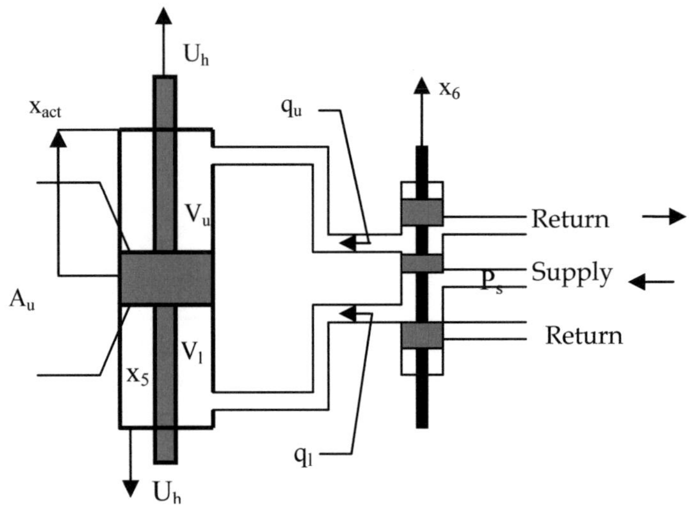

Fig. 6. Hydraulic valve and cylinder

5. Controller design

The design of controller is given by,

$$U_c = K_p e(t) + \frac{K_p}{T_i} \int_0^t e(t)dt + K_p T_d \frac{de(t)}{dt} \quad (3)$$

U_c is the current input from the controller, K_p is the proportional gain, T_i and T_d is the integral and derivative time constant of the PID controller respectively.

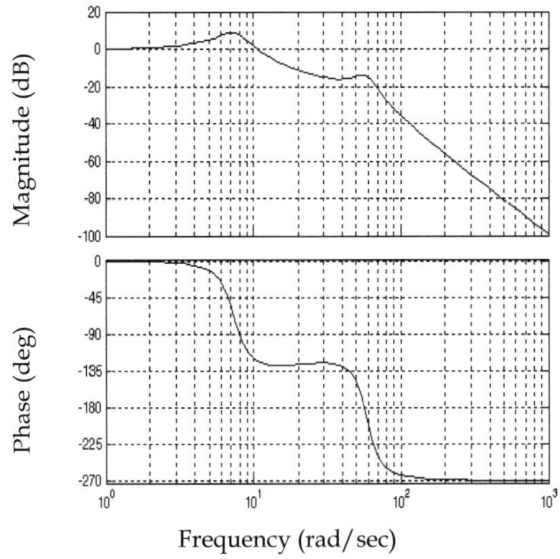

Fig. 7. Bode plot of passive suspension system

The values of gain margin and phase margin obtained from the frequency response plot of car body displacement of the passive suspension system shown in Figure 7 are used to determine the tuning parameters of the PID controller for the active quarter car model. The Ziegler-Nichols tuning rules are used to determine proportional gain, reset rate and derivative time of PID controller.

5.1 Tuning of PID controller

The process of selecting the controller parameters to meet given performance specification is known as controller tuning. Zeigler and Nichols suggested rules for tuning PID controllers (meaning to set values K_p, T_i, T_d) based on experimental step responses or based on the values of K_p that results in marginal stability when only proportional control action is used. Ziegler-Nichols rules, which are briefly presented in the section 5.1.1 are very much useful. Such rules suggest a set of values of K_p, T_i, and T_d that will give a stable operation of the system. However, the resulting system may exhibit a large maximum overshoot in the step response, which is unacceptable. In such a case we need a series of fine tunings until an acceptable results is obtained. In fact, the Zeigler-Nichols tuning rules give an educated guess for the parameter values and provide a starting point for fine tuning, rather than giving the final settings for K_p, T_i, and T_d in a single shot.

5.1.1 Zeigler-Nichols rules for tuning PID controllers

Zeigler and Nichols proposed rules for determining the proportional gain K_p, integral time T_i, and derivative time T_d based on the transient response characteristics of a given system. In this method, we first set $T_i = \infty$ and $T_d = 0$. Using the proportional control action only, increase K_p from 0 to a critical value K_{cr} at which the output first exhibits sustained oscillations. Thus the critical gain K_{cr} and the corresponding period P_{cr} are experimentally determined. Zeigler and Nichols suggested that we set the values of the parameter K_p, T_i, and T_d according to the formula shown in Table 1.

Type of Controller	Kp	Ti	Td
P	$0.5\ K_{cr}$	∞	0
PI	$0.45\ K_{cr}$	$0.83\ P_{cr}$	0
PID	$0.6\ K_{cr}$	$0.5P_{cr}$	$0.125\ P_{cr}$

Table 1. Zeigler-Nichols tuning rules

It can be noted that if the system has a known mathematical model, then we can use the root-locus method to find the critical gain K_{cr} and the frequency of the sustained oscillations ω_{cr}, where $2\pi/\omega_{cr} = P_{cr}$. These values can be found from the crossing points of the root locus branches with the $j\omega$ axis. The passive suspension (open loop) system of the quarter car model is analyzed and the bandwidth and gain margin of the system are found to be 1.92 Hz and -13.9 db respectively as shown in Figure 8. Gain margin is the gain, at which the active suspension (closed loop) system goes to the verge of instability; (Gain margin is the gain in db at which the phase shift of the system is -180^0). The gain margin of the system is found to be 4.91. It is the value of the gain, which makes the active suspension (closed loop) system to exhibit sustained oscillation (the vibration of car body of the active suspension (closed loop) system is maximum for this value of gain).

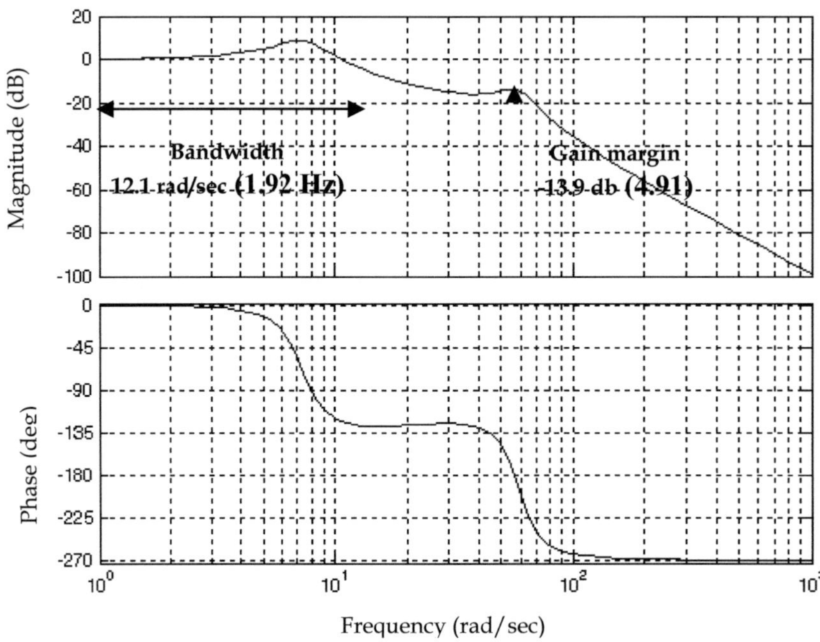

Fig. 8. Open loop unit step response

When the gain of the system is increased beyond 4.915 the response (vibration of car body displacement) of the active suspension (closed loop) system is increased instead of being reduced. The system becomes unstable when the gain of the system is increased beyond 4.915 which is shown in Figure 9.

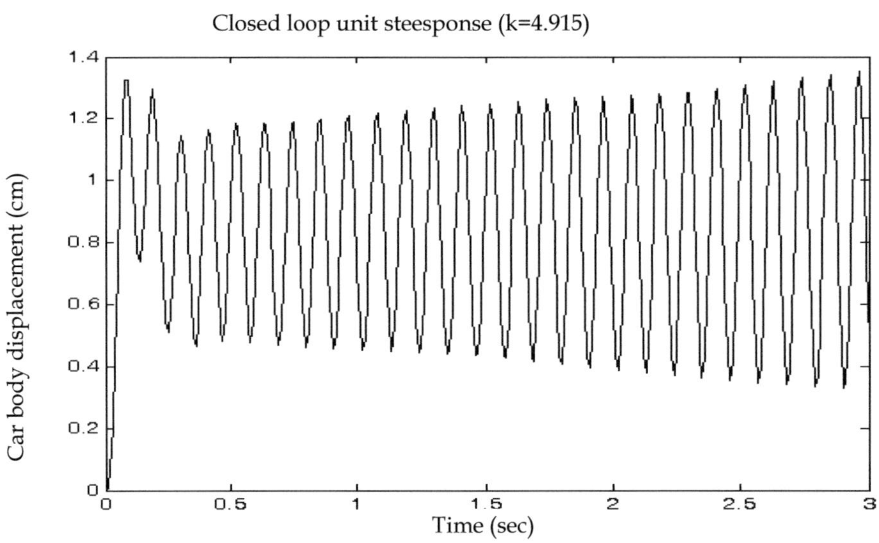

Fig. 9. Closed loop unit step response (k>4.915)

The response of the active suspension (closed loop) system of the quarter car model for the critical gain value (K_{cr} = 4.915) is as shown in Figure 10 and the time period of the sustained oscillation for this value of critical gain K_{cr} is called critical period P_{cr}, which is determined from the step response of the closed loop system and is found to be P_{cr} = 0.115 sec.

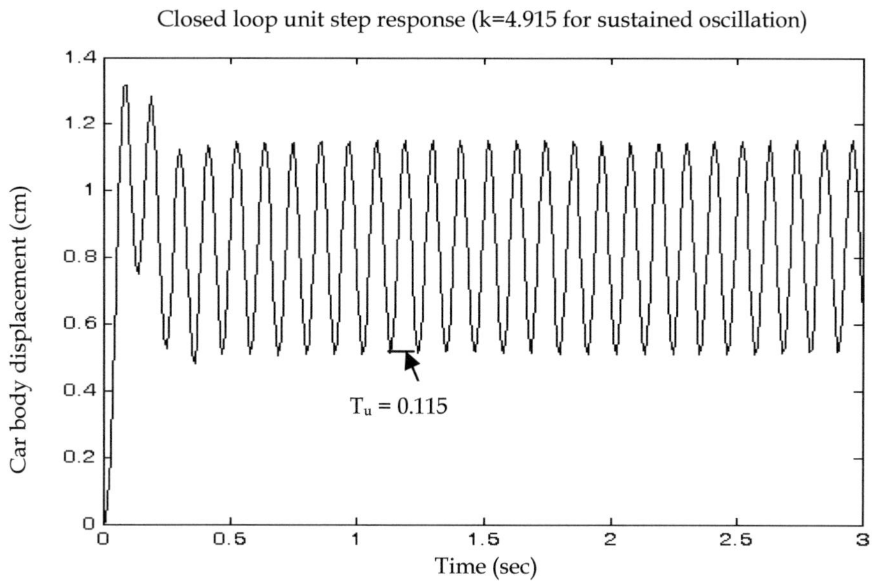

Fig. 10. Closed loop unit step response (k=4.915) for sustained oscillation

The critical gain (K_{cr}) and critical time period (P_{cr}), determined above are used to set the tuning rules for the quarter car model using the Zeigler-Nichols tuning rules. As discussed, the values of the P, PI and PID controller are obtained and are tabulated in Table 2.

Type of Controller	Kp	Ti	Td
P	2.4575	∞	0
PI	2.212	0.096	0
PID	2.95	0.0575	0.014

Table 2. Zeigler-Nichols tuning values

5.2 Hydraulic dynamics

Three-land four-way valve-piston system as shown in Figure 6 is used in the hydraulic controller design. The force U_h from the actuator is given by,

$$U_h = AP_L \tag{4}$$

where A (=A_u, Area of upper chamber; = A_l, Area of lower chamber) is the piston area and P_L is the pressure drop across the piston. Following Merritt (1967), the derivative of P_L is given by

$$\frac{V_t}{4\beta}\dot{P}_L = Q - C_{tp}P_L - A(x_2 - x_4) \tag{5}$$

where V_t is the total actuator volume, β is the effective bulk modulus of the fluid, Q is the hydraulic load flow ($Q = q_u+q_l$), where q_u and q_l are the flows in the upper and lower chamber respectively and C_{tp} is the total leakage coefficient of the piston. In addition, the valve load flow equation is given by

$$Q = C_d \omega x_6 \sqrt{\frac{1}{\rho}\left[P_s - sgn(x_6)x_5\right]} \tag{6}$$

where C_d is the discharge coefficient, ω is the spool valve area gradient, x_5 is the pressure inside the chamber of hydraulic piston and $x_6 = x_{sp}$ is the valve displacement from its closed position, ρ is the hydraulic fluid density and P_s is the supply pressure, Since, the term $\left[P_s - sgn(x_6)x_5\right]$ may become negative, Equation (6) is replaced with the corrected flow equation as,

$$Q = sgn\left[P_s - sgn(x_6)x_5\right]C_d\omega x_6\sqrt{\frac{1}{\rho}\left[P_s - sgn(x_6)x_5\right]} \tag{7}$$

Finally, the spool valve displacement is controlled by the input to the valve U_c, described by Equation (3), which could be a current or voltage signal. Equations (2) to (7) used to derive the equation of the active suspension system, including the hydraulic dynamics are rewritten as Equation (8).

$$\begin{aligned}
\dot{x}_1 &= x_2 \\
\dot{x}_2 &= \frac{K_s}{M_s}x_1 + \frac{C_a}{M_s}x_2 + \frac{K_s}{M_s}x_3 + \frac{C_a}{M_s}x_4 + \frac{A_l}{M_s}x_5 \\
\dot{x}_3 &= x_4 \\
\dot{x}_4 &= \frac{K_s}{M_{us}}x_1 + \frac{C_a}{M_{us}}x_2 + \frac{K_s+K_t}{M_{us}}x_3 + \frac{C_a}{M_{us}}x_4 + \frac{A_l}{M_{us}}x_5 + \frac{K_t}{M_{us}}r \\
\dot{x}_5 &= \beta x_5 + A(x_2 - x_4) + x_6\omega_3 \\
\dot{x}_6 &= \frac{x_6}{\tau} + U_c
\end{aligned} \tag{8}$$

where, $\omega_3 = sgn\left[P_s - sgn(x_6)x_5\right]\sqrt{\left|P_s - sgn(x_6)x_5\right|}$

Thus Equation (8) becomes state feedback model of active suspension system including hydraulic dynamics. Figure 11 represents the Simulink model of active suspension system.

6. Simulation

To ensure that our controller design achieves the desired objective, the open loop passive and closed loop active suspension system are simulated with the following values.

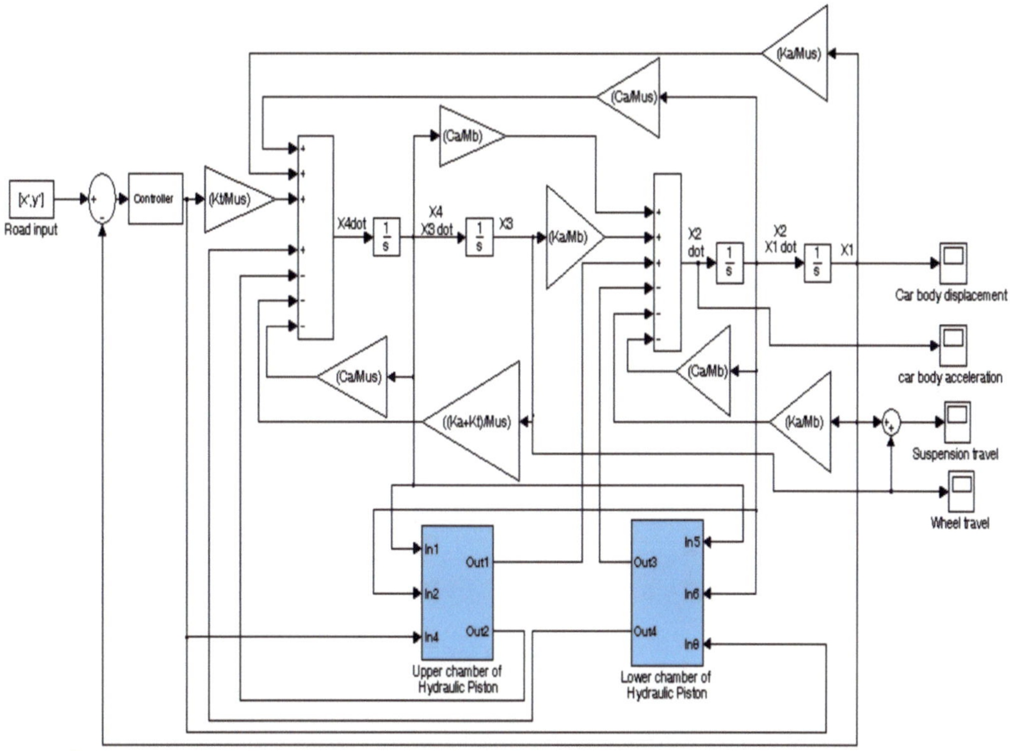

Fig. 11. Simulink model of active suspension system

M_b = 300 Kg	K_t = 190000 N/m
M_{us} = 60 Kg	β = 1 sec^{-1}
K_a = 16850 N/m	P_s = 10.55 MPa
C_a = 1000 N/(m/sec)	

6.1 Bumpy road (sinusoidal input)

A single bump road input, Z_r as described by (Jung-Shan Lin 1997), is used to simulate the road to verify the developed control system. The road input described by Equation (9) is shown in Figure 12.

$$Zr = \begin{cases} a(1 - \cos\omega t) & 0.5 \le t \le 0.75 \\ 0, & otherwise, \end{cases} \tag{9}$$

In Equation (9) of road disturbance, 'a' is set to 0.02 m to achieve a bump height of 4 cm. All the simulations are carried out by MATLAB software. The following assumptions are also made in running the simulation.

a. Suspension travel limits: $\pm$ 8 cm
b. Spool valve displacement $\pm$ 1 cm

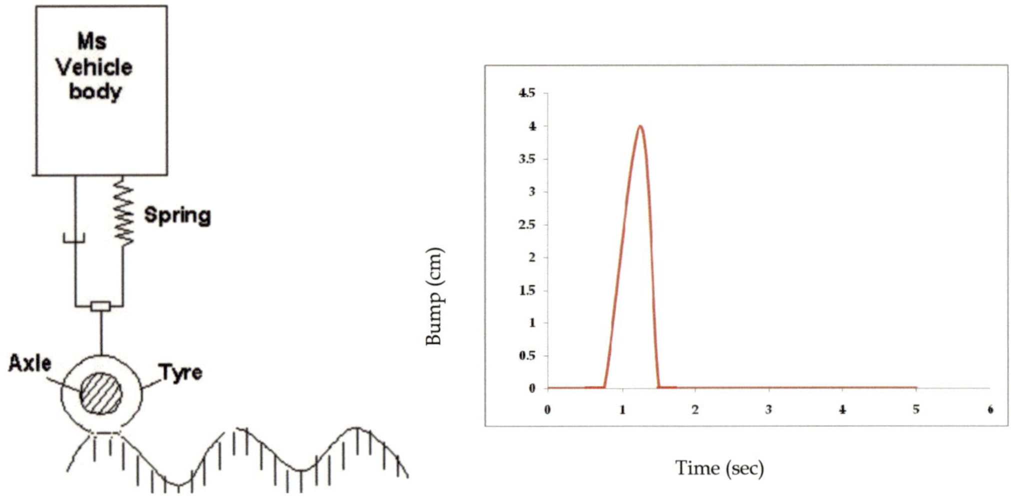

(a) Actual bumpy road (b) Bumpy road input

Fig. 12. Road input disturbance

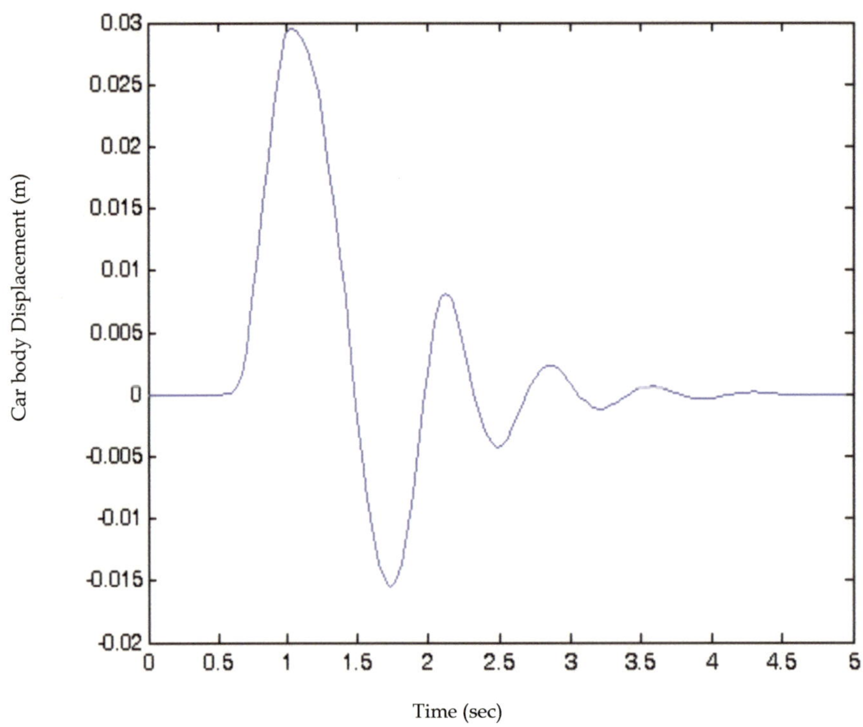

Fig. 13. Car body displacement of passive suspension system

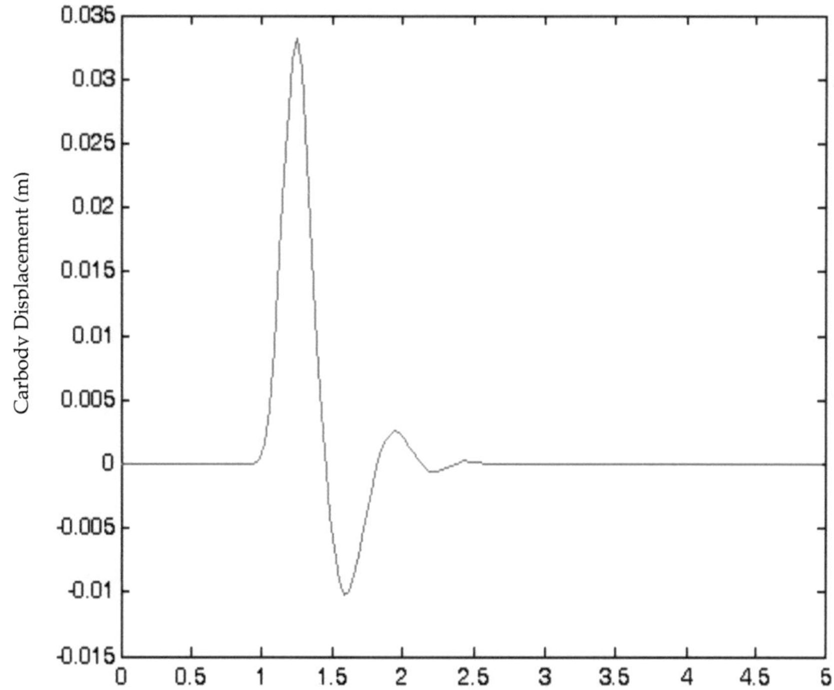

Fig. 14. Car body displacement of active suspension system

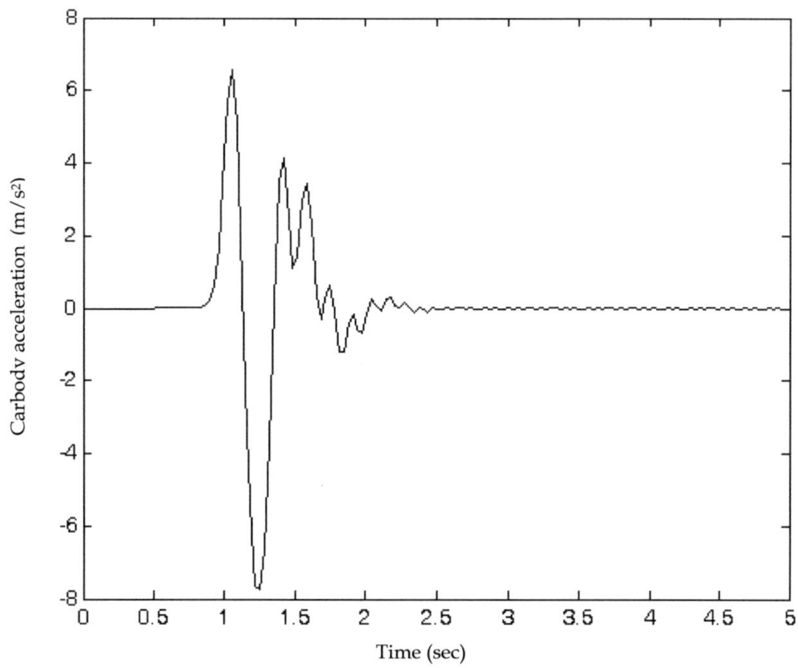

Fig. 15. Car body acceleration of passive suspension system

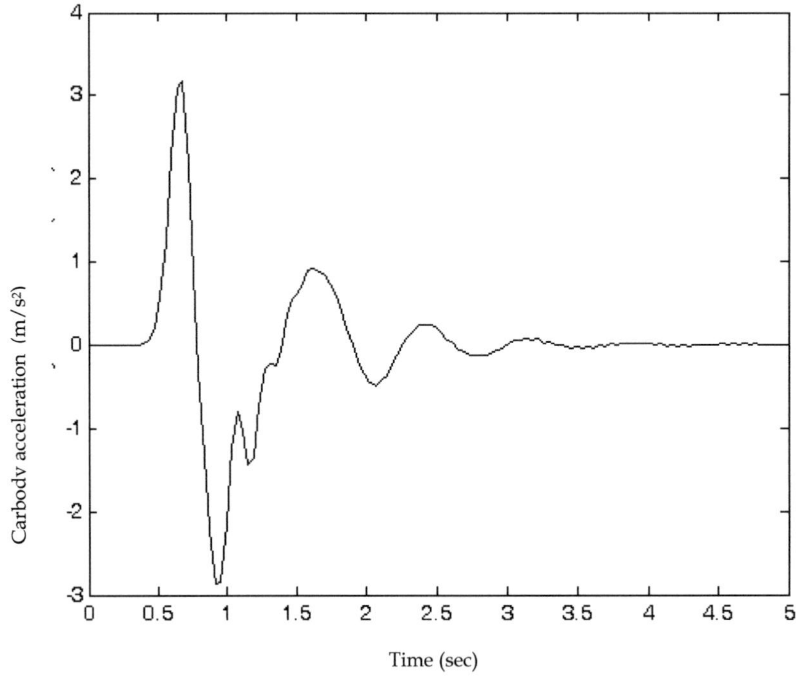

Fig. 16. Car body acceleration of active suspension system

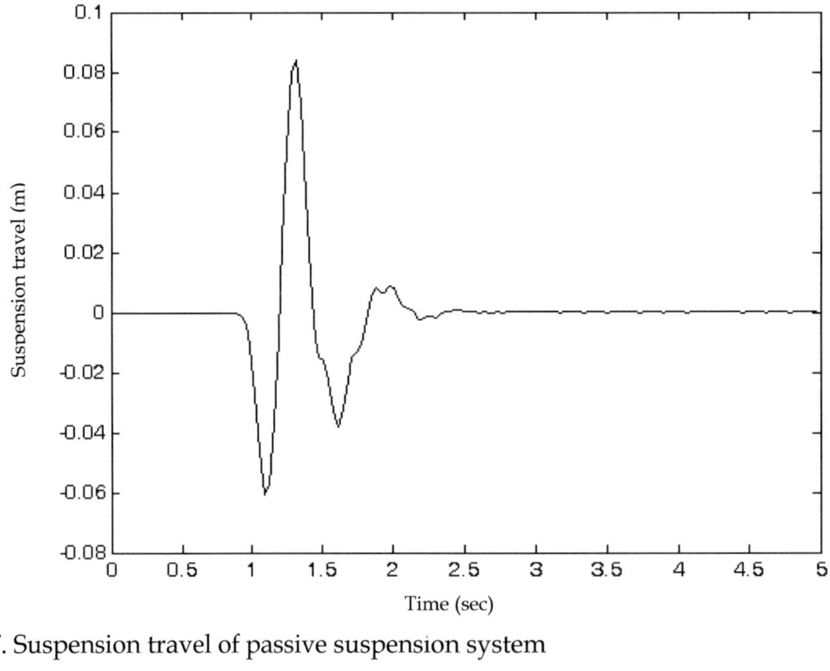

Fig. 17. Suspension travel of passive suspension system

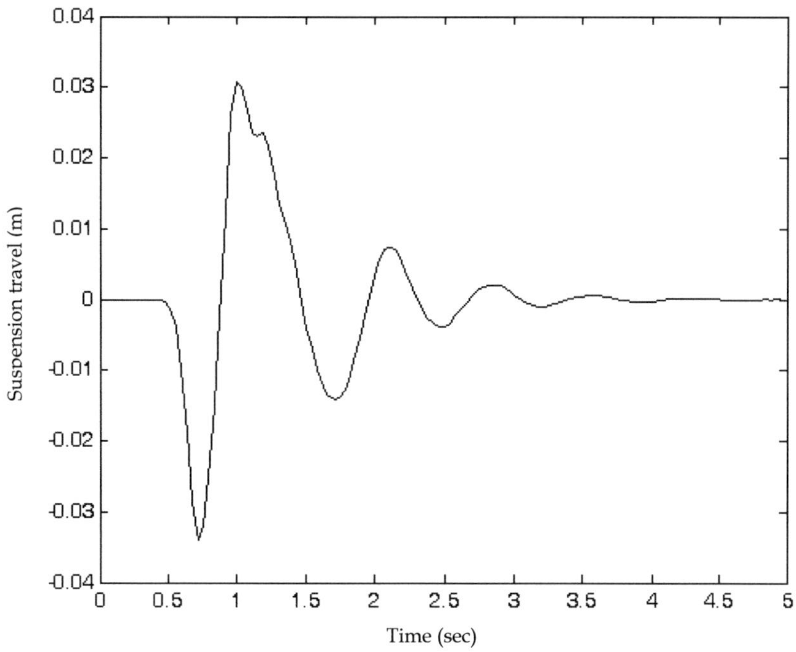

Fig. 18. Suspension travel of active suspension system

Figures 13-18 represent the time response plots of car body displacement, car body acceleration and suspension travel of both passive and active suspension system without tuning of controller parameters respectively. The PID controller designed produces a large spike (0.0325 m) in the transient portion of the car body displacement response of active suspension system as shown in Figure 14, compared to the response (0.03 m) of passive suspension system shown in Figure 13. The spike is due to the quick force applied by the actuator in response to the signal from the controller. Even though there is a slight penalty in the initial stage of transient vibration in terms of increased amplitude of displacement, the vibrations are settled out faster as it takes only 2.5 sec against 4.5 sec taken by the passive suspension system as found from Figure 14.

The force applied between sprung mass and unsprung mass would not produce an uncomfortable acceleration for the passengers of the vehicle, which is depicted in Figure 15 ($3.1\ m/s^2$ in active system), in comparison with the acceleration ($6.7\ m/s^2$) of passenger experienced in passive system as shown in Figure 15. Also, it is found that the suspension travel (0.031 m) is very much less as seen in Figure 18 compared with suspension travel (0.081 m) of passive suspension system as seen in Figure 17. Therefore rattle space utilization is very much reduced in active suspension system when compared with passive suspension system in which suspension travel limit of 8 cm is almost used.

Figures 19-22 represent the behavior of both active suspension systems with tuned parameters.

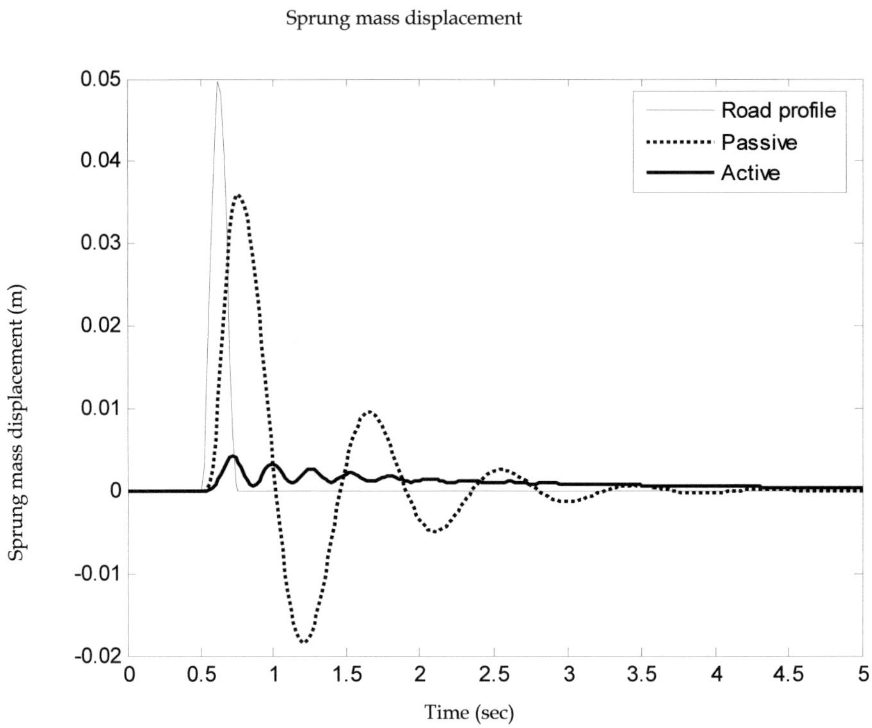

Fig. 19. Sprung mass displacement Vs time (Bumpy road)

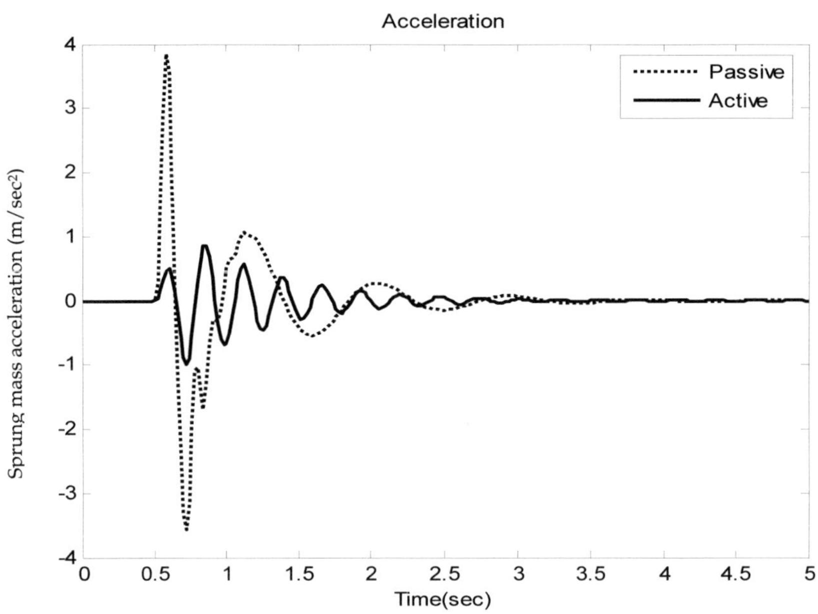

Fig. 20. Sprung mass acceleration Vs time (Bumpy road)

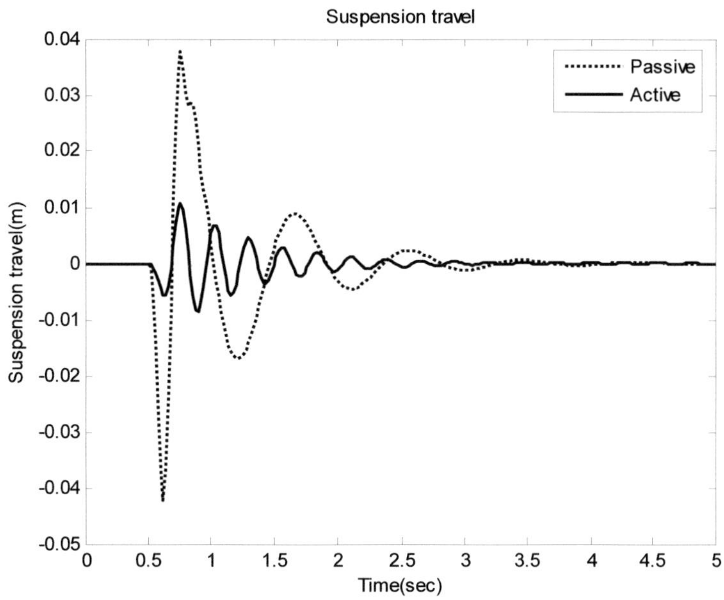

Fig. 21. Suspension travel Vs time (Bumpy road)

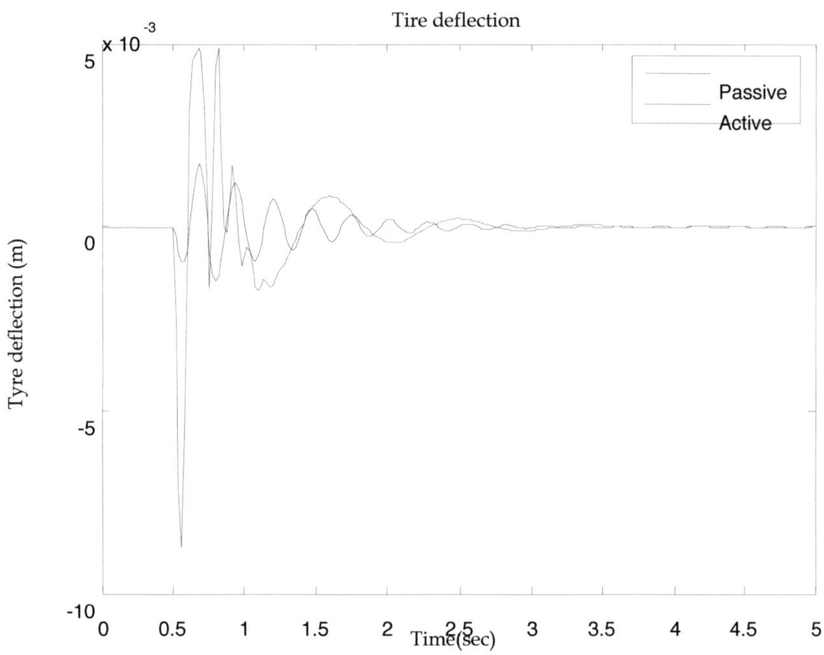

Fig. 22. Tyre deflection Vs time (Bumpy road)

Figures 19–22 illustrate that both peak values and settling time have been reduced by the active system compared to the passive system for all the parameters of sprung mass displacement, sprung mass acceleration (ride comfort), suspension travel and tyre deflection (road holding). Table 3 gives the percentage reduction in peak values of the various parameters for the sinusoidal bumpy road input.

Parameter	Passive	Active	% Reduction
Sprung Mass Acceleration	3.847 m/s^2	0.845m/s^2	78.03
Suspension Travel	0.038 m	0.011 m	71.05
Tyre Deflection	0.005 m	0.002 m	60.00

Table 3. Reduction in peak values different parameters (Bumpy road)

6.2 Pot-hole (step input)

The step input characterizes a vehicle coming out of a pothole. The pothole has been represented in the following form.

$$Z_r = \begin{cases} 0 & \rightarrow \ t \leq 1 \\ 0.05 & \rightarrow \ t > 1 \end{cases} \tag{10}$$

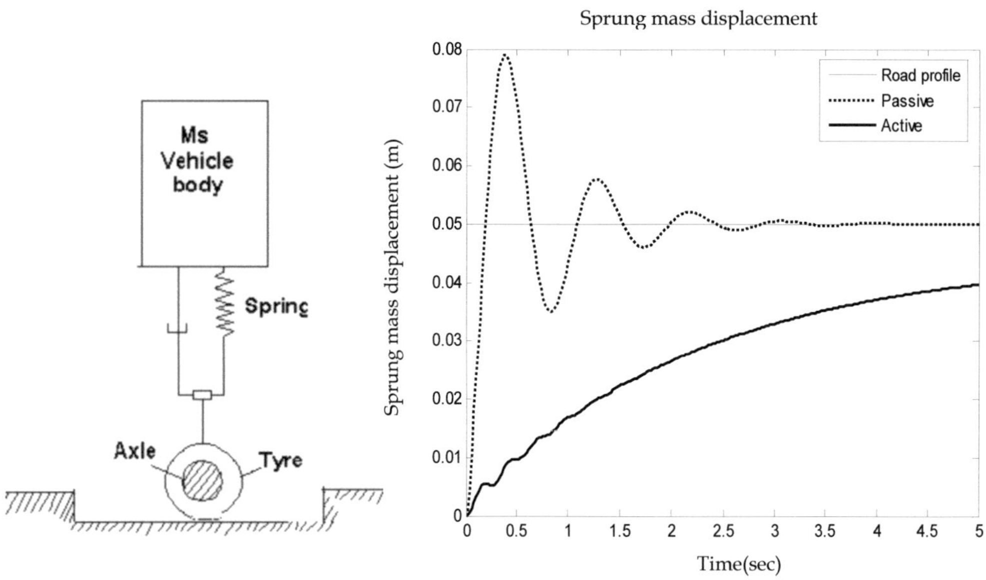

(a) Pot hole (b) Sprung mass displacement Vs time

Fig. 23. Sprung mass displacement (Pot hole)

Figures 23-26 illustrates the performance comparison between passive and active suspension system for the vehicle coming out of a pot-hole of height 0.05 m.

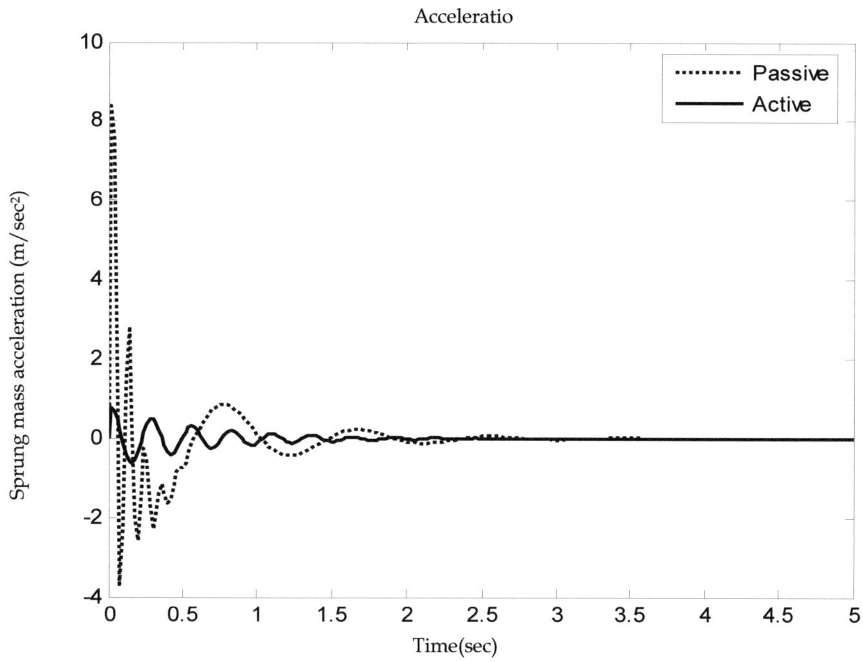

Fig. 24. Sprung mass acceleration Vs time (Pot hole)

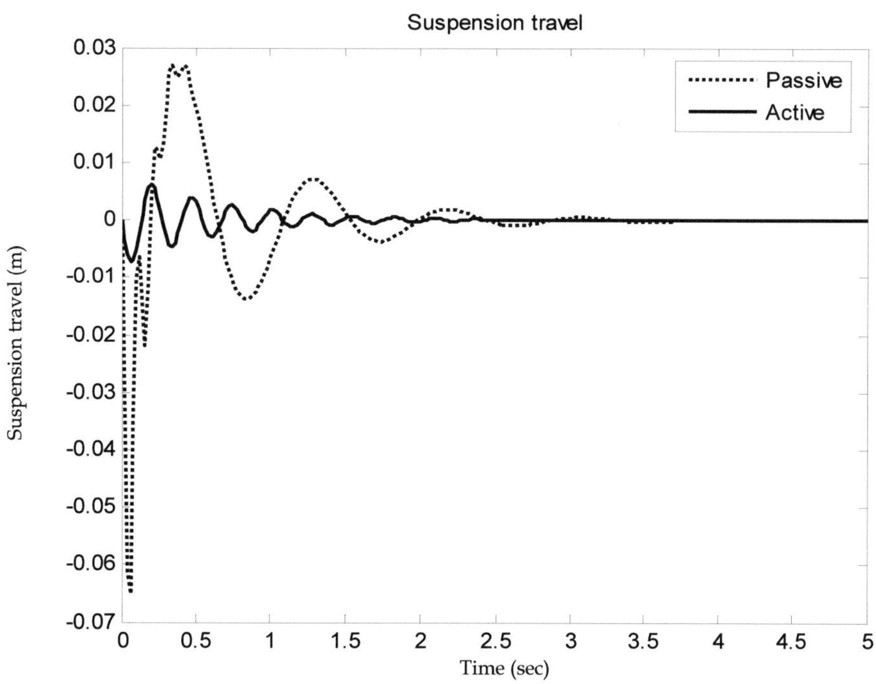

Fig. 25. Suspension travel Vs time (Pot hole)

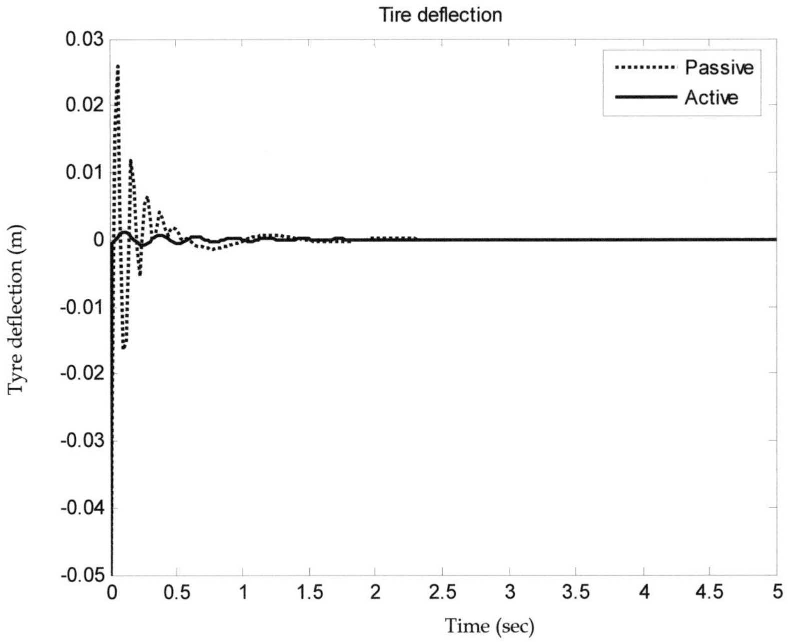

Fig. 26. Tyre deflection Vs time (Pot hole)

Parameter	Passive	Active	% Reduction
Sprung Mass Acceleration	8.3793 m/s^2	0.7535 m/s^2	91.01
Suspension Travel	0.06510 m	0.00732 m	88.75
Tyre Deflection	0.02597 m	0.00102 m	96.04

Table 4. Reduction in peak values of different parameters (Pot-hole)

From Figures 23-26, it could be observed that both peak overshoot and settling time have been reduced by the active system compared to the passive system for sprung mass acceleration, suspension travel, and tyre deflection. Table 4 shows the percentage reduction in various parameters which guarantees the improved performance by active suspension system.

6.3 Random road

Apart from sinusoidal bumpy and pot-hole type of roads, a real road surface taken as a random exciting function is used as input to the vehicle. It is noted that the main characteristic of a random function is uncertainty. That is, there is no method to predict an exact value at a future time. The function should be described in terms of probability statements as statistical averages, rather than explicit equations. In road model power spectral density has been used to describe the basic properties of random data.

Several attempts have been made to classify the roughness of a road surface. In this work, classifications are based on the International Organization for Standardization (ISO). The ISO has proposed road roughness classification (classes A-H) based on the power spectral density values is as shown in Figure 27.

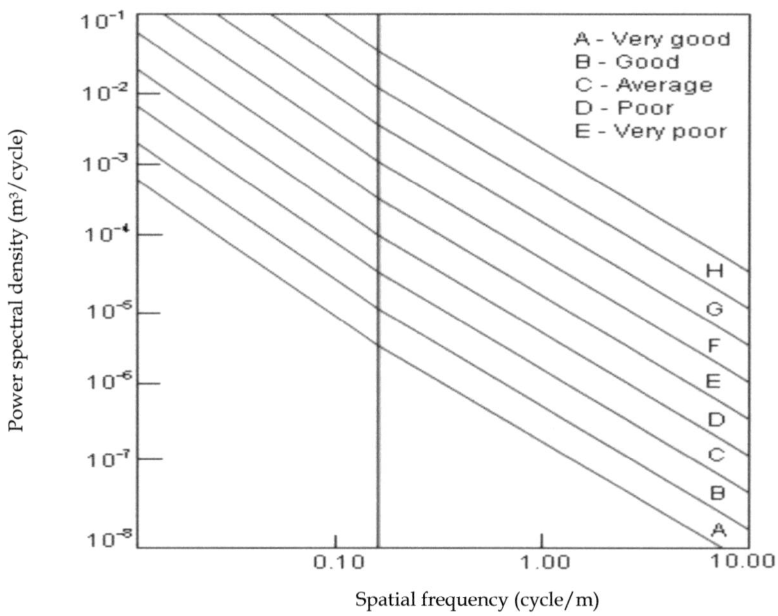

Fig. 27. Road roughness classification by ISO (2000)

6.3.1 Sinusoidal approximation

A random profile of a single track can be approximated by a superposition of $N \rightarrow \infty$ sine waves

$$Z_r(s) = \sum_{i=1}^{N} A_i \sin(\Omega_i s - \psi_i) \tag{11}$$

where each sine wave is determined by its amplitude A_i and its wave number Ω_i. By different sets of uniformly distributed phase angles ψ_i, i = 1(1) N in the range between 0 and 2π different profiles can be generated which are similar in the general appearance but different in details.

A realization of the class E road is shown in Figure 28. According to Equation (11) the profile z = z(s) was generated by N = 10 sine waves in the frequency range from 0.1cycle/m (0.628rad/m) to 1cycle/m (6.283rad/m). The amplitudes A_i, i = 1(1)N were calculated and the MATLAB function 'rand' was used to produce uniformly distributed random phase angles in the range between 0 and 2π. Figure 28 shows road profile input in time domain.

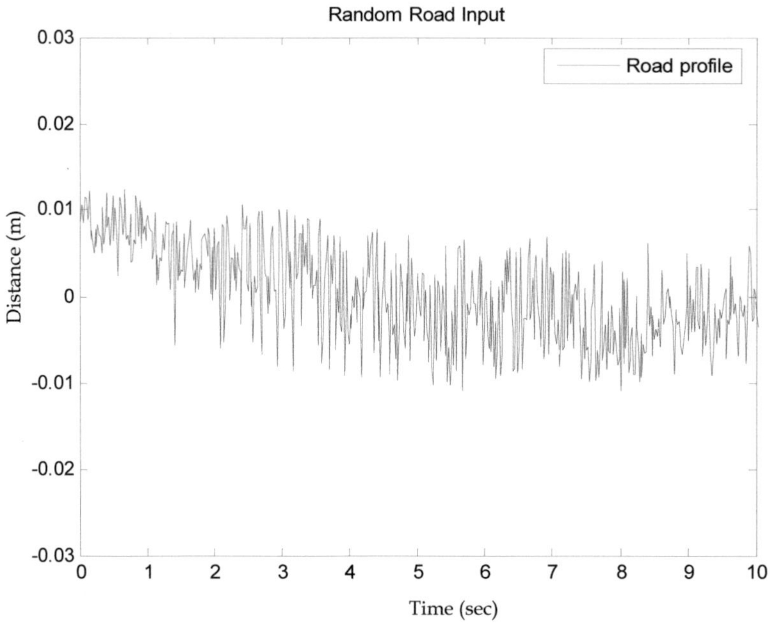

Fig. 28. Road disturbance Vs time (Random)

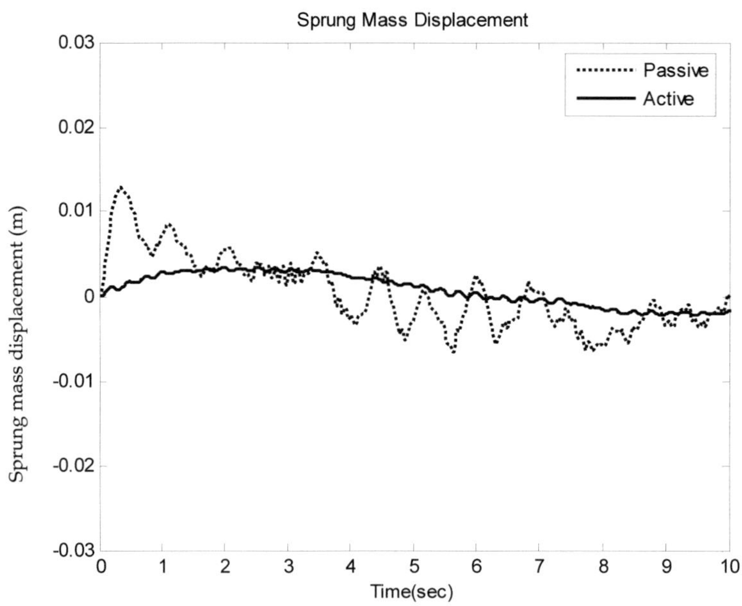

Fig. 29. Sprung mass displacement Vs time (Random road)

Figures 29-32 represent the behaviour of both passive and active suspension systems subjected to random road profile. Table 5 shows the percentage reduction in peak values of suspension parameters.

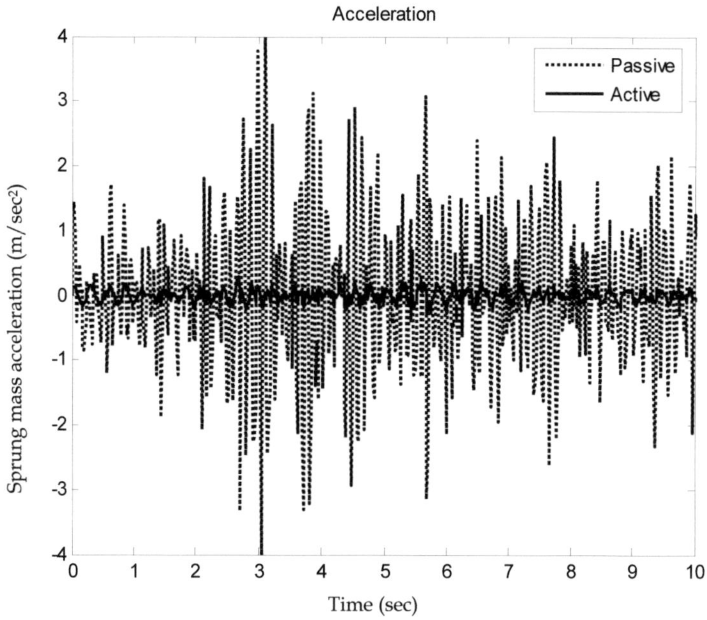

Fig. 30. Sprung mass acceleration Vs time (Random road)

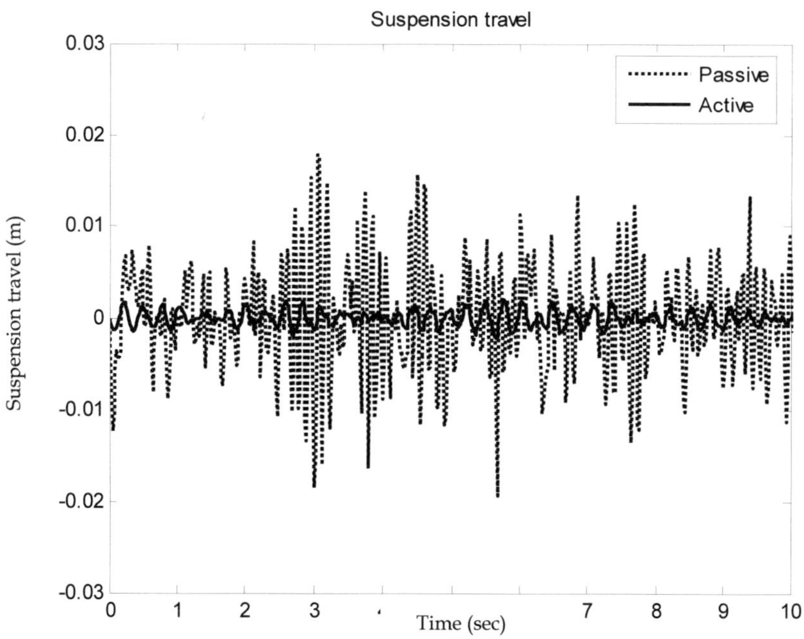

Fig. 31. Suspension travel Vs time (Random road)

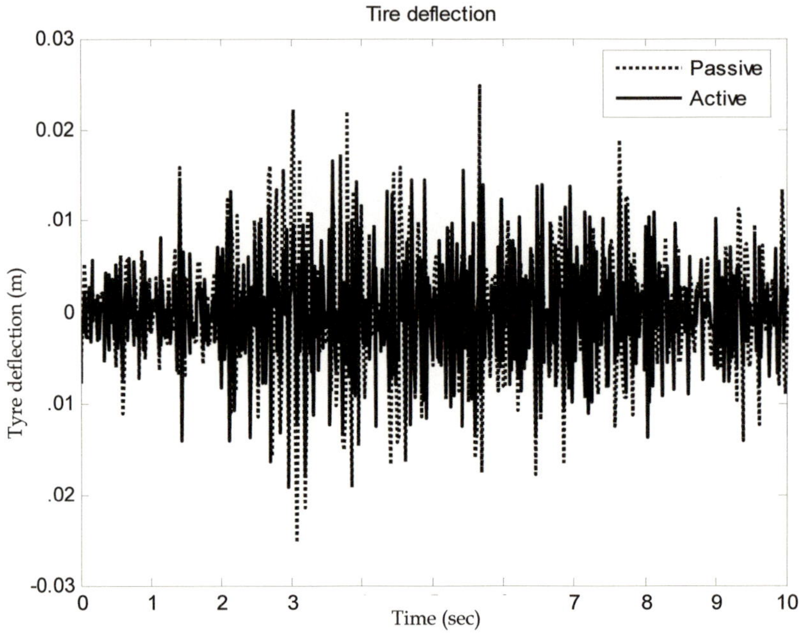

Fig. 32. Tyre deflection Vs time (Random road)

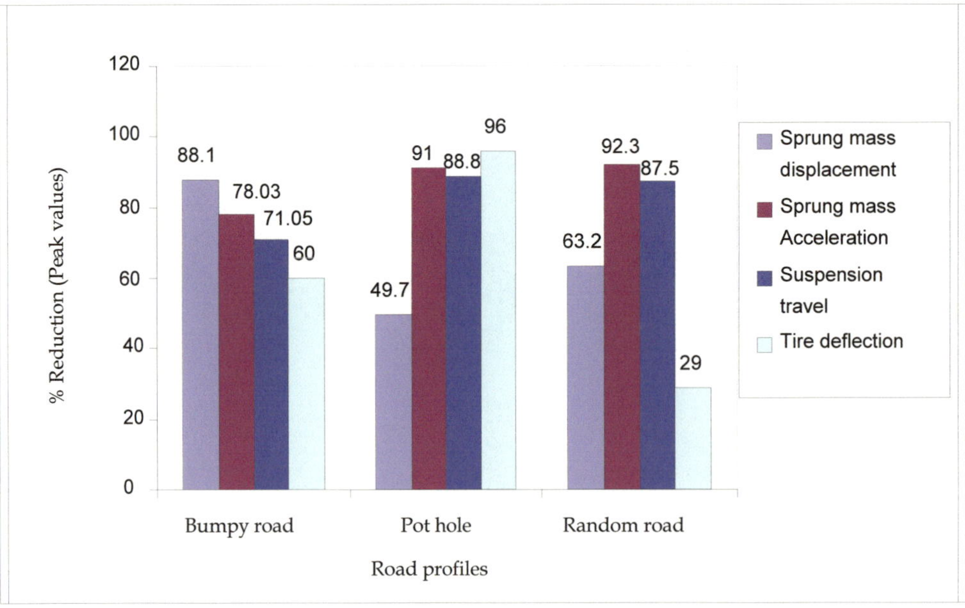

Fig. 33. Active system performance for various road inputs

Parameter	Passive	Active	% Reduction
Sprung Mass Acceleration	4.1551 m/s^2	0.3203 m/s^2	92.29
Suspension Travel	0.0195 m	0.0024 m	87.45
Tyre Deflection	0.0254 m	0.018 m	29.02

Table 5. Reduction in peak values of different parameters (Random road)

It is illustrated that both peak overshoot and settling time have been reduced by the active system compared to the passive system for all the parameters of sprung mass displacement, sprung mass acceleration (ride comfort), suspension travel. Moreover, there is no significant decrease in tyre deflection, but still it is lesser than the static spring deflection. The reason for no improvement in the tyre deflection is the wheel oscillations due to sudden variations of road profile due to randomness.

Figure 33 shows the percentage reduction of the peak values of sprung mass displacement, sprung mass acceleration, suspension travel and tyre deflection for active system for three road inputs of sinusoidal bump, step and random road profiles. The peak values of sprung mass acceleration have reduced for all the road profiles, which show the improved ride performance of active suspension system. The peak values of sprung mass displacement and suspension travel have also reduced significantly. As the ride comfort and road holding are mutually contradicting parameters the tyre deflection peak value has reduced only by 29% for random road profile.

7. Concluding remarks

The PID controller is designed for active suspension system. A quarter car vehicle model with two-degrees-of-freedom has been modeled. Hydraulic dynamics is also considered while simulated. Ziegler-Nichols tuning rules are used to determine proportional gain, reset rate and derivative time of PID controllers. The system is developed for bumpy road, pothole and random road inputs. The simulated results prove that, active suspension system with PID control improves ride comfort. At the same time, it needs only less rattle space. However, there is no significant improvement in road holding ability observed especially for random road surface. Besides its relative simplicity in design and the availability of well-known standard hardware, the viability of PID controller as an effective tool in developing active suspension system has been proved.

8. Acknowledgement

This work is carried out as part of research project supported by All India Council for Technical Education (AICTE), India. Author is grateful for the financial support provided by DRDO.

9. References

Aldair, A. &. Wang, W. (2010). Design of Fractional order Controller Based on Evolutionary Algorithm for a Full Vehicle Nonlinear Active Suspension System, *International journal of Control and Automation (IJCA)*, 3(4). pp. 33-46.

Aldair, A & Wang, W. J. (2011). Design an Intelligent Controller for Full Vehicle Nonlinear Active Suspension Systems, *International Journal on Smart Sensing and Intelligent Systems*, vol. 4, no. 2, 224-243

Alexandru, C. & Alexandru, P. (2011). *Control Strategy for an Active Suspension System*, World Academy of Science, Engineering and Technology, 79, 126-131.

Elbeheiry, E.M. Bode, O. & Cho, D. (1995). Advanced ground vehicle suspension systems, *Vehicle System Dynamics*, Vol. 24, pp. 231-258, ISSN 0042-3114

Elbeheiry, E.M. (2000). Effect of small travel speed variation on active vibrations on active vibration control in modern vehicles, *Journal of Sound and Vibration*, Vol. 232, No. 5, pp. 857-875, ISSN 0022-460X

Fatemeh Jamshidi. & Afshin Shaabany. (2011). Robust Control of an Active Suspension System Using H_2 & H_∞ Control Methods, *Journal of American Science*, 7(5), pp.1-5

Hedrick, J.K. & Wormely, D.N. (1975). Active suspension for ground support transportation, *ASME AMD*, Vol. 15, pp. 21-40, ISSN 0160-8835

Hrovat, D. (1997). Survey of advanced suspension development and related optimal control application, *Automatica*, Vol. 30, No. 10, pp. 1781-1817, ISSN 0005-1098

Hrovat, D. (1988). Influence of unsprung weight on vehicle ride quality, *Journal of Sound and Vibration*, Vol. 124, pp. 497-516, ISSN 0022-460X

Hrovat, D. (1997). Survey of advanced suspension development and related optimal control application, *Automatica*, Vol. 30, No. 10, pp. 1781-1817, ISSN 0005-1098

Jung-Shan Lin & Ioannis Kanellakopoulos. (1995). Nonlinear design of active suspensions, *34th IEEE Conference on Decision and Control*, ISBN 0-1803-2685-7, New Orleans, LA pp. 11-13,

Karnopp, D. (1986). Theoretical limitations in active vehicle suspensions, *Vehicle System Dynamics*, Vol. 15, pp. 41-54, ISSN 0042-3114

Karnopp, D. (1992). Power requirements for vehicle suspension systems, *Vehicle System Dynamics*, Vol. 21, No.2, pp. 65-72, ISSN 0042-3114

Karnopp, D.C. (1995). Active and semi-active vibration isolation, *Journal of Mechanical Design*, Vol.117, pp. 177-185., ISSN 0738-0666

Lin, J. & Lian, R.J. (2011). Intelligent control of active suspension systems, *IEEE Transactions on Industrial Electronics*, vol. 58, pp. 618–628

Nemat, C. & Modjtaba, R. (2011), Comparing PID and fuzzy logic control of a quarter car suspension system, *The Journal of Mathematics and Computer Science* , Vol.2 No.3, pp.559-564, ISSN 2008-949X.

Purdy,D.J. & Bulman, D.N. (1993). An experimental and theoretical investigation into the design of an active suspension system for a racing car, *Proc. Instn. Mech. Engrs. - Part D*, Vol. 211, pp. 161-173, ISSN 3-540-76045-8

Pilbeam, C. & Sharp, R.S. (1996). Performance potential and power consumption of slow – active suspension systems with preview, *Vehicle system Dynamics*, Vol. 25, pp. 169 183, ISSN 0042-3114

Senthilkumar, M. & Vijayarangan, S. (2007). Analytical and experimental studies on active suspension system of light passenger vehicle to improve ride comfort, *Mechanika*, Vol.3, No.65,pp.34-41, ISSN 1392-1207

Sharp, R.S. & Crolla, D.A. (1987). Road vehicle suspension system design, *Vehicle System Dynamics*, Vol. 16, No. 3, pp. 167-192, ISSN 0042-3114

Yoshimura, T. & Teramura, I. (2005). Active suspension control of a one-wheel car model using single input rule modules fuzzy reasoning and a disturbance observer, *Journal of Zhejiang University SCIENCE*, Vol.6A, No.4, pp.251-256, ISSN 1009-3095

Part 2

Intelligent Control and Genetic Algorithms Approach

5

Fuzzy PID Supervision for an Automotive Application: Design and Implementation

R. Sehab and B. Barbedette
Ecole Supérieure des Techniques Aéronautiques et de Construction Automobile
France

1. Introduction

In the control of plants with good performances, engineers are often faced to design controllers in order to improve static and dynamic behavior of plants. Usually the improvement of performances is observed on the system responses. For illustration, an example of a DC machine is chosen. Two cases of study are presented:

1. In open loop, the velocity response depends on the mechanical time constant of the DC machine (time response) and the value of the power supply. Indeed, for each value of power supply, a velocity value is reached in steady state. Therefore the DC machine can reach any value of velocity which depends only of the power supply. In this case no possibility to improve performances.
2. For a specific need, the open loop control is not sufficient. Engineers are faced to a problem of control in order to reach a desired velocity response according to defined specifications such as disturbance rejection, insensitivity to the variation of the plant parameters, stability for any operation point, fast rise-time, minimum Settling time, minimum overshoot and a steady state error null. Also, the designed control is related to other constraints related to the cost, computation complexity, manufacturability, reliability, adaptability, understandability and politics (Passino & Yurkovich, 1998).

In general the design of the control needs to identify the dynamic behavior of the system. Therefore a dynamic model of the plant is developed in order to reproduce the real response in open loop. Developing a model for a plant is a complex task which needs time and an intuitive understanding of the plant's dynamics. Usually, on the basis of some assumptions to choose, a simplified model is developed and the physical parameters of the established model are identified using some experimental responses. If the model is nonlinear, we need to linearize the model around a steady state point in order to get a simplified linear model. Therefore a linear controller is designed with techniques from classical control such as pole placement or frequency domain methods. Using the mathematical model and the designed controller, a simulation in closed loop is carried out in order to study and to analyze its performances. This step of study consists to adjust controller parameters until performances are reached for a given set point. In the last step, the designed controller is implemented via, for example, a microprocessor, and evaluating the performance of the closed-loop system (again, possibly leading to redesign).

In industry most time engineers are interested by linear controllers such as proportional-integral-derivative (PID) control or state controllers. Over 90% of the controllers in operation today are PID controllers (or at least some form of PID controller like a P or PI controller). This approach is often viewed as simple, reliable, and easy to understand and to implement on PLCs. Also performances of the plant are on-line improved by adjusting only gains. In spite of the advantages of PID controllers, the process performances are never reached. This is due mainly to the accuracy of the model used to design controller and not properly to the controller.

In the development of analytical models, variation of physical parameters, operation conditions, disturbances are not taken into account. This is due to the difficulty to identify all the physical phenomena in the process and to find an appropriate model for each. Therefore the closed loop specifications of the process are not maintained. Engineers are however in the need to adjust permanently PID controllers even it's a heavy task. For these different reasons, a new approach of control is proposed taking into account the constraints related to the process, the parameter variation and disturbances. This type of control, named fuzzy control, is designed from the operator experience on the process over many years. Under different operation conditions, linguistic rules are established taking into account constraints and environment process. In this case, modeling the process is not necessary and the designed fuzzy controller is sufficient to ensure the desired performances. In other cases, if the model is known with a good accuracy, industrials prefer to implement a fuzzy PID supervisor to their existing PID controllers where gains are on-line adjusted taking into account different operation conditions, variation of plant parameters and disturbances. Using this type of control, performances are better in comparison to PID controllers and no need to adjust on-line the PID parameters. Currently, Fuzzy control has been used in a wide variety of applications in engineering, science, business, medicine, psychology, and other fields (Passino & Yurkovich, 1998). For instance, in engineering some potential application areas include the following:

1. Aircraft/spacecraft: Flight control, engine control, avionics systems, failure diagnosis, navigation, and satellite attitude control.
2. Automated highway systems: Automatic steering, braking, and throttle control for vehicles.
3. Automobiles: Brakes, transmission, suspension, and engine control.
4. Autonomous vehicles: Ground and underwater.
5. Manufacturing systems: Scheduling and deposition process control.
6. Power industry: Motor control, power control/distribution, and load estimation.
7. Process control: Temperature, pressure, and level control, failure diagnosis, distillation column control, and desalination processes.
8. Robotics: Position control and path planning.

In this chapter, a brief description of fuzzy control is given in order to understand how to design a fuzzy controller. Among the different type of fuzzy controllers, a Fuzzy PID supervision associated to a PID control is presented using different approaches. Also a new approach based a non linear model is proposed to design a fuzzy PID supervisor where performances are a priori defined and taken into. Finally, an automotive application is chosen for illustration and validation of the proposed approach. For a steer-by-wire a fuzzy

PD supervisor is designed and a comparison of performances is carried out with a PID control.

2. Fuzzy control

Fuzzy control is useful in some cases where the control processes are too complex to analyze by conventional quantitative techniques. Fuzzy control design is very interesting for industrial processes where modeling is not easy to make or conception of nonlinear controllers for industrial processes with models. The available sources of information of a process are interpreted qualitatively, inexactly or uncertainly. The main advantages of fuzzy logic control remains in (Passino & Yurkovich, 1998):

1. Parallel or distributed multiple fuzzy rules -complex nonlinear
2. Linguistic control, linguistic terms -human knowledge
3. Robust control

3. Fuzzy control system design

Figure 1 gives the fuzzy controller block diagram, where we show a fuzzy controller embedded in a closed-loop control system. The plant outputs are denoted by y(t), its inputs are denoted by u(t), and the reference input to the fuzzy controller is denoted by r(t). The design of fuzzy logic controller is based on four main components (Passino & Yurkovich, 1998):

1. The fuzzification interface which transforms input crisp values to fuzzy values
2. The knowledge base which contains a knowledge of the application domain and the control objectives
3. The decision-making logic which performs inference for fuzzy control actions
4. The defuzzification interface which provides the control signal to the process

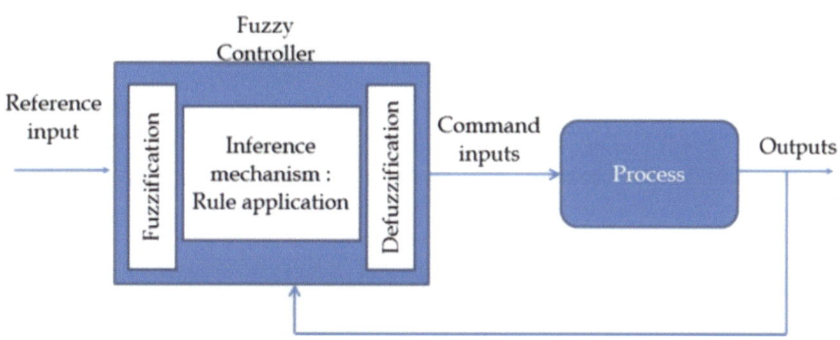

Fig. 1. Fuzzy controller diagram

3.1 Fuzzification

The fuzzification block contains generally preliminary data which are obtained from:

- Conversion of measured variables with analog/digital converters
- Preprocessing of the measured variables in order to get the state, error, state error derivation and state error integral of the variables to control (output variables or other state variables).

- Choice of membership functions for the input and output variables namely the shape, the number and distribution. Usually three to five triangular or Gaussian membership functions are used with a uniform distribution presenting 50% of overlapping. More than seven membership functions, the algorithm processing becomes long and presents a drawback for fast industrial processes.

3.2 Inference mechanism and rule–base

The Inference block is used to link the input variables to the output variable denoted X_R and considered as a linguistic variable given by a set of rules:

X_R = (**IF** (condition 1), **THEN** (consequence 1) **OR**
IF (condition 2), **THEN** (consequence 2) **OR**
....
.... **OR**
IF (condition **n**), THEN (consequence **n**).

n corresponds to the product of the number of membership functions of each input variable of the fuzzy logic controller.

In these rules, the fuzzy operators **AND, OR** link the input variables in the "condition" while the fuzzy operator **OR** links the different rules. The choice of these operators for inference depends obviously on the static and dynamic behaviors of the system to control. The numerical processing of the inference is carried out by three methods (Bühler, 1994; Godjevac, 1997):

1. max-prod inference method
2. max-min inference method
3. sum-prod inference method

3.3 Defuzzification

The numerical processing of the three methods provides a resultant membership function, $\mu_{RES}(x_R)$ for the output variable which is a fuzzy information. It's however necessary to convert the fuzzy information to an output signal x_R^* which is well defined. This conversion is named defuzzification and we have mainly four methods to get the output signal x_R^* (Bühler, 1994; Godjevac, 1997):

- centre of gravity
- maximum value
- centre of sums
- Height method or weight average

Also, the determined output signal x_R^* is converted to a control signal noted u_{cm} . This analog signal is provided to the power amplifier (power stage) of the process to control.

3.4 Different types of fuzzy logic controllers

On the basis of the consequence of rules given above, different types of fuzzy logic controllers are presented.

1. If the consequence is a membership function or a fuzzy set, the fuzzy controller is Mamdani type. In this case, the processing of inference uses often the max-min or max-prod inference method while for defuzzification, the center of gravity method is often used and in some cases we use the maximum value method if fast control is needed.
2. If the consequence is a linear combination of the input variables of the fuzzy logic controller. Indeed each rule corresponds to a local linear controller around a steady state. Consequently, the set of the established rules correspond to a nonlinear controller. In this case, we use max-min or max-prod inference method and for deffuzification, we often use the weight average method.

Also, it exists other types of fuzzy logic controllers such as Larsen or Tsukamoto (Driankov et al., 1993). Most of time Mamadani and TSK controllers are used in the design of controllers for nonlinear systems with or without models (Bühler, 1994; Godjevac, 1997; Nguyen, 1997).

The advantages of the design of a fuzzy logic controller using Mamdani type are an intuitive method, used at a big scale and well suited for translation of human experience on linguistic rules.

On the other hand, the advantages of a fuzzy logic controller using a Takagi-Sugeno type are:

1. Good operation with linear techniques (the consequence of a rule is linear)
2. Good operation with optimization techniques and parameters adaptation of a controller
3. Continuous transfer characteristics
4. very suited for systems with a model
5. fast processing of information

3.5 Discussion

The different steps followed in the processing of the input variables of the fuzzy logic controller namely fuzzification, inference and defuzzification, allows to get a non linear characteristic. Indeed it's an advantage when compared to the classical control. The nonlinearity of this characteristic depends on some parameters. For example the number, the type and the distribution of membership functions. Also, other parameters can be considered such as the number of rules and inference methods. Finally, the nonlinearity can be more or less pronounced depending on all these parameters.

In this case we consider the fuzzy logic controller as a nonlinear controller. Another possibility to get a non linear controller is to design and to add a fuzzy supervision to a PID controller. Industrials are motivated to keep PID controllers which are well known and to add a fuzzy supervisor which modifies on-line PID parameters in order to reach and to maintain high performances whatever the parameters change and operations conditions maybe. In the design of the fuzzy supervision, the outputs are the PID parameters to provide on-line to the PID controller.

4. Fuzzy supervisory control

Fuzzy Supervisory controller is a multilayer (hierarchical) controller with the supervisor at the highest level, as shown in Figure 2. The fuzzy supervisor can use any available data from the control system to characterize the system's current behavior so that it knows how to change the controller and ultimately achieve the desired specifications. In addition, the supervisor can be used to integrate other information into the control decision-making process.

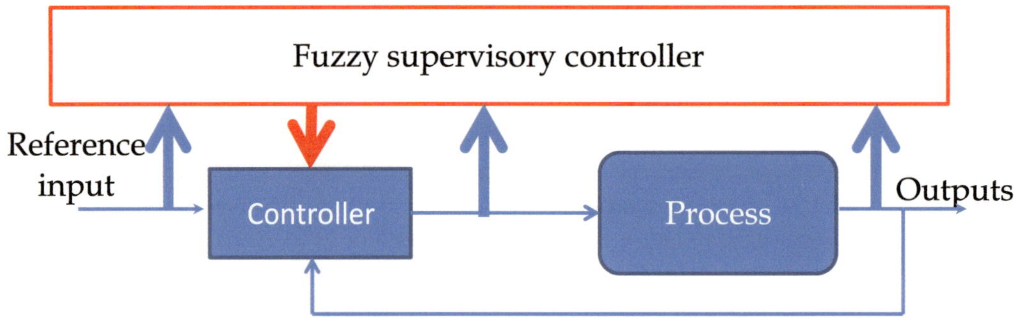

Fig. 2. Fuzzy Supervisory controller

Conceptually, the design of the supervisory controller can then proceed in the same manner as it did for direct fuzzy controllers (fuzzification, inference and defuzzification): either via the gathering of heuristic control knowledge or via training data that we gather from an experiment. The form of the knowledge or data is, however, somewhat different than in the simple fuzzy control problem. For instance, the type of heuristic knowledge that is used in a supervisor may take one of the following two forms:

1. Information from a human control system operator who observes the behavior of an existing control system (often a conventional control system) and knows how this controller should be tuned under various operating conditions.
2. Information gathered by a control engineer who knows that under different operating conditions controller parameters should be tuned according to certain rules.

Fuzzy supervisor is characterized by:

1. The outputs which are not control signals to provide to the control system but they are parameters to provide to the controller in order to compute the appropriate control.
2. Fuzzy supervision associated to the controller can be considered as an adaptatif controller
3. Fuzzy supervisor can integrate different types of information to resolve problems of control.

4.1 Supervision of conventional controllers

Most controllers in operation today have been developed using conventional control methods. There are, however, many situations where these controllers are not properly tuned and there is heuristic knowledge available on how to tune them while they are in operation. There is then the opportunity to utilize fuzzy control methods as the supervisor that tunes or coordinates the application of conventional controllers. In this part, supervision of conventional controllers concerns only PID controllers and how the supervisor can act as a gain scheduler

4.2 Fuzzy tuning of PID controllers

Over 90% of the controllers in operation today are PID controllers. This is because PID controllers are easy to understand, easy to explain to others, and easy to implement. Moreover, they are often available at little extra cost since they are often incorporated into the programmable logic controllers (PLCs) that are used to control many industrial processes. Unfortunately, many of the PID loops that are in operation today are in continual need of monitoring and adjustment since they can easily become improperly tuned.

Because PID controllers are often not properly tuned (e.g., due to plant parameter variations or operating condition changes), there is a significant need to develop methods for the automatic tuning of PID controllers for nonlinear systems where the model is not well known. In this method, the fuzzy supervisor knows, from a response time, when the controller is not well tuned and acts by adjusting the controller gains in order to improve system performances. The principle scheme of the fuzzy PID auto tuner (Passino & Yurkovich, 1998) is given by figure 3.

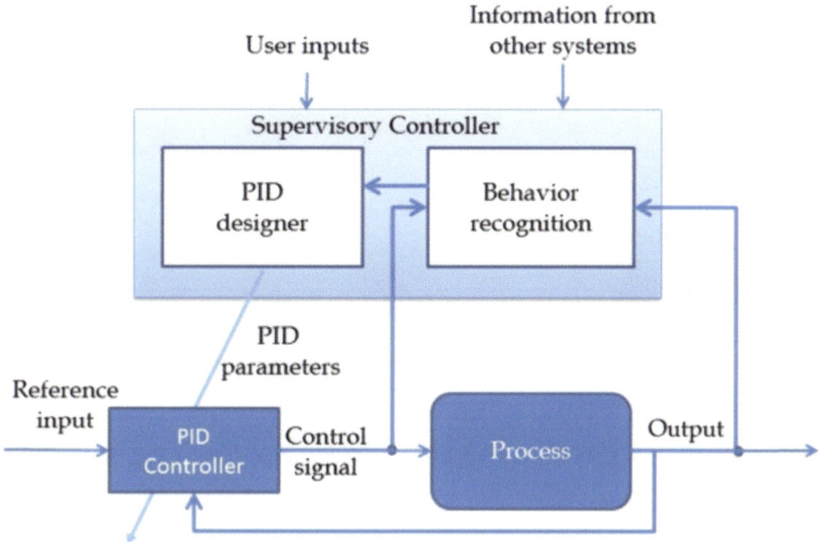

Fig. 3. Fuzzy PID auto-tuner

The block "Behavior Recognition" is used to characterize and analyze the current response of the system and provides information to the "PID Designer" in order to determine the new parameters of the PID controllers and to improve performances. The basic form of a PID controller is given by:

$$u(t)=K_P e(t)+K_I \int_0^t e(\tau)d\tau+K_D \frac{d}{dt} e(t) \tag{1}$$

Where u is the control signal provided by the PID controller to the plant.
e: the error deuced from the reference input r and the plant output y.
K_p is the proportional gain, K_i is the integral gain, and K_d is the derivative gain.

In this case, the adjustment of PID parameters is carried out by some candidate rules as follows:

- **If** steady-state error is large **Then** increase the proportional gain.
- **If** the response is oscillatory **Then** increase the derivative gain.
- **If** the response is sluggish **Then** increase the proportional gain.
- **If** the steady-state error is too big **Then** adjust the integral gain.
- **If** the overshoot is too big **then** decrease the proportional gain.

In these rules conditions are deal with the block «Behavior Recognition" and consequences are evaluated by the block "PID Designer" of the fuzzy supervisor. In some applications controller gains are quantified according to different types of responses a priori identified from experiments on the real process and implemented on the block "Behavior Recognition" (Passino & Yurkovich, 1998).

4.3 Fuzzy gain scheduling

Conventional gain scheduling involves using extra information from the plant, environment, or users to tune (via "schedules") the gains of a controller. The overall scheme is shown in Figure 4. A gain schedule is simply an interpolator that takes as inputs the operating condition and provides values of the gains as its outputs. One way to construct this interpolator is to view the data associations between operating conditions and controller gains.

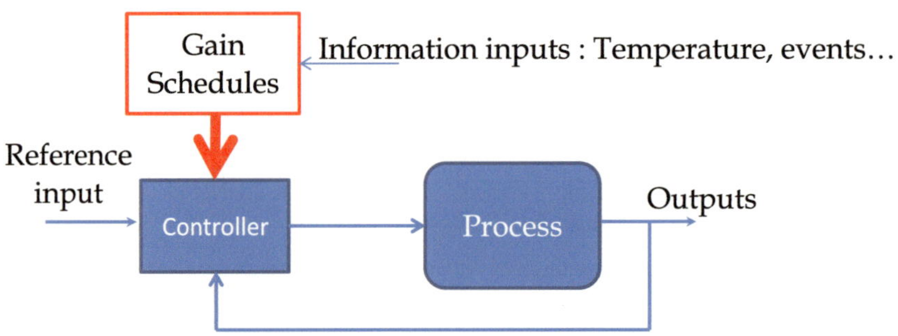

Fig. 4. Conventional fuzzy gain scheduler

The controller gains are established on the basis of information collected from the plant to control, the operator or the environment. Three approaches are proposed for the construction of the fuzzy gain scheduling (Passino & Yurkovich, 1998) :

- Heuristic Gain Schedule Construction
- Construction of gain schedule by fuzzy identification
- Construction of gain schedule using the PDC method (Parallel Distributed Compensation method)

4.3.1 Construction of an heuristic schedule gain

This method is applied for plants with specific particularities not involved in the design of classical controllers. The PID parameters are deduced intuitively and the rules used for the adjustment of parameters are heuristic. This is for example the case of a tank with an oval shape (figure 5). In the heuristic rules, the condition corresponds to the water levels and the consequence corresponds to the values of the controller gain (Passino & Yurkovich, 1998).

Each rule covers a set of water levels taking into account the tank section. For low levels, the gain is higher in order to get high flow rates and for high water levels, the gain is small in order to get small flow rates. This approach is very useful for systems without models.

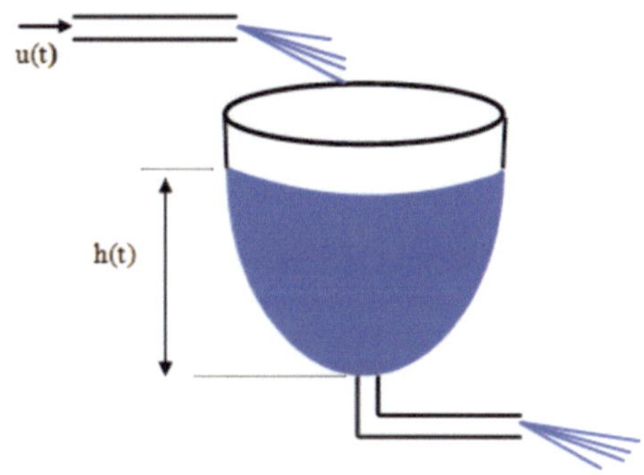

Fig. 5. Tank

4.3.2 Construction of a schedule gain by fuzzy identification

This approach is useful for plants where we know a priori how to adjust the controller gains under different operation conditions (Passino & Yurkovich, 1998). For example if a control engineer knows how to adjust gain controller according to certain rules, he can represent this data by a fuzzy model of Mamdani or TSK type. Indeed it's the equivalent of a set of controllers which are active in terms of the operation points. Also, the gain controllers are deduced on-line by the inference mechanism between controllers for any operation point. Indeed it's a soft transition from controller to another one.

4.3.3 Construction of a gain schedule using the PDC method

This approach is applied particularly for processes that can be modelled. Most of time, the established models are nonlinear. In this case, the nonlinear model is replaced by a sum of linearized models around different operation points (Passino & Yurkovich, 1998), (Vermeiren, 1998). For each linearised model, a linear controller is designed (figure 6). These linear controllers could be PI, PID, PD ou state controllers. The set of the designed controllers is finally a non linear controller which is a fuzzy controller.

In this approach the n linearized models and the n corresponding controllers are rules which are active simultaneously two by two since the condition is similar for both, thus the name of the method "**Parallel Distributed Compensation**".

In all the approaches presented above, performances are not used directly when designing controllers. Also non linearities, disturbances and variation parameters of the plant are not taken into account in the systems with models.

In some cases stability is not ensured particularly when a change set point occurs or when a disturbance is present. In the case of the PDC approach, local and global stabilities are checked using Lyapunov theory (Passino & Yurkovich, 1998). For the other approaches, stability is checked when implementing fuzzy supervision for classical controllers.

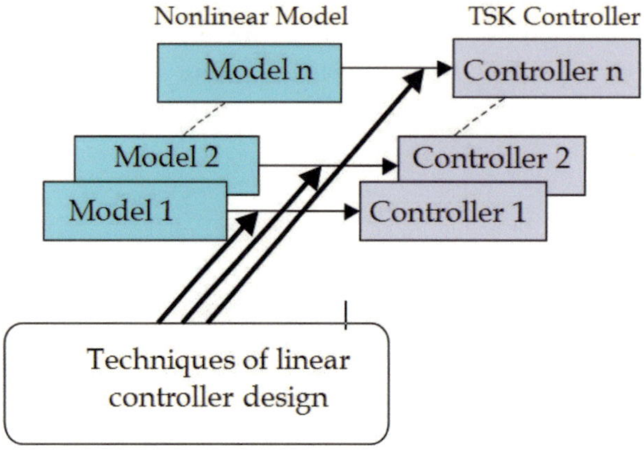

Fig. 6. PDC Concept

Usually specifications and performances in closed loop are a priori defined. Therefore, it's more interesting to use them and to design controllers ensuring stability and same performances in closed loop whatever the operation conditions maybe.

5. The proposed approach

In the proposed approach (Sehab, 2007), Fuzzy PID Supervision is designed for systems with models. PID parameters are designed taking into all the nonlinearities of the system in closed loop where noise measurement is considered. Also the desired output performances are fixed using the Simulink Response Optimization for the complete range of operation. Choosing a set of set points, the optimal PID parameters are computed according to the desired performances. With the collected data, fuzzy blocks are designed and implemented on the Simulink model in order to modify on-line the PID parameters in terms of the set point and the output variable of the process and other parameters.

The advantages of the proposed approach are:

- All the model nonlinearities are taken into account in the design of the controller gains.
- The commutation from PID controller to another is soft when a set point change occurs. This is due to the nonlinear interpolation of two controller parameters (defuzzifation).
- The designed fuzzy blocks provide on-line the appropriate control signal to the plant for any set point where the output performances are all the time maintained.
- Stability is ensured for any set point or physical parameter changes.
- The control is robust since the performances are maintained for any operation conditions.

The proposed approach is applied and validated on a test bench of a three tank system (Sehab et al, 2001). Different tests are carried out varying the configuration, modifying the set point profile during operation (Sehab, 2007). The obtained results confirm the robustness of the fuzzy PID supervision implemented on this application. In this chapter, a second application is chosen in order to design a fuzzy PID controller according to the same proposed approach.

In automotive, many systems are integrated in vehicle in order to ensure continuously the driver comfort and the safety during driving. Among them, the active suspension, antibraking system (ABS), electronic stability program (EPS) and steer by wire SBW). In our case, the steer by wire is chosen for this study. Some works on "steering by wire" using fuzzy control were realized. The interest of using fuzzy techniques consisted to treat the global non-linearity of this device. For example, fuzzy controller improved the vehicle's stability at different speeds (Shu et al., 2011). An hybrid-fuzzy controller combining a conventional PID and a fuzzy controller together were developed (Qu et al. 2010). This controller gave results showing quick responses and little overshoot like conventional steering device.

6. System description

Steer-By-Wire (SBW) System (Figure 7) is a new design which eliminates the mechanical link between the handwheel and the roadwheel. The measured handwheel position controls the roadwheel position using an electrical motor (road motor). However, the driver has to feel the feedback road force as in the case of a classical steering. For sensing this force, a sensor torque is added in order to control a second electrical motor (Feedback motor). The associated inputs and outputs of motors are managed by an Electronic Control Unit (ECU). The road motor and the feedback motor are respectively controlled in angular position and torque by P and PI controllers. Both are implemented in the ECU.

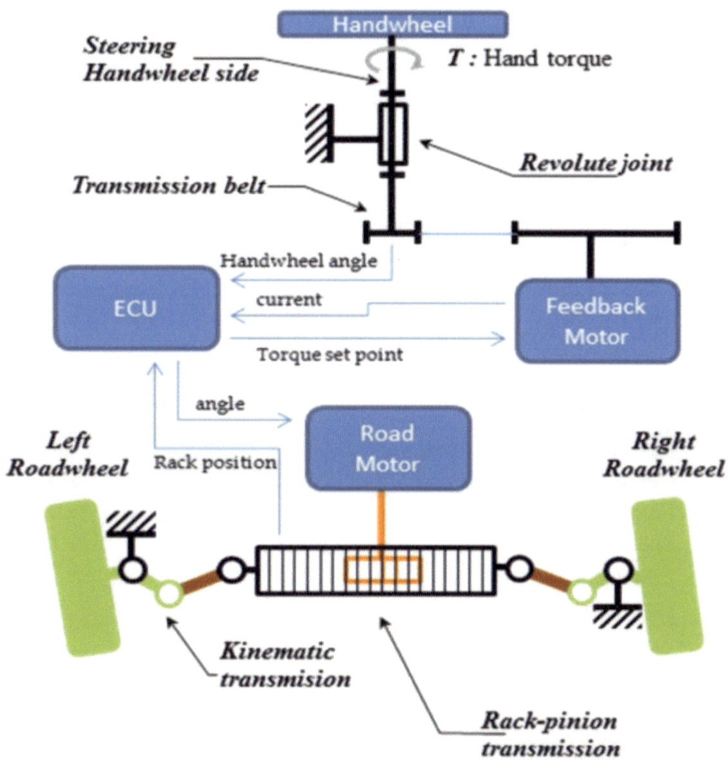

Fig. 7. Steer-by-Wire System

7. System modeling

In the system given by figure 7, different subsystems are considered. The modeling of each subsystem of the steer by wire is described in the following paragraphs.

7.1 Load model (Vehicle)

The appropriate physical load used for testing the architecture is the well-known "bicycle model" (Brossard, 2006) given by figure 8. The choice of this load allows simulating all technological components, such as angle and torque sensors, and actuators. This model load is a planar model (longitudinal, lateral and yaw motions) where wheels are in contact with the ground by a rigid beam. The front wheel has one rotation degree of freedom and is perpendicular to the plane. The rear wheel is blocked.

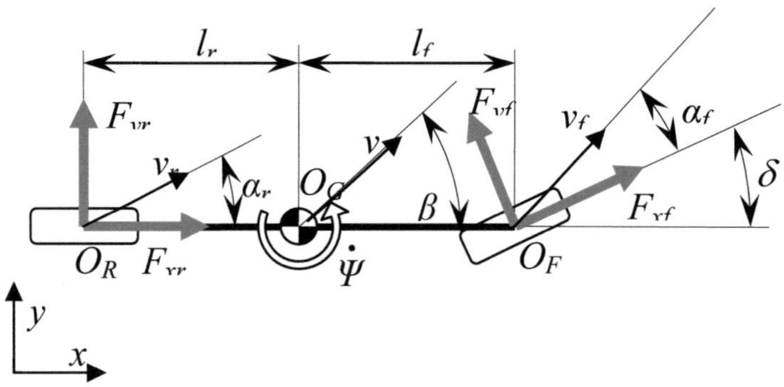

Fig. 8. Description of the bicycle model (load)

O_G, O_R, O_F	Point marked the center of gravity, the center of rotation for the rear wheel and the center of rotation for the front wheel
l_f (l_r)	Distance between O_G *and* O_F (O_G *and* O_R)
m	Total mass of the vehicle
Iz	Inertia along z axis
F_{yr} (F_{yf})	Rear (or front) lateral wheel force
F_{xf}	Front longitudinal wheel force
	Front wheel steering angle
v_x (v_y)	Cartesian component of vehicle velocity vector
α_f (α_r)	Front (or rear) tire side slip angle
β	Slip angle at the vehicle center of gravity
C_f (C_r)	Lateral stiffness of front (or rear) tire
ψ	yaw angle vehicle
ψ	yaw rate

Notice: All the variables associated to the load model are given in the MKSA system.

In the modeling of the load some assumptions are considered:

- small variations are taken into account

- the longitudinal velocity is constant
- the vehicle doesn't slip

According to figure 8, the mechanical dynamic equations are given by:

$$F_{yf}\cos\delta - F_{xf}\sin\delta + F_{yr} = m\left(\dot{v}_y + v_x\dot{\psi}\right) \tag{2}$$

$$l_f\ F_{yf}\cos\delta - l_r\left(F_{xf}\sin\delta + F_{yr}\right) = I_z\ddot{\psi} \tag{3}$$

Including the non-holomic kinematic constraints, the slip angles of the tires are given as:

$$\alpha_f = \tan^{-1}\left(\frac{v_y + l_f\dot{\psi}}{v_x}\right) - \delta \tag{4}$$

$$\alpha_r = \tan^{-1}\left(\frac{v_y - l_r\dot{\psi}}{v_x}\right) \tag{5}$$

Considering the force generated by the wheels as linearly proportional to the slip angle, the lateral forces are defined as:

$$F_{yf} = -C_f\alpha_f \tag{6}$$

$$F_{yr} = -C_r\alpha_r \tag{7}$$

The whole of those equations gives the model of the vehicle where the front wheel steering angle is the input and the yaw rate is the primary output. With this output, the lateral forces are deduced.

7.2 DC motor models

The DC motor model is valid for feedback and road motors of the Steer by Wire system given by figure 1.The electrical equation is given by:

$$U = R.i + L\frac{di}{dt} + e \tag{8}$$

where:

R, L are respectively the armature resistance (Ω) and inductance (H)
i: armature current (A)
e: back e.m.f (V)

Also, the mechanical equation is given by:

$$J_m.\dot{\omega} = T_m - T_{friction} - T_{load} \tag{9}$$

J_m: motor inertia (kgm^2)

ω: motor rotary velocity(rd/s)
T_m: motor torque (Nm)
$T_{friction}$: friction torque (Nm)
T_{load}: load torque (Nm)

The friction torque is defined by the resultant of viscous and dry frictions:

$$J_m.\dot{\omega} = T_m - T_{friction} - T_{load} \tag{10}$$

Also, motor torque and back e.m.f are given by:

$$e = k.\omega \tag{11}$$

$$T_m = k.i \tag{12}$$

k: the back e.m.f coefficient (Nm/A)

7.3 Rack/pinion model

The rack and pinion converts a linear displacement to an angle or a torque to a force. This transformation respects a ratio corresponding to the primitive radius *R* of the pinion. As the frictions (no backlash and no slide) are neglected, the corresponding equations are given by:

$$V = R.\omega \tag{13}$$

$$T_m = R.F \tag{14}$$

7.4 Simulink model description

Using the DC motor model for the wheel and feedback motors, load and Rack/Pinion models and the corresponding physical parameters, an implementation is carried out in the environment of Simulink. Figure 9 describes the whole architecture of the Steer-by-Wire model in closed loop. Also the PI and P controllers of the ECU are designed and implemented.

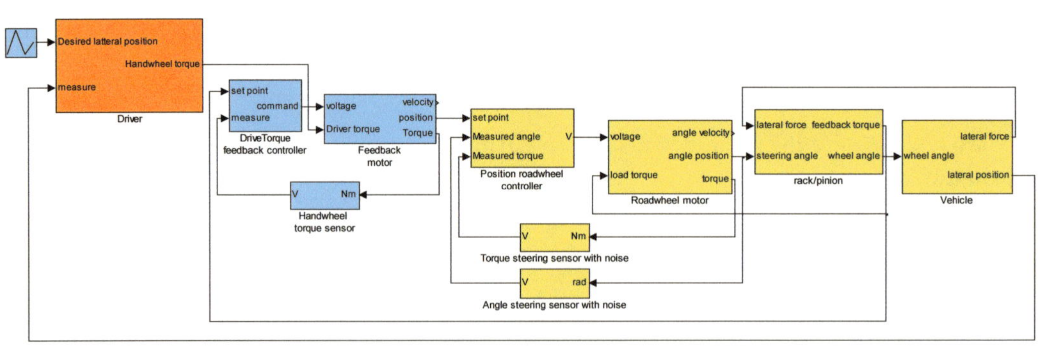

Fig. 9. Simulink Model of the Steer by Wire in closed loop

In order to reach the desired lateral position, the driver provides the appropriate handwheel torque according to the vehicle speed. Therefore, a driver model is proposed and implemented in the Simulink model of the Steer-by-Wire of figure 9. A proportional derivative controller is however chosen to reproduce the dynamic behavior of the driver. Indeed no need of the integral part in the controller since in the model, rack/pignon has an integral behavior.

Also random disturbances are added in the Simulink model for the measured armature currents (torque sensors) of the handwheed and the roadwheel motors and measured angular steering as shown in figure 9.

On the basis of the complete Simulink model, the PD parameters are determined for a chosen lateral position and a given vehicle speed where vehicle reaches the lateral position with minimum overshoot and time response.

Using the designed controller, a simulation is carried out for other operation points defined by the lateral position and the vehicle speed chosen by the driver. In all the studied cases, the performances are not maintained and in other cases system is instable.

Indeed, all the nonlinearities presented in the model involve, in some cases, instability and in others, degradation of system performances even responses, for some operation points, are stable. For this reason, it's interesting to design a Fuzzy PID Supervisor in order to modify on-line the PD parameters ensuring permanently good performances and stability whatever the operation conditions maybe.

8. Design procedure of fuzzy PD supervision

According to the proposed approach, the first step to follow is to design PD controller for each operation point where performances are imposed using optimization response time toolbox of Matlab. In this case, the chosen performances are minimum response time and overshoot of the lateral position followed by the vehicle according the lateral position of the trajectory followed by the driver. The operation ranges are defined by [1.5 - 7] m for the lateral position and by [0 - 72] km/h for the vehicle speed. These values correspond to the normal driving ensuring the driver safety. Choosing different set points from the defined operation ranges, optimal PD parameters are computed. For each case, stability is ensured but the obtained responses do not satisfy the desired performances at a high level.

On the basis of the collected data (K_p and T_d) two fuzzy blocks are conceived using Mamdani approach of fuzzy logic toolbox of Matlab. The first block provides K_p in terms of the error on the lateral position and the second provides T_d in terms also of the error on the lateral position. Figure 10 gives the designed fuzzy PD Supervisor which varies on-line the parameters K_p and K_d of the PD controller. Indeed the Fuzzy PD supervisor associated to the PD controller is also considered as a fuzzy logic controller. Also, the control signal noted U_c corresponds to the resistive torque to apply on the hanwheel taking into account the vehicle speed and the desired lateral position. For a given vehicle speed, the values of K_p and T_d are computed by defuzzification (nonlinear interpolation) from the measured lateral position (Y) and the chosen set point ($Y_{Set\ point}$).

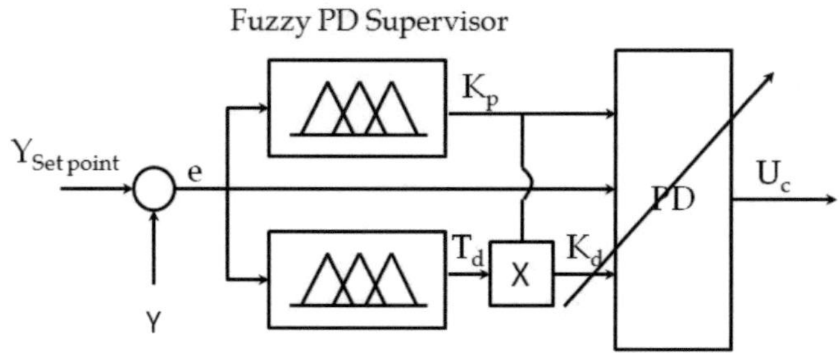

Fig. 10. Fuzzy Controller

9. Implementation and simulation

In the second step of this study, the fuzzy PD supervisor of figure 10 is implemented on the simulator of the Steer-by-Wire given by figure 9. Choosing a vehicle speed and lateral position set points from the operation ranges, a simulation is carried out in order to make a comparison with classical control using the corresponding PD controller.

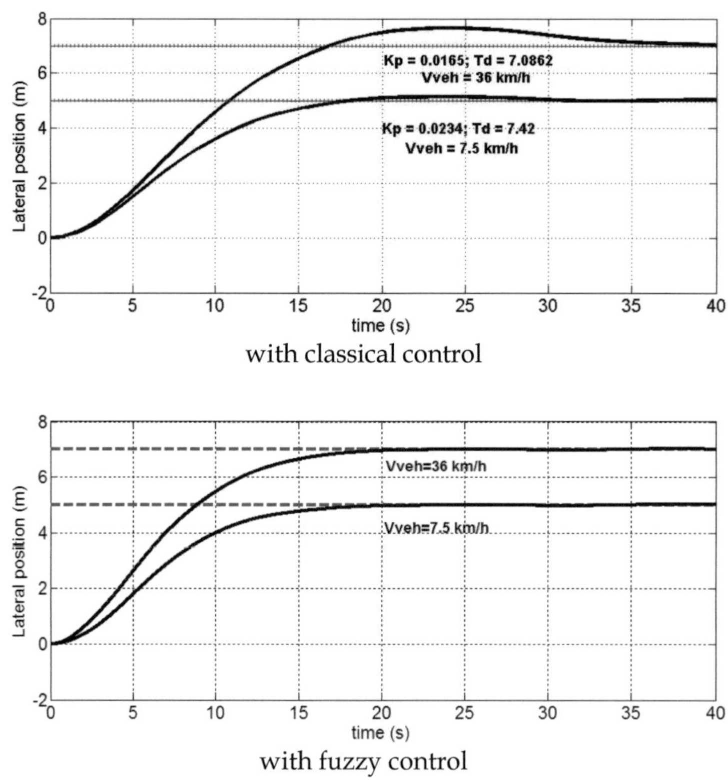

Fig. 11. Lateral position

For a lateral position set point and a vehicle speed, figure 11, gives the evolution of the lateral position in case of classical control (figure 11-a) and in case of fuzzy logic control (figure 11-b). Indeed with the fuzzy control the steady state is reached, without overshoot in

20s, while with classical control, the steady state is reached, with an overshoot, in 35s. Indeed the performances are better with fuzzy control in comparison to the classical control.

Also, for the chosen lateral position y = 7m and a vehicle speed Vveh= 36 km/h, the evolution of torques of the feedback motor are shown in figure 12. Indeed the rack torque is well compensated by the load torque applied by the driver on the handwheel even with fuzzy or classical control. Also, the steady state torque responses are reached in 15 s (Fig. 12-b) when compared to the classical control (Fig.12-a). Indeed this is due to the high value of the load torque applied by the driver in case of fuzzy control during transient time.

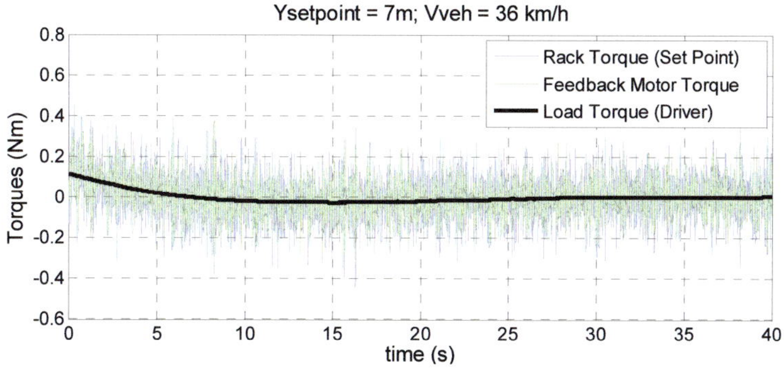

-a- with classical control

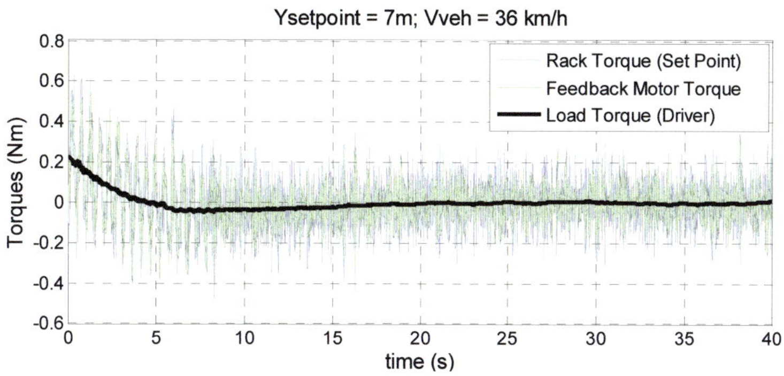

-b- with fuzzy control

Fig. 12. Torques of the feedback motor

Also in figure 13, the rack angular position is reached with a response time of 15 s (Fig.13-b) in comparison to the classical control (Fig.13-a). Also, during transient time, the rack angular set point provided by the handwheel motor, in case of fuzzy control, presents some oscillations when compared to the same response with classical control. Indeed it's a drawback due to the nonlinearity of the vehicle dynamics and eventually variations of PD parameters.

Also for the wheel angle, the steady state is reached in 15s with the same oscillations which are involved by the rack angular position during the transient time (Fig.14-b) while with the classical control, the steady state is reached in 25 s with small oscillations during transient time (Fig.14-a).

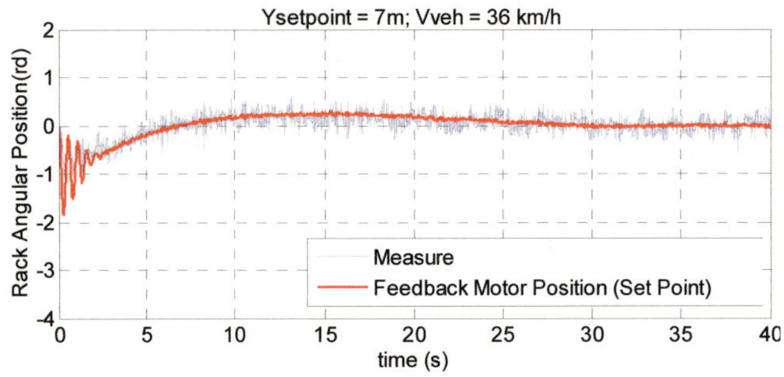

a- with classical control

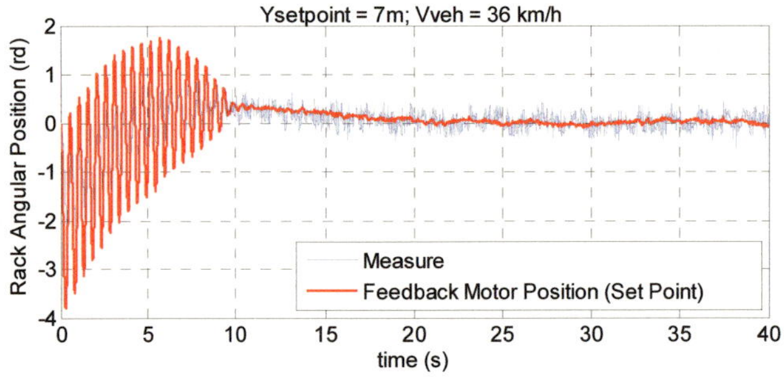

b- with fuzzy control

Fig. 13. Rack angle position

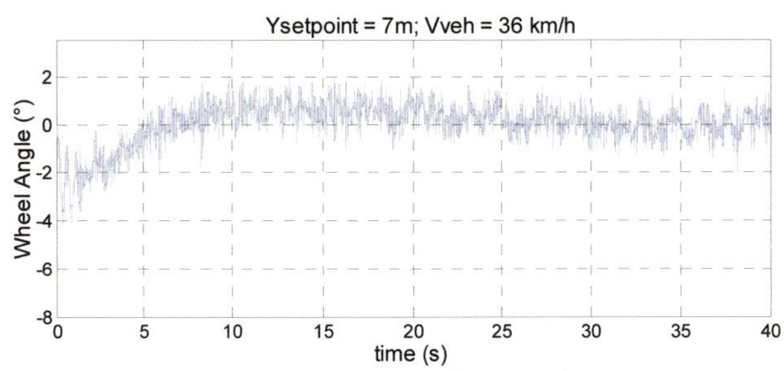

-a- with classical control

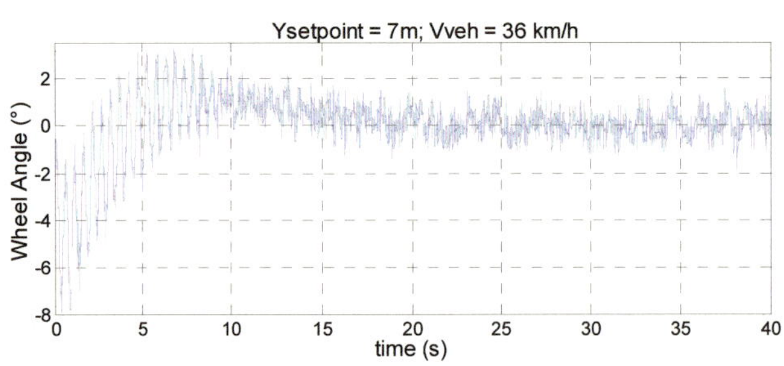

with fuzzy control

Fig. 14. Wheel angle

According the different responses, it's clear that performances are better with the fuzzy control (Fuzzy PD supervisor with a PD controller) in comparison to the classical control. Indeed, the designed Fuzzy PD supervisor provides on-line the PD parameters Kp and Td allowing to reach the desired lateral position with good performances. Also, for the chosen lateral position, the evolution of the PD parameters Kp and Td is given by figure 15. For both, Kp and Td, have steady state values while for the classical control, the PD parameters are constant during all the operation time. In this case, we consider that performances are better because the PD parameters vary in order to reach the desired lateral position very quickly and without overshoot.

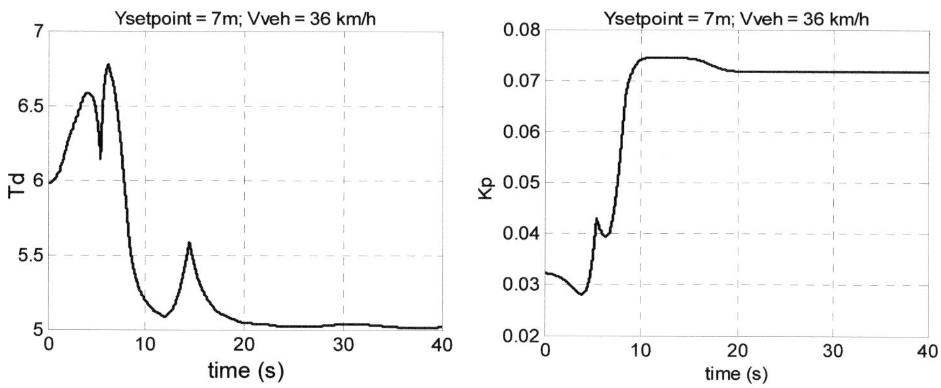

a) Derivative time constant b) Proportional gain

Fig. 15. PD Parameters

Also, in order to validate the designed fuzzy controller, other operation points and profiles of lateral position are simulated for different vehicle speeds. For each case, performances are maintained and the steady state is reached with a minimum response time and without overshoot when compared to the classical control.

10. Conclusion

In the control of a nonlinear process, classical control is robust but not optimal for the complete range of operation conditions. Indeed, the design of one controller is not sufficient to ensure good performances and stability for all the operation set points. Also, the variation of physical parameters of a process over time affects the performances. Therefore a continuous adjustment of controller gains is necessary to improve and eventually to maintain performances. On the basis of the proposed approach, performances are used a priori in the design of the fuzzy PID supervision taking into account the variation of parameters and operation conditions. Indeed in terms of both, the designed fuzzy PID supervision provides on-line the appropriate gains to the PID controllers ensuring the same performances whatever the operation conditions maybe. For the chosen application, the designed fuzzy PD controller ensures for any value of vehicle speed and lateral position, a good response where performances are maintained. In this application, the designed fuzzy PD supervisor associated to the PD controller is used to provide the resistive torque to apply on the handwheel. The wheel resistive torque is however compensated according to the desired lateral position by the steer-by-wire. For safety reasons, the proposed Fuzzy controller could be used in the vehicle as assistance during driving.

11. References

Babuska, B. (1997). *Fuzzy Modelling and Identification*, Ph.D thesis, University of Delft, ND.

Bongards, M. (1996). Application of Fuzzy Adapted PID Controllers in the Closed Loop Control of Industrial Compressed-Air System," *Proceedings of EUFIT'96*, Aachen, Germany, Vol 2, September 2-5, pp. 1258-1265.

Brossard, J.P. (2006) *Dynamique du véhicule. Modélisation des systèmes complexes*, Presses Polytechniques et Universitaires Romandes,. chap. 9.

Bühler, H. (1994). *Réglage par Logique Floue.* Presses Polytechniques et Universitaires Romandes.

Driankov, D., Hellendoorn, H. & Reinfrank, M. (1993) *An Introduction to Fuzzy Control*, Springer Verlag, Berlin Heidelberg.

Godjevac, J. (1997). *Neuro-Fuzzy controller: Design and Application.* Presses Polytechniques et Universitaires Romandes.

Nguyen, H. & Sugeno, M. (1998). *Fuzzy Systems: Modelling and Control* Library of Congress Cataloging-in-Publication Data.

Passino, K.M. & Yurkovich, S. (1998). *Fuzzy control*, Addison-Wesley-Longman, Menlo Park, CA.

Qu, A.Q., Wang, T.J & Liu X.H. (2010). Study on Hybrid-Fuzzy Control Algorithm of Wheel Loader Steer by Wire System, 2nd International Conference on Industrial and Information Systems, Dalian, China.

Sehab, R., Remy, M. & Renotte, CH. (2001). An Approach to Design Fuzzy PI Supervisor For a Nonlinear System, *Proceedings of North American Fuzzy Information Processing Society, NAFIPS'2001*, July 25-28, Vancouver, Canada, pp 894-899.

Sehab, R.(2007). Fuzzy PID Supervision for a Non linear System: Design and Implementation, *Proceedings of NAFIPS'2007, the North American Fuzzy Information Processing Society Conference, pp 894-899*, June 24-27, San Diego, California, USA.

Shu, F.G., Qing, F. P., Li, F.W. (2011), The Research of Steer by Wire System Based on Fuzzy Logic Control, *Applied Mechanics and Materials (Volumes 55 - 57)*, pp. 780-784.

Tanaka, K. (1998). *Stability of Fuzzy Controllers*, Library of Congress Cataloging-in-Publication Data. U.S.A.

Van Nauta Lemke, H.R. & Wang D. Z. (1985). Fuzzy PID Supervisor, *Proceedings of 24th IEEE Conference on Decision and Control*, Ft. Lauderdale, FL, USA December, pp. 602-608.

Verbruggen, H.B. & Babuska B. (1999) *Fuzzy Logic Control advances in application* World Scientific series in Robotics and Intelligent Systems. Volume 23, April.

Vermeiren, L. (1998), *Proposition de lois de commande pour la stabilisation de Modèles Flous*, Thèse de Doctorat, Université de Valenciennes et du Hainaut-Cambrésis, France.

White, D.A., Sofge, D.A. (1992). *Neural,Fuzzy, and Adaptative Approaches.* Handbook of Intelligent Control. U.S.A.

Xie, X.Q., Zhou, D.H. & Jin, Y.H. (1999). Strong tracking filter based adaptive generic model control, *Journal of Process Control*, N° 9, pp 337-350.

6

Conceptual Model Development for a Knowledge Base of PID Controllers Tuning in Closed Loop

José Luis Calvo-Rolle[1], Héctor Quintián-Pardo[1], Antonio Couce Casanova[1] and Héctor Alaiz-Moreton[2]
[1]*University of Coruña*
[2]*University of León*
Spain

1. Introduction

In the area of control engineering work must be constant to obtain new methods of regulation, to alleviate the deficiencies in the already existing ones, or to find alternative improvements to the ones used previously. This huge demand of control applications is due to the wide range of possibilities developed to this day.

Regardless of this increasing rhythm of discovery of different possibilities, it has been impossible at this moment to oust relatively popular techniques, as can be the 'traditional' PID control. Since the discovery of this type of regulators by Nicholas Minorsky Mindell (2004) Bennett (1984) in 1922 to this day, many works have been carried out about this controller. In this period of time there was an initial stage, in which the resolution of the problem was done analogically and in it the advances were not as remarkable as have been since the introduction of the computer, which allows to implement the known structure of direct digital control Auslander et al. (1978), illustrated in figure 1.

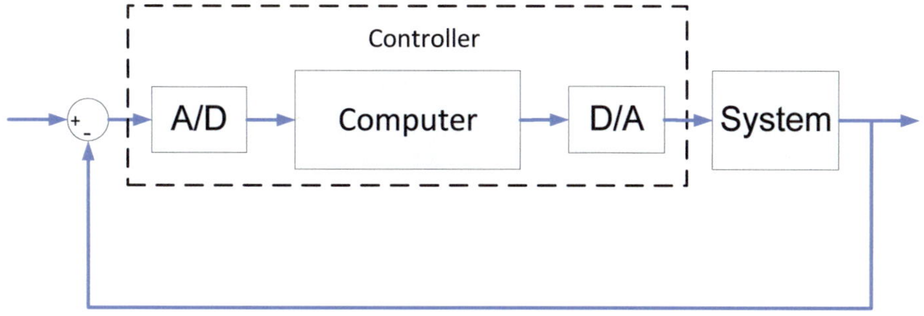

Fig. 1. Structure of direct digital control

Since then, regulators have passed from being implemented in an analogous way to develop its control algorithm digitally, by signal digital processors. As well as carrying out the classic PID control in digital form, its development based on computer allows adding features to the regulator with difficulty could have been obtained analogically.

We must say that there exist usual control techniques for the processes in any area, in which innovations have been introduced. But nevertheless, the vast majority of these techniques in their implementation employ PID traditional controllers, although in an improved way, increasing the percentage of use around 95% Astrom & Hagglund (2006). Its use is still very high due to various reasons like: robustness, reliability, relative simplicity, fault, etc.

The great problem of the PID control is the adjustment of the parameters that it incorporates. Above all in its conventional topology Astrom & Hagglund (2006) Feng & Tan (1998), as a consequence of the investigations carried out in the area, many contributions have been made by specialists, existing among them many methods to obtain the parameters that define this regulator, achieved through different ways, and working conditions of the plant to control. It must be highlighted that the methods developed to obtain the terms, which in occasions are empiric, are always directed to optimize defined specifications; the negative thing is that frequently when some specifications are improved others get worse.

It is necessary to highlight that empirical methods have been the first to be discovered and they are often the ones first learnt in the training of technicians for this discipline. In this sense the parameters obtained in this way through the application of formulas of different authors, are a starting point of adjustment of the regulator, being necessary, normally, a later fine adjustment by trial and error.

Regardless of what has been said, in practice there is a wide variety of regulators working in the industry with an adjustment far from what can be considered optimum Astrom & Hagglund (2006). This fact is originated among other reasons due to a lack of adjustment techniques by the users.

This fact creates the necessity to use intelligent systems, due to the demand of a better performance and resolution of complex problems both for men as well as for the machines. Gradually time restrictions imposed in the decision making are stronger and the knowledge has turned out to be an important strategic resource to help the people handling the information, with the complexity that this involves. In the industry world, intelligent systems are used in the optimization of processes and systems related with control, diagnosis and repair of problems. One of the techniques employed nowadays are knowledge based systems, which are one of the streams of artificial intelligence.

The development of knowledge based systems is very useful for certain knowledge domains, and also indispensable in others. Some of the most important advantages offered by knowledge based systems are the following:

- Permanence: Unlike a human expert, a knowledge based system does not grow old, and so it does not suffer loss of faculties with the pass of time.
- Duplication: Once a knowledge based system is programmed we can duplicate it countless times, which reduces the costs.
- Fastness: A knowledge based system can obtain information from a data base and can make numeric calculations quicker than any human being.
- Low cost: Although the initial cost can be high, thanks to the duplication capacity the final cost is low.
- Dangerous environments: A knowledge based system can work in dangerous or harmful environments for the human being.

- Reliability: A knowledge based system is not affected by external conditions, as a human being is (tiredness, pressure, etc).
- Reasoning explanation: It helps justify the exits in the case of problematic or critical domain. This quality can be used to train not qualified personnel in the area of the application.

Up to now the existing knowledge based systems for resolution of control systems have reduced features Pang (1991) Wilson (2005) Zhou & Li (2005) Epshtein (2000) Pang et al. (1994) Pang (1993), summarizing, in application of the method known as "Gain Schedulling" Astrom & Wittenmark (1989a), which is based in programming the profits of the regulator with reference to the states variables of the process. For the cases in which the number of control capacities have increased, the knowledge based system, is applicable to specific problems. There is the possibility to implement knowledge based systems programming them in the devices, but without taking advantage of the existing specific tools of Knowledge Engineering Calvo-Rolle & Corchado (n.d.) Calvo-Rolle (2007).

According to what has been said, the development of a PID conceptual model is described in this document to obtain the parameters of a PID regulator with the empirical adjustment method in a closed loop; feasible in the great majority of cases in which such method is applicable. The model has been developed for six groups of different expressions with highly satisfactory results, and of course expandable to more following the same methodology.

The present document is structured starting with a brief introduction of PID regulator topology employed, along with the traditional technique of which the conceptual method is derived, an explanation of the proposed method that is divided in three parts: In the first part the tests done to representative systems are explained, in the second part how the rules have been obtained and in the third one how the knowledge has been organized. Concluding with the validation of the proposed technique.

2. PID controller

There are multiple forms of representation of PID regulator, but perhaps the most extended and studied one is the one given by equation 1.

$$u(t) = K\left[e(t) + \frac{1}{T_i}\int_0^t e(t)dt + T_d\frac{de(t)}{dt}\right] \tag{1}$$

where u is the control variable and e is the error of control given by $e = y_{SP} - y$ (difference between the specified reference by the entry and exit measured of the process). Therefore, the control variable is the sum of three different terms: P which is proportional to the error, I which is proportional to the integral of the error and D which is proportional to the derivative of the error (expression 2). The parameters of the controller are: the proportional gain K, the integral time T_i and the derivative time T_d. If the function transfer of the controller is obtained and a representation of the complex variable is done, the form is the one illustrated in expression 2.

$$G_C(s) = \frac{U(s)}{E(s)} = K\left(1 + \frac{1}{T_i \cdot s} + T_d \cdot s\right) \tag{2}$$

There are several ways for the representation of a PID regulator, but for the implementation of the PID regulator used, defined in the previous formula and more commonly known as the

Standard format Astrom & Hagglund (2006) Feng & Tan (1998), shown in the form of blocks in figure 2.

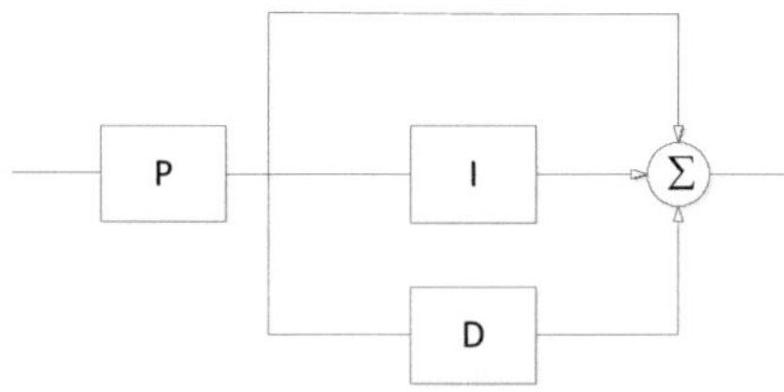

Fig. 2. PID regulator in standard topology

The industrial processes whose normal function is not adequate for certain applications are infinite. This problem, in many cases, is solved through the employment of this regulator, with which defined specifications are obtained in the control of processes leading to optimum values for what was being done. The adjustment of this controller is carried out varying the proportional gain and the integral and derivative times.

3. Empirical adjustment in closed loop of PID regulators

It is true that this day there are analytical methodologies to obtain the parameters of a PID regulator, in order to achieve improved one or various specifications. From a chronological point of view, the empirical procedures were born before the obtaining of the parameters, and currently they are still used for various reasons: the parameters are obtained in an empiric way, they are simple techniques, a given characteristic is optimized, good results are obtained in many cases, there is usually always a rule for the case that is trying to be controlled, etc.

3.1 Steps to obtain the parameters

The empiric techniques are based on the following steps:

1. Experimental establishment of certain characteristics of the response of the process that can be carried out with the plant working either in open or closed loop.
2. Application of formulas depending on the data previously obtained, to get the parameters of the regulator, with the aim that the operation of the plant with the controller is within certain desired specifications.

3.2 Measurement of the characteristics of the response of the process

In the first of the steps listed above for obtaining the PID controller parameters, the goal is to measure the characteristics of the responses of the process, it can be done in different ways, obtaining identical results in theory, although with small variations feasible in reality.

3.2.1 Sustained oscillation method

A fundamental method in the tuning of PID controllers is a closed loop method proposed by Ziegler and Nichols in 1942, whose best-known name is "the method of sustained oscillation".

It is an algorithm based on the frequency response of the process. The features to determine are:

- Critical proportional gain (K_c).- It is the gain of a proportional controller only, which causes the system to be oscillatory (critically stable).
- Sustained oscillation period (T_c).- It is the period of oscillation is achieved with the critical gain.

The procedure for obtaining these data is described below:

1. Closed loop system is located with a regulator that is only proportional (figure 3).

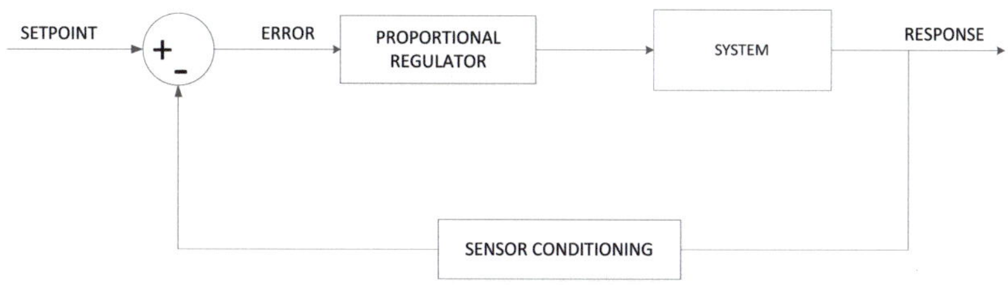

Fig. 3. System regulated with a proportional controller

2. Set any value of the proportional gain of the regulator and cause abrupt changes in the set point, then watching as the system response.
3. Increasing or decreasing the proportional gain of the regulator as necessary (if the system response is stabilized at a value, increase and if the output takes values without periodicity, decrease) until the system oscillates with constant amplitude and frequency as in figure 4. At that moment, writing down the value of the proportional gain applied to the regulator to achieve this state, this value corresponds to the gain of the system Kc, and also measure the period of oscillation of the output in these conditions, which is the period of sustained oscillation system T_c.

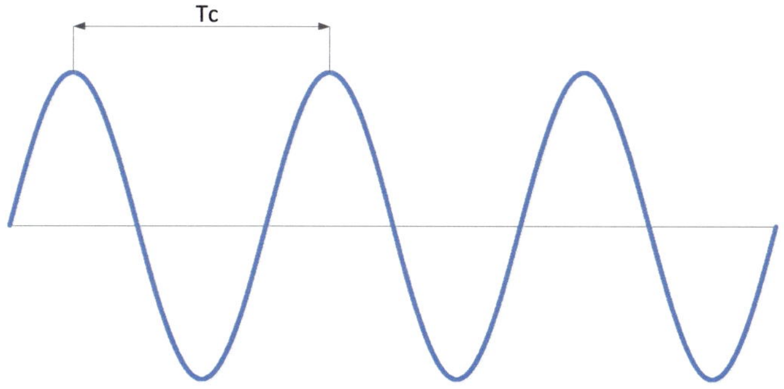

Fig. 4. Shape of the output in sustained oscillation state.

3.2.2 Relay-feedback method

The above method for obtaining parameters in closed loop of a regulator is a method that cannot be often used, since what is actually being done by increasing the proportional gain is to bring the system to a border zone of stability (oscillation), so that is possible to pass to the unstable region with relative ease. Sometimes without taking the system to instability, and just placing it in an area of sustained oscillation, the system is in a prohibited area, which could not operate by what might happen in the plant to be controlled. Therefore the application of this technique is only valid in certain specific cases where it can be passed to the oscillation or instability without major consequences.

An alternative way to locate the empirical critical gain (K_C) and the sustained oscillation period (T_C) of the system is by using the Relay method (Relay Feedback) developed by Aström and Hägglud, which is to bring the system to state of oscillation with the addition of a relay as shown in figure 5.

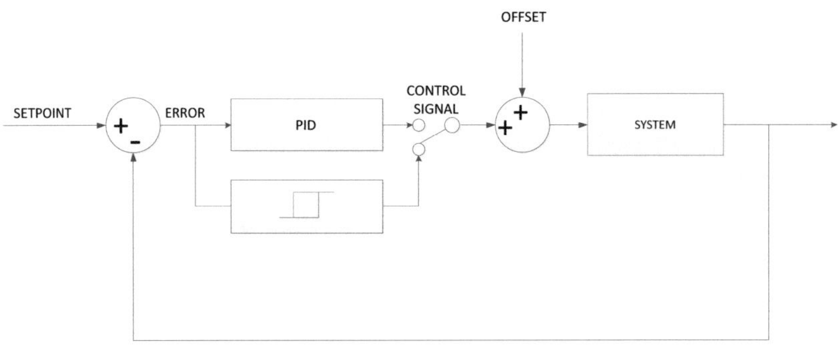

Fig. 5. Scheme for implementation of Relay-Feedback

The period of this achieved oscillation is approximately the same value as the sustained oscillation period T_C. In the experiment it is convenient to use a relay with hysteresis whose characteristics are shown in figure 6, an amplitude d and a window width of hysteresis h.

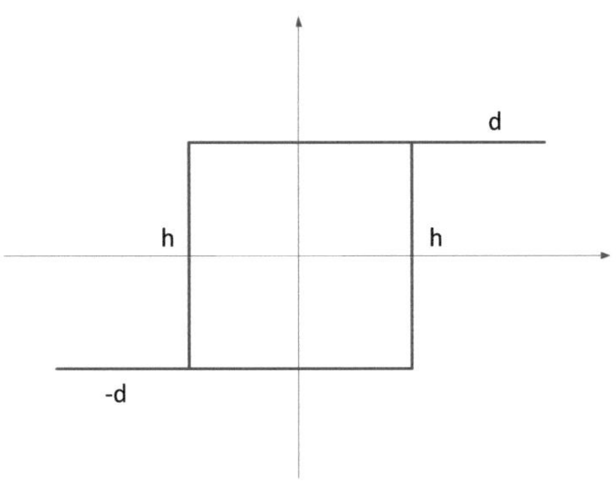

Fig. 6. Relay hysteresis used in the Relay-Feedback

Once the assembly, proceed as follows to obtain the mentioned parameters:

1. Bring the process to a steady state mode with the system regulated by PID controller, with any parameters that allow to achieve the aforementioned state. The values of the control signal (controller output) and the process output in the above conditions shall be recorded.
2. Then, control is closed with the relay instead of the PID controller. As a set point, the value read from the process output in the previous step is given. The value of the control signal taken in the previous section needed to bring the process in steady state is introduced into the entry shown in Figure 5 as Offset.
3. The process is put into operation with the indications made int the previous section, and wait for the output becomes periodic (in practice it can be considered to have achieved this state when the maximum value of the output repeats the same value in at least two periods in a row).
4. Two parameters must be recorded as shown in Figure 7, where T_c is the period of sustained oscillation.

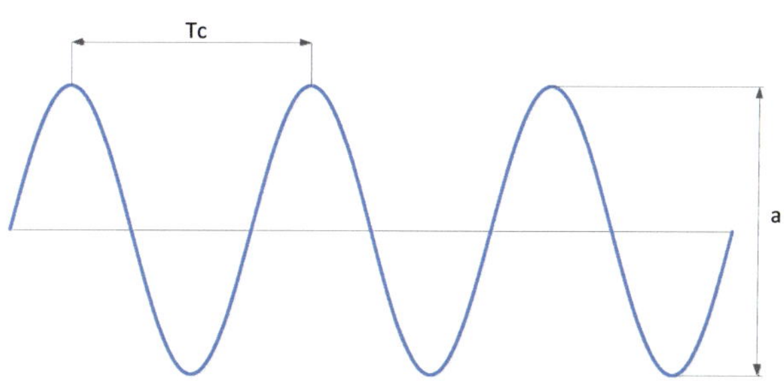

Fig. 7. System output with Relay-Feedback

5. The critical gain of the process is determined by the expression 3.

$$K_c = \frac{4d}{\pi\sqrt{a^2 - h^2}} \tag{3}$$

The Relay Feedback has the advantage that the adjustment can be made on the setpoint and can be carried out at any time. However, it has the disadvantage that, for the tuning process must overcome several occasions the setpoint and can be cases in which this is inadvisable because of the damage that can caused in the process.

3.2.3 Measurement of the characteristics of the response of the process from frecuency response example of bode diagrams

As previously indicated, the method of adjustment in closed loop is applicable to all those systems whose root locus cuts with the imaginary axis. In other words, occurs when increasing a proportional gain can bring the plant into a state of oscillation and subsequent instability.

This has an immediate translation in the field frequency on the Bode curves (figure 8), and consists of increasing the gain commented, causing a rise in the curve of modules to match

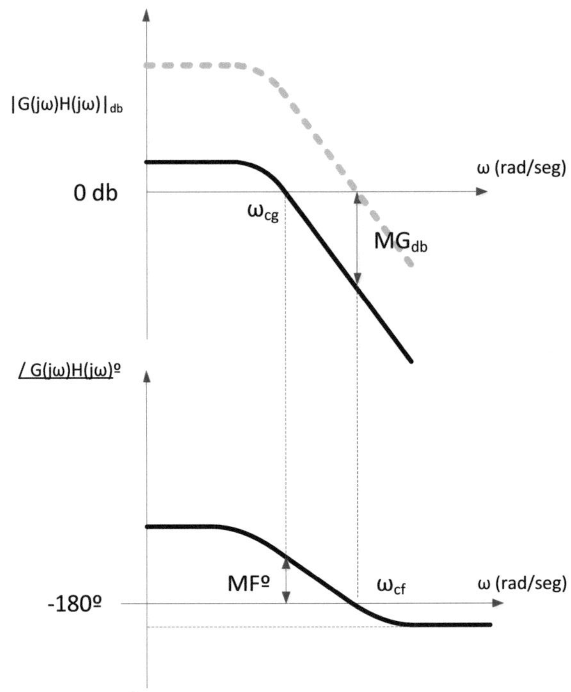

Fig. 8. Effect on Bode curves by increasing the proportional gain

the gain crossover frequency to the phase crossover frequency, state in which the system is oscillating (oscillates at the gain crossover frequency or phase crossover frequency with period T_c). The value of the gain that must be introduced to reach this state is the gain margin expressed in units (critical gain K_c).

This method is applicable in systems in which to practice a frequency analysis is possible. The parameters to introduce in the expressions of the terms of the controller can be seen on the results.

3.3 Parameters calculation through application of formulas

Once the characteristics of the response of the process have been measured and know what specification wants to be optimized, the following is to apply formulas developed to fulfill the description sought, bearing in mind the scopes of application for which they were obtained. The application range for the case of empiric adjustment in a closed loop comes defined usually by the product of the critical gain K_c and the process gain k.

Different authors propose expressions, depending on the characteristics of the response measured, for the achievement of the parameters of the regulator. It must be highlighted that there are multiple expressions given that work in an adequate way in certain cases for which they were developed. It is frequent also that the manufacturers of controllers deduce their own expressions that work satisfactorily above all with the products that they manufacture and especially for those applications to which are destined. It must be highlighted that there are no general equations that always work well, because of this, it will be necessary to select the expressions that best adjust in each specific case to the control intended.

In this study, the most known and usual Ziegler (1942) Astrom & HÃd'gglund (1995) McCormack & Godfrey (1998) Tyreus & Luyben (1992) that are employed in the achievement of the parameters of the PID regulators have been compiled, even though the methodology followed can be used for any case. In table 1 the different expressions used in the present study are shown, together with the scope of application in each case Astrom & Wittenmark (1989b).

Method	K_p	T_i	T_d	Application range
Ziegler-Nichols	$0.6 \cdot K_c$	$0.5 \cdot T_c$	$0.125 \cdot T_c$	$2 < k \cdot K_c < 20$
Ziegler-Nichols modified (little overshoot)	$0.33 \cdot K_c$	$\frac{T_c}{2}$	$\frac{T_c}{3}$	$2 < k \cdot K_c < 20$
Ziegler-Nichols modified (without overshoot)	$0.2 \cdot K_c$	T_c	$\frac{T_c}{3}$	$2 < k \cdot K_c < 20$
Tyreus-Luyben	$0.45 \cdot K_c$	$2.2 \cdot T_c$	$\frac{T_c}{6.3}$	$2 < k \cdot K_c < 20$

Table 1. Expressions of parameters of authors and scopes of application

4. Design rules of PID regulators in closed loop

In the first part of this section a sweep at the different expressions of achievement of parameters of the PID regulator previously mentioned is made, in which the systems are controlled with this type of regulator, with the aim of obtaining some generic design rules, or in their case particular rules for certain types of systems.

Due to the general character of the rules it will be necessary to use significant systems for this. In this aspect it has been opted to use a known source in this scope, which is the Benchmark of systems to control PID developed by Äström y Hägglund Astrom (2000). In this source a collection of systems is presented which: are usually employed in the testing of PID controllers. These systems are based in countless sources of importance and also the immense majority of the existing systems adapt to some of those included in this source.

4.1 Benchmark systems to which closed loop empiric adjustment is not applicable

There are a set of systems included in the Benchmarking to which the empiric adjustment in closed loop is not applicable. If for instance there is a system whose transfer function is the expression 4, which deals with a system of first order.

$$G(s) = \frac{1}{s+1} \tag{4}$$

The root locus for this transfer function is shown in figure 9, where one can see that does not cut the imaginary axis, and therefore cannot get the setting parameters of the PID closed-loop for this case.

If there is a system with a transfer function as the one in the expression 5, which is an unstable system, when introducing a step type input, will not be able to apply this method to regulate it.

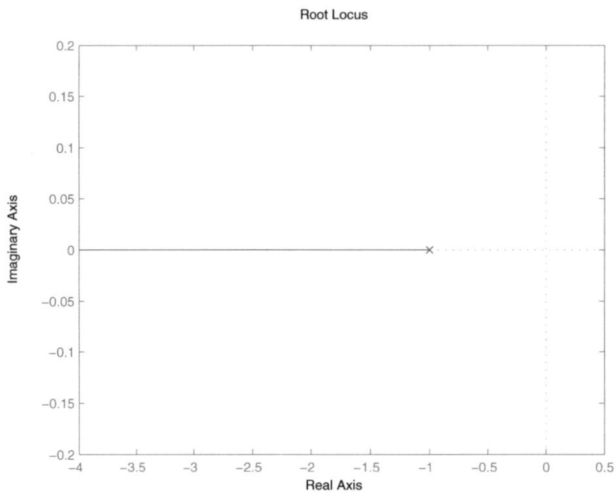

Fig. 9. Root locus for 1st order system with pole at -1

$$G(s) = \frac{1}{s^2 - 1} \tag{5}$$

Another possibility within the contemplated functions in the Benchmark is the systems that possess a function like the one in expression 6.

$$G(s) = \frac{1}{(s+1)^2} \tag{6}$$

The root locus in this case is shown in figure 10, where one can see that it does not cut the imaginary axis, and therefore cannot get the setting parameters of the PID controller for this case either.

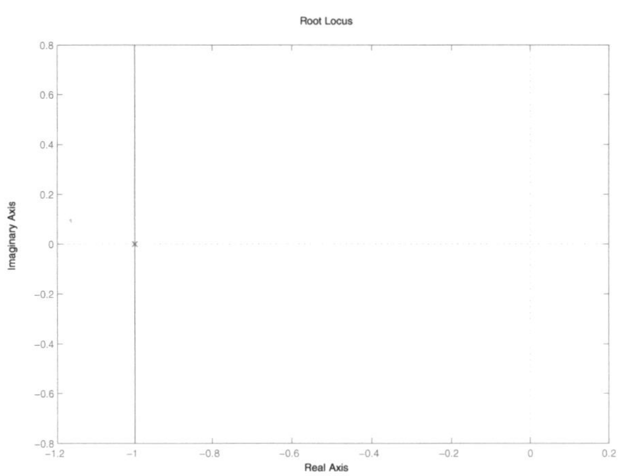

Fig. 10. Root locus for a system with two poles at -1

4.2 Benchmark systems to which empiric adjustment in closed loop is applicable

Apart from the types of systems found in some of the examples in the previous section, the rest can be regulated by a controller PID applying the empiric adjustment in closed loop to obtain its parameters. If there is a transfer function like the one in expression 7, and whose root locus is the one of figure 11, it will be possible to adjust the PID controller in closed loop.

$$G(s) = \frac{1}{(s+1)^3} \tag{7}$$

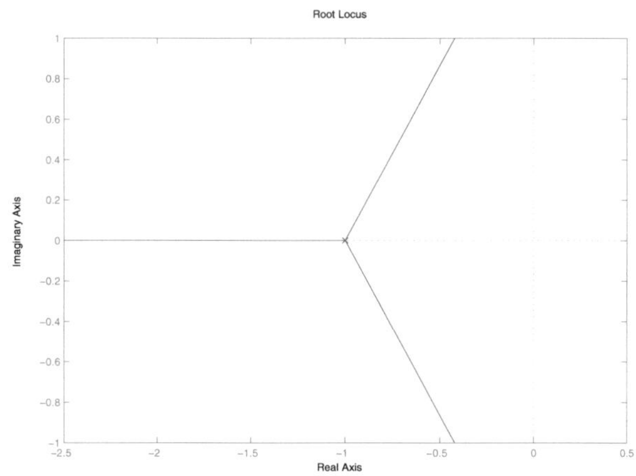

Fig. 11. Root locus for a system with three poles at -1

The response characteristics of the process in closed loop are determined obtaining values of k = 1, K_c = 8.0011, T_c = 3.6274. The range of application is given by the product k x K_c taking a value in this case 8.0011, which, as indicated in table 1 applies to the four methods outlined.

4.3 Analysis of the methods applied to obtain the rules

Having measured the response characteristics of the system, regulating it with the different expressions in the case study proceeds, extracting significant specifications like: response time, peak time, overshoot and settle time.

All the tests will be carried out on all the systems proposed by Ästrom in Benchmark in which they are applicable, to check the results and be able to extract conclusions from which rules will be obtained. If system of the expression 7 is regulated, the results obtained are illustrated in figure 12.

5. PID controller conceptual modeling

The conceptual model of a domain consists of the organization as strict as possible of the knowledge from the perspective of the human brain. In this sense for the domain that is being dealt with in this case study, a general summarized model is proposed and shown in figure 13.

As can be observed it is divided in three blocks:

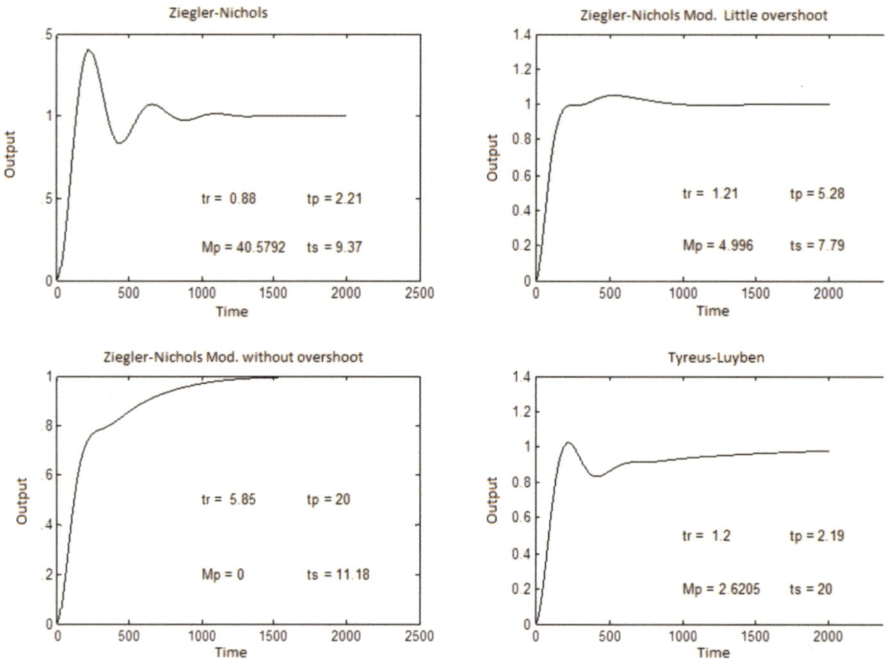

Fig. 12. Response of the system regulated by the expressions of Ziegler-Nichols (initial and modified) and Tyreus-Luyben

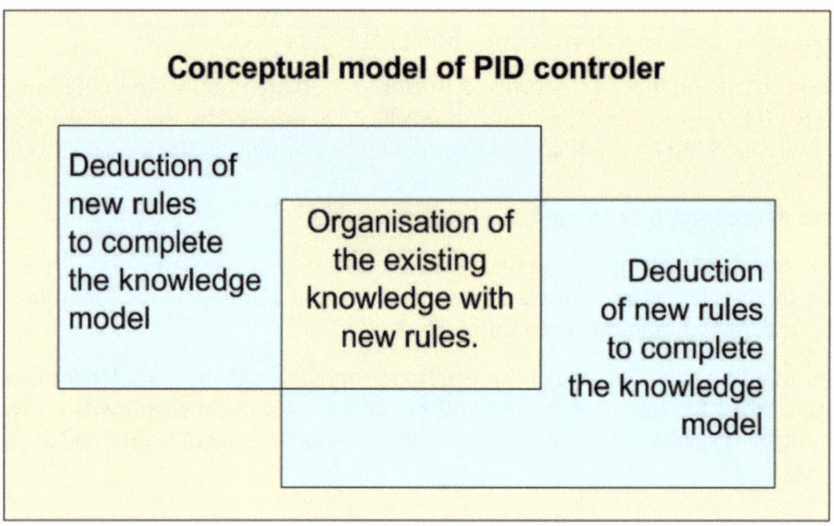

Fig. 13. General schema summarized from the conceptual model of empiric adjustment of PID regulators in closed loop

- Organization of the existing rules: In this block the aim is to organise the existing rules of the types of expressions, scopes of application range, change criteria in the load disturbance or follow up of the setpoint control criterion, etc.

- Organization of existing knowledge with new rules: This block is the meeting point between the other two, and it aims to organise the existing knowledge in an adequate way for which it will be necessary to create new rules.
- Deduction of new rules to complete the knowledge model: In this part it has been detected the necessity to deduce new rules to make a complete knowledge model, from the system itself and the desired specifications, to the final derivation of the parameters of the controller in a reasoned way.

5.1 General diagram of knowledge in the closed loop

In accordance with the steps deducted by the elaboration of the conceptual model, a general diagram of the knowledge for the adjustment of PID controllers in closed loop shown in figure 14 is obtained.

Following the above a more detailed description of the knowledge schema is done, in different figures with their corresponding explanation. It starts with the corresponding part in the top right corner of the general diagram, detailed in figure 15.

In this first part it checks whether the system can be brought to the sustained oscillation by any existing methods. On one hand are the methods of sustained oscillation, and the frequentials methods with Bode as an example, and on the another hand is the method of Relay-Feedback. This is because in some systems empirical adjustment parameters in closed loop can be deducted by the latter procedure in a more feasible, for example, processes in which the transfer function is unknown, it is more viable its use that any of the other two. After selecting the method, in all cases checks if it have been able to achieve sustained oscillation. If the answer is negative in the case of Relay-Feedback is concluded that no empirical adjustment will be applied in closed loop system. If it is negative in the case of sustained oscillation or frequential methods, the user is given the possibility to check if is achieved by the method of Relay-Feedback, if that is not desired, it also concludes that it is not possible to apply empirical adjustment in closed loop system.

If by any method sustained oscillation is reached, it may be possible to adjust the PID closed loop. To check it first calculates the parameters k, K_c and T_c, and then checks if the product k x K_c is within the range of application of expressions.

In figure 16 (continuation of figure 15) check that the measured parameters are within range, if the answer is affirmative then Benchmark systems that do not meet the range are discarded, and the scheme ends in the rule rg. 6. Otherwise, it requires the user to specify if he/she wishes to apply the expressions, although not being within the range. Otherwise, it requires the user to specify whether the terms are applied, although not being within the range. If the answer is negative adjustment in closed loop cannot be done. If the answer is affirmative, a group with generic characteristics to which the system belongs will be determined depending on the product k x K_c. For this, it first checks whether its value is infinite. If not, the scheme applies the rule rg.5, if it were, have to check whether the system is unstable because it can happen that it has a pole at the origin, and in this case, the product is infinite. If the system is unstable and the product is infinite, rule rg.5.1 will be applied and, if unstable, it is concluded that the adjustment in closed loop cannot be used.

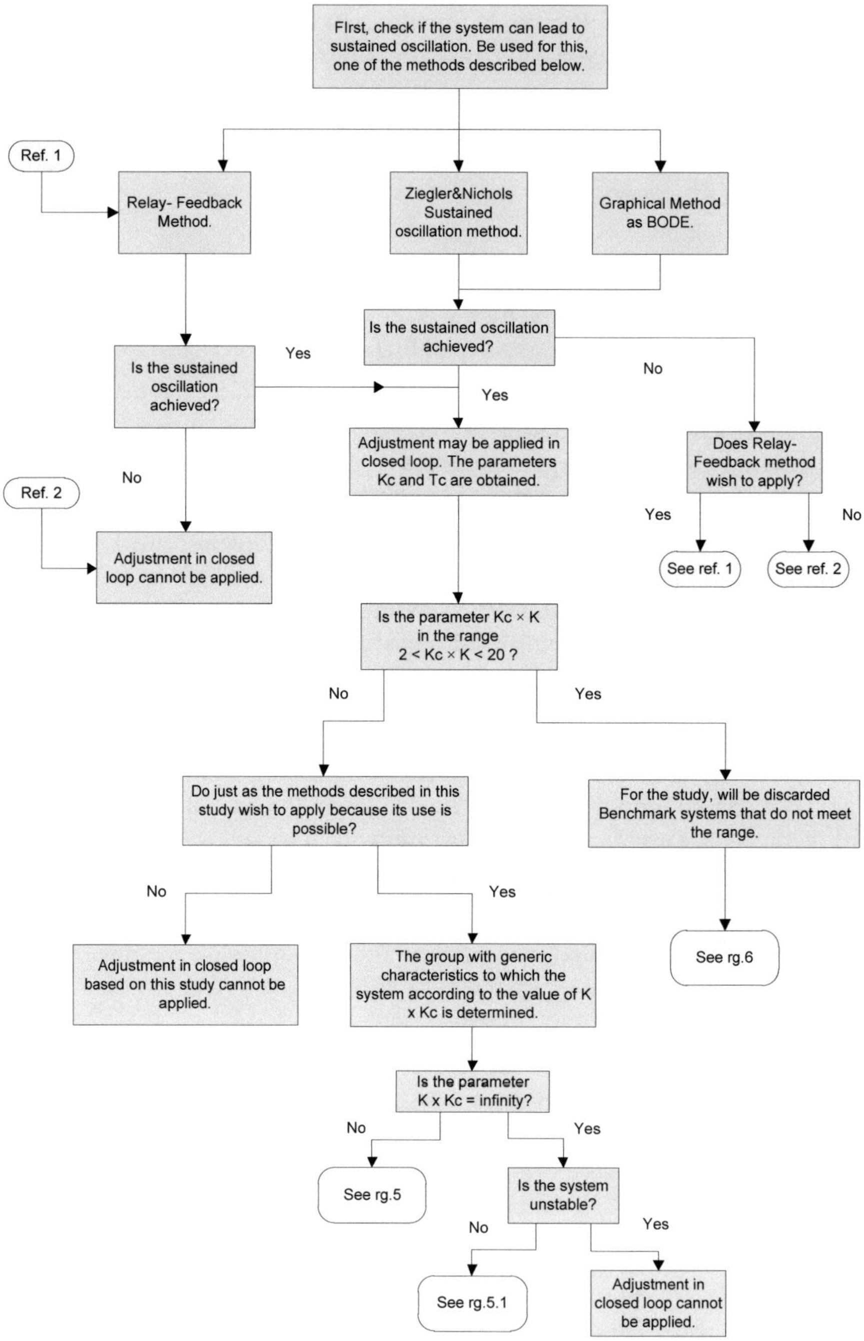

Fig. 14. General diagram of knowledge for closed loop empirical tuning of PID controllers

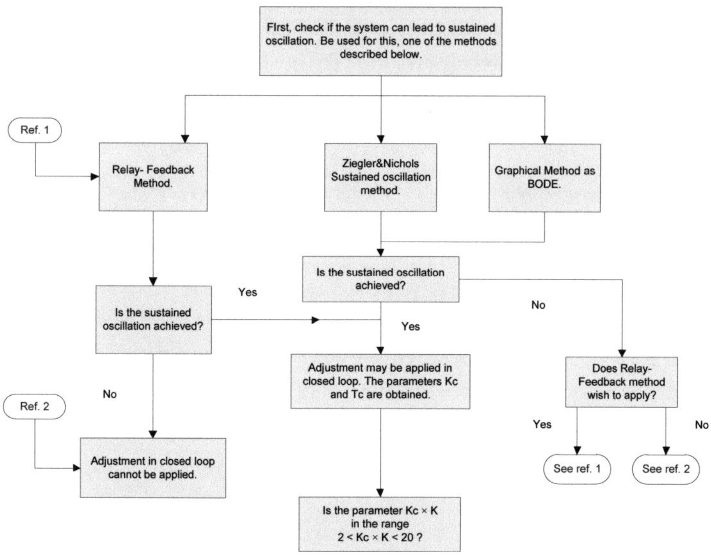

Fig. 15. Area 1 of the diagram

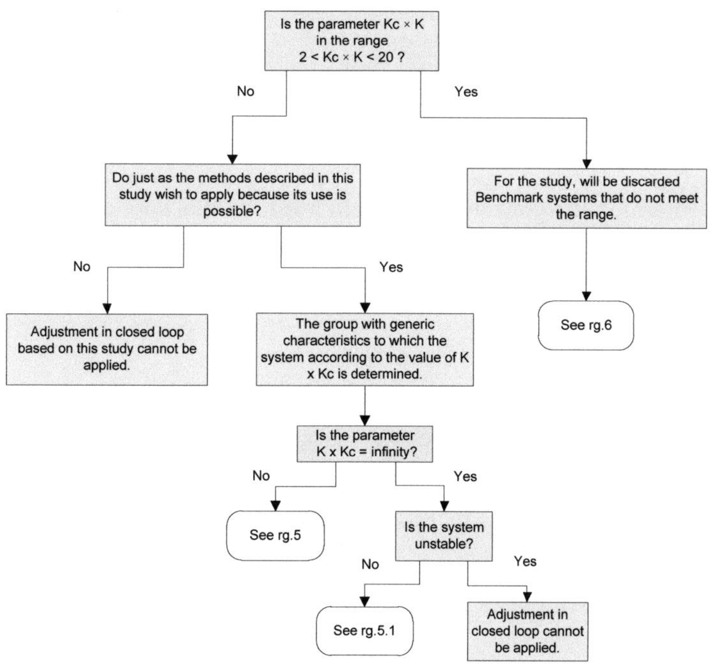

Fig. 16. Area 2 of the diagram

5.2 Deduction of rules to complete the knowledge model

As has been commented in the general summarized schema of knowledge, it is necessary to draw new rules to complete the knowledge model: In this part the need to do a model of complete knowledge model has been detected, from the system itself and the specifications desired, up to the final obtaining of the parameters of the controller. In this sense two examples are shown in which the two possibilities of deduction of the rules are clarified.

5.2.1 Deduction of the rules rg.5

This rule as shown in figure 16, is applied in the worst case, where the product k x K_c is not within the application range of expressions.

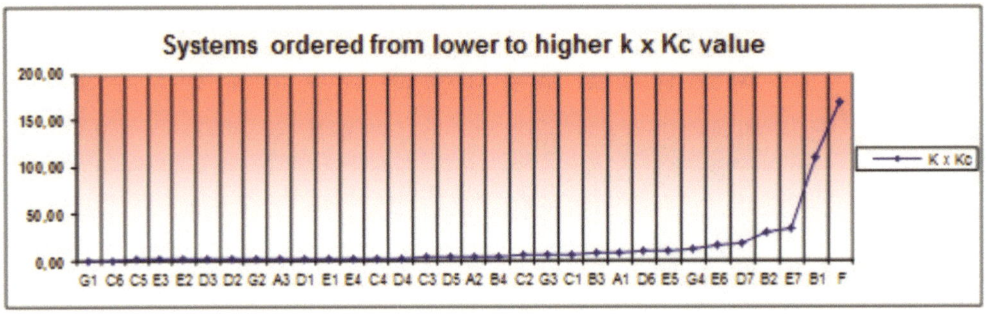

Fig. 17. Systems of Benchmark ordered from lower to higher k x K_c value

In order to create the groups with generic characteristics, different systems are sorted from lowest to highest value of the ratio k x K_c (figure 17 and table 2). In this case is done in a table (table 3) as it intends to have generic groups in all specifications.

In the table the values of the specification in each case have been indicated, alongside the expressions for obtaining the parameters used to improve this specification. Next, a division in groups is made in which the systems with groups of equal expressions are concentrated. Having this in mind, for instance systems G1 to A1, with the condition that $0 < L/T \leq 8.0011$, and establish the following rules:

- To minimize T_r, apply the method Ziegler&Nichols.
- To minimize T_s, apply the method Ziegler&Nichols.
- For the lowest percentage of M_p, apply the method of Ziegler&Nichols modified (without overshoot).
- To minimize Tp, apply again the method Ziegler&Nichols.

Table 3 shows that there are several exceptions which correspond to areas of the range where would be the systems G2, A1 and C6 of Benchmark. If the values obtained in the specifications after applying the above rules are checked, concludes that in the case of A1 and G2, the difference is very small. For the C6 system, these value differences are greater and the rule would not be entirely valid, but it is the only exception and therefore is generalized, despite making a small mistake.

System	k x K_c
G1	0.44
C6	0.5
C5	1.1429
E3	1.5377
E2	1.6422
D3	1.7071
D2	1.7972
G2	1.8812
A3	1.884
D1	1.9052
E1	1.9052
E4	1.9317
C4	2
D4	2
C3	3.2
D5	3.8891
A2	4
B4	4
C2	5
G3	5.24
C1	6.1585
B3	6.7510
A1	8.0011
D6	8.8672
E5	9.7963
G4	11.2414
E6	16.818
D7	17.5578
B2	30.2375
E7	35.1592
B1	110.1
F	167.7135

Table 2. Values of the parameters K x K_c of each system

6. Validation

A validation of the conceptual model proposed is carried out. This will not be done on the cases in which the transfer function is known, and it is exactly adapted to one of the systems referred in the Benchmark, but it will be carried out on the worst case, when the transfer function is not known or if it is known and it does not adapt to any of the systems.

The validation is done on 9 systems not contemplated in the Benchmark and it is checked for each one of the specifications that the model has been developed. There are a total of 36 checking cases, in which the results shown in table 4 are obtained.

Therefore it is considered that the model proposed has a satisfactory functioning, and that overall, the results are the following:

SYSTEM	T_r	T_s	M_p	T_p
G1	24.45 (Z&N)	48.18 (Z&N)	0 (Z&N s M_p)	131.21 (Z&N)
C6	0.44 (Z&N p M_p)	48.95 (Z&N)	0 (Z&N s M_p)	2.81 (Z&N p M_p)
C5	7.92 (Z&N)	19.03 (Z&N)	0 (Z&N s M_p)	110 (Z&N)
E3	0.76 (Z&N)	7.25 (Z&N)	0 (Z&N s M_p)	2.02 (Z&N)
E2	0.72 (Z&N)	6.79 (Z&N)	0 (Z&N s M_p)	1.94 (Z&N)
D3	0.74 (Z&N)	6.53 (Z&N)	0 (Z&N s M_p)	1.99 (Z&N)
D2	0.77 (Z&N)	5.58 (Z&N)	0 (Z&N s M_p)	2.04 (Z&N)
G2	1.05 (Z&N)	9.29 (Z&N p M_p)	0 (Z&N s M_p)	4.71 (Z&N)
A3	4.01 (Z&N)	32.75 (Z&N)	0 (Z&N s M_p)	9.75 (Z&N)
D1	0.84 (Z&N)	5.42 (Z&N)	0 (Z&N s M_p)	2.21 (Z&N)
E1	0.84 (Z&N)	5.42 (Z&N)	0 (Z&N s M_p)	2.21 (Z&N)
E4	0.85 (Z&N)	7.56 (Z&N)	0 (Z&N s M_p)	2.25 (Z&N)
C4	1.32 (Z&N)	8.51 (Z&N)	0 (Z&N s M_p)	3.79 (Z&N)
D4	0.8 (Z&N)	7.25 (Z&N)	0 (Z&N s M_p)	2.13 (Z&N)
C3	1.14 (Z&N)	7.39 (Z&N)	0 (Z&N s M_p)	3.31 (Z&N)
D5	0.77 (Z&N)	6.17 (Z&N)	0 (Z&N s M_p)	2.39 (Z&N)
A2	1.62 (Z&N)	11.76 (Z&N)	0 (Z&N s M_p)	3.94 (Z&N)
B4	1.62 (Z&N)	11.76 (Z&N)	0 (Z&N s M_p)	3.94 (Z&N)
C2	1.02 (Z&N)	8.69 (Z&N)	0 (Z&N s M_p)	2.76 (Z&N)
G3	0.34 (Z&N)	2.88 (Z&N)	0 (Z&N s M_p)	0.78 (Z&N)
C1	0.96 (Z&N)	8.27 (Z&N)	0 (Z&N s M_p)	2.5 (Z&N)
B3	0.52 (Z&N)	4.47 (Z&N)	0 (Z&N s M_p)	1.37 (Z&N)
A1	0.88 (Z&N)	7.79 (Z&N p M_p)	0 (Z&N s M_p)	2.19 (T&L)
D6	0.71 (Z&N)	7.98 (Z&N)	0 (T&L)	2.46 (Z&N)
E5	0.84 (Z&N)	8.95 (Z&N s M_p)	0.31574 (Z&N s M_p)	2.59 (Z&N)
G4	0.16 (Z&N)	1.46 (Z&N)	0 (T&L)	0.4 (Z&N)
E6	1.52 (Z&N)	19.92 (Z&N s M_p)	3.4625 (Z&N s M_p)	4.77 (Z&N)
D7	0.67 (Z&N)	3.98 (T&L)	1.7451 (T&L)	2.46 (Z&N)
B2	0.12 (Z&N)	0.9 (T&L)	10.062 (Z&N s M_p)	0.34 (Z&N)
E7	2.05 (Z&N)	19.75 (T&L)	14.3172 (Z&N s M_p)	6.34 (Z&N)
B1	0.04 (Z&N)	0.33 (T&L)	20.9154 (Z&N s M_p)	0.12 (Z&N)
F	0.13 (Z&N)	1.29 (T&L)	13.2937 (Z&N s M_p)	0.36 (Z&N)

Table 3. Groups rg.5 for changes in the load

- The scores are $36/36 = 100\%$
- The misses are $0/36 = 0\%$

Comment	Number	Percent overall experiments
The expression indicated by the rule coincides with the one that has to actually be used.	30 *cases*	83.4%
The expression indicated by the rule does not coincide with the one that has to actually be used, but the deviation is very small.	6 *cases*	16.6%
The expressions indicated by the rule makes the system unstable	0 *cases*	0%
The expressions indicated by the rule does not coincide with the one that has actually to be used so the deviation is considerable.	0 *case*	0%

Table 4. Results of the validation

7. Conclusions

The task of selection of the adjustment expression to be used has been solved with the proposed technique in the present paper, thus through the follow up of the rules procedure the adjustment expressions can be selected for the case disposed and also choose among them if more than one is applicable.

Having selected the expression or expressions to obtain the parameters, the calculation of these is carried out following the procedure for the case that has been chosen previously in a structured way. And so the possible paths to be followed are solved with rules, including those to reach a balance between specifications that do not improve in one same path.

When carrying out the conceptual modeling two relevant contributions have been obtained. First, clarity has been added in various stages of the adjustment of a PID. Second, some contradictions have been manifested between different expressions that have been solved with it.

The procedure in real plants whose function transfer is different to the ones mentioned in the Benchmark, has been validated for the more restrictive cases of the deduced rules. The results obtained and presented in the corresponding section to validating satisfy the initial objectives when verifying the functioning of the rules in the plants used.

8. References

Astrom, K. & Hagglund, T. (2006). *PID controllers: Theory, Desing and Tuning*, Research Triangle Park, USA.

Astrom, K. & HÃd'gglund, T. (1995). *Adaptive Control*, IEEE/CRC Press, Sweden.

Astrom, K. & Wittenmark, B. (1989a). *Adaptive Control*, Addison Wesley Publishing Company, Sweden.

Astrom, K. & Wittenmark, B. (1989b). Adaptive control, *Addison Wesley USA* pp. 332–336.

Astrom, K.J. Hagglund, T. (2000). Benchmark systems for pid control, *Preprints IFAC Workshop on Digital Control. Past, present and future of PID Control*, Elsevier Science and Technology, Terrasa, Spain, pp. 181 –182.

Auslander, D., Takahashi, Y. & Tomizuka, M. (1978). Direct digital process control: Practice and algorithms for microprocessor application, *Proceedings of the IEEE* 66(2): 199 – 208.

Bennett, S. (1984). Nicolas minorsky and the automatic steering of ships, *Control System Magazine* Vol. 4(No. 4): 10–15.
URL: *10.1109/MCS.1984.1104827*

Calvo-Rolle, J. & Corchado, E. (n.d.). A bio-inspired robust controller for a refinery plant process, *Logic Journal of IGPL* .
URL: *http://jigpal.oxfordjournals.org/content/early/2011/02/04/jigpal.jzr010.abstract*

Calvo-Rolle, J.L. Alonso-Alvarez, A. F.-G. R. (2007). Using knowledge engineering in a pid regulator in non linear process control, *Ingenieria Quimica* 32: 21 – 28.

Epshtein, V. (2000). Hypertext knowledge base for the control theory, *Automation and Remote Control* 61(11): 1928–1933.

Feng, Y. & Tan, K. (1998). Pideasytm and automated generation of optimal pid controllers, *Third Asia-Pacific Conference on Control&Measurement*, Aviation Industry Press, Dunhuang, China, pp. 29–33.

McCormack, A. S. & Godfrey, K. R. (1998). Rule-based autotuning based on frequency domain identification, *IEEE Transactions on Control Systems Technology* 6(1).

Mindell, D. (2004). *Between human and machine: Feedback, Control, and Computing before Cybernetics*, Johns Hopkings Paperbacks edition, London.

Pang, G. (1991). An expert adaptive control scheme in an intelligent process control system, *Proceedings of the IEEE International Symposium on the intelligent Control*, IEEE Press, Arlington, Virginia, pp. 13–18.

Pang, G. (1993). Implementation of a knowledge-based controller for hybrid systems, *Decision and Control, 1993., Proceedings of the 32nd IEEE Conference on*, IEEE Press, San Antonio, TX , USA, pp. 2315 –2316 vol.3.

Pang, G., Bacakoglu, H., Ho, M., Hwu, Y., Robertson, B. & Shahrrava, B. (1994). A knowledge-based system for control system design using medal, *Computer-Aided Control System Design, 1994. Proceedings., IEEE/IFAC Joint Symposium on*, IEEE Press, Tucson, AZ , USA, pp. 187 –196.

Tyreus, B. & Luyben, W. (1992). Industrial engineering chemistry research, *IEEE Transactions on Control Systems Technology* pp. 2625–2628.

Wilson, D. (2005). Towards intelligence in embedded pid controllers, *Proceedings of the Eight IASTED International Conference on Intelligent Systems and Control*, ACTA Press, Cambridge, USA, pp. 25–30.

Zhou, L. Li, X. H. T. & Li, H. (2005). Development of high-precision power supply based on expert self-tuning control, *ICMIT 2005: Control Systems and Robotics*, SPIE-The International Society for Optical Engineering, Wuhan, China, pp. 60421T.1–60421T.6.

Ziegler, J. Nichols, N. R. N. (1942). Optimum settings for automatic controllers, *Transactions of ASME* 64: 759 – 768.

7

Stabilizing PID Controllers for a Class of Time Delay Systems

Karim Saadaoui, Sami Elmadssia and Mohamed Benrejeb
UR LARA-Automatique, Ecole Nationale d'Ingénieurs de Tunis,
University of Tunis, ElManar
Tunisia

1. Introduction

It is well known that delay exists in many practical systems such as industrial systems, biological systems, engineering systems, population dynamic models, etc… Time delay may appear as part of the internal dynamics or may result because of the actuators and sensors used. Communication delays between different parts of a system are also sources of this phenomenon. Existence of time delays may cause oscillations and even instability; moreover it complicates the analysis and control of these systems (Zhong, 2006; Normey-Rico & Camacho, 2007). For all these reasons there is an extensive literature on stability analysis, stabilization and control of time delay processes, which continue to be active areas of research.

Recently, there has been a great interest in stabilizing high order systems by low fixed order controllers. The main motivation for using low order controllers comes from their simplicity and practicality, which explains the desire to reduce controller complexity and to determine as low order a controller as possible for a given high order or complex plant.

Computational methods for determining the set of all stabilizing controllers, of a given order and structure, for linear time invariant delay free systems are reported in the literature (Saadaoui & Ozguler, 2005; Silva, 2005). In this line of research, the main objective is to compute the stabilizing regions in the parameter space of simple controllers since stabilization is a first and essential step in any design problem. Once the set of all controllers of a given order and structure is determined, further design criteria can be added. To this end, several approaches are employed, among which extensions of the Hermite-Biehler theorem and the D-decomposition method seem to be the most appropriate. Stabilization of special classes of time delay systems by fixed-order controllers was also addressed. Using a generalization of the Hermite-Biehler theorem the set of all stabilizing PI and PID controllers are determined for first order systems with dead time (Silva et al., 2001; Silva et al., 2002). Graphical methods were used to determine all stabilizing parameters of a PID controller applied to second order plants with dead time (Wang, 2007b).

In this chapter, we consider determining all stabilizing PID controllers for n-th order all poles systems with time delay. These systems can be used to model many physical examples. This includes the representation of a ship positioning an underwater vehicle through a long cable (Zhong, 2006). It can be used to represent the dynamic behavior of temperature control in a mix process (Normey-Rico & Camacho, 2007). One of the main contributions of this work is the determination of admissible values of one of the controller's parameters. This is done using two approaches, the first approach is based on the use of an extension of the Hermite-Biehler theorem and the second approach is characterized by the application of a necessary condition which is a consequence of Kharitonov's lemma. The D-decomposition method is then applied to find stabilizing regions in the parameter space of the remaining two parameters. By sweeping over the admissible ranges of the first parameter, the complete set of stabilizing PID controllers for this class of systems is obtained. It can be easily shown that the proposed method is applicable to first order controllers, being phase lead or phase lag controllers, by carrying out the appropriate modifications.

Stabilization being the most basic requirement in most controller design problems, determining the set of all stabilizing PID controllers for this class of systems is a first step in searching, among such controllers, those that satisfy further performance criteria, such as those imposed on the unit step response of the closed loop system. Once this first step is done, genetic algorithms are used to minimize several performance measures of the closed loop system. Genetic algorithms (GAs) (Chen et al., 2009; Chen & Chang, 2006; Jan et al., 2008 ; Goldberg, 1989; Haupt & Haupt, 2004; Melanie, 1998) are stochastic optimization methods. Unlike classical optimization methods, they are not gradient based which makes GAs suitable to minimize performance measures such as maximum percent overshoot, rise time, settling time and integral square error. Moreover, genetic algorithms explore the entire admissible space to search the optimal solution. This is another reason for using GAs after finding the complete set of stabilizing PID controllers which form the set of admissible solutions. In this work, the ranges of proportional, derivative and integral gains are not set arbitrary but they are set within the stabilizing regions determined, so that we are searching among the stabilizing values of the PID controller.

The chapter is organized as follows. In Section 2, a constant gain stabilizing algorithm is used to determine the admissible ranges of one of the controller's parameters. Next, the D-decomposition method is introduced and a necessary condition is used to compute the admissible ranges of proportional and derivative gains. In Section 3, the stabilizing regions in the space of the controller's parameters are determined and the genetic algorithm method is used to minimize several performance measures of the step response of the closed loop system. In Section 4, illustrative examples are given. Finally, Section 6 contains some concluding remarks.

2. Determining controller's stabilizing parameters

In this work, we consider determining the stabilizing regions in the parameter space of a PID controller,

$$C(s) = \frac{k_d s^2 + k_p s + k_i}{s} \tag{1}$$

Applied to an n-th order all poles system with delay,

$$G(s) = \frac{e^{-Ls}}{Q(s)} = \frac{e^{-Ls}}{q_n s^n + \ldots + q_0} \tag{2}$$

Where $L > 0$ represents the time delay. In this section, the admissible ranges of the parameters (k_p, k_d) are found, where k_p denotes the proportional gain and k_d the derivative gain. The exact range of stabilizing (k_p, k_d) values is difficult to determine analytically. In fact, the problem of determining the exact range of stabilizing k_p values analytically is solved for the case of a first order plant with dead time (Silva et al., 2002). Therefore, instead of determining the exact range of k_p (or k_d) a necessary condition will be used to get an estimate of the stabilizing range.

Let us first fix the notation used in this work. Let R denote the set of real numbers and C denote the set of complex numbers and let R_-, R_0 and R_+ denote the points in the open left-half, jw -axis and the open right-half of the complex plane, respectively. The derivative of a polynomial or a quasi-polynomial Ψ is denoted by Ψ'. The set H of stable polynomials or quasi-polynomials are,

$$H = \{\Psi(s) \in R[s] : \Psi(s) = 0 \Rightarrow s \in C_-\}$$

Given a set of polynomials $\Psi_1, \cdots, \Psi_l \in R[s]$ not all zero and $l > 1$ their greatest common divisor is unique and it is denoted by $\gcd\{\Psi_1, \cdots, \Psi_l\}$; If $\gcd\{\Psi_1, \cdots, \Psi_l\} = 1$ then we say $\{\Psi_1, \cdots, \Psi_l\}$ is coprime. The signature $\sigma(\Psi)$ of a polynomial $\Psi \in R[s]$ is the difference between the number of its C_- roots and C_+ roots. Given $\Psi \in R[s]$ *the even-odd components* (a,b) *of* $\Psi(s)$ are the unique polynomials $a, b \in R[s^2]$ such that

$$\Psi(s) = a(s^2) + sb(s^2)$$

It is possible to state a necessary and sufficient condition for the Hurwitz stability of $\Psi(s)$ in terms of its even-odd components (a,b). Stability is characterized by the interlacing property of the real, negative, and distinct roots of the even and odd parts. This result is known as the Hermite-Biehler theorem. Below is a generalization of the Hermite-Biehler theorem applicable to not necessarily Hurwitz stable polynomials. Let us define the *signum function* $S : R \to \{-1, 0, 1\}$ *by*

$$Su = \begin{cases} -1 & if\ u < 0 \\ 0 & if\ u = 0 \\ 1 & if\ u > 0 \end{cases}$$

Lemma 1. (Saadaoui & Ozguler, 2003) *Let a nonzero polynomial* $\Psi \in R[s]$ *has the even-odd components* (a,b) *Suppose* $b \neq 0$ *and* (a,b) *is coprime. Then,* $S(\Psi) = r$ *if and only if at the real negative roots of odd multiplicities* $v_1 > v_2 > \cdots > v_l$ *of b the following holds:*

$$r = \begin{cases} Sb(0_-)[Sa(0) - 2Sa(v_1) + 2S\,a(v_1) + \cdots + (-1)^l 2Sa(v_l)] & if \deg\Psi\ odd \\ Sb(0_-)[Sa(0) - 2Sa(v_1) + 2S\,a(v_1) + \cdots + (-1)^{l+1} Sa(-\infty)] & if \deg\Psi\ even \end{cases}$$

where $Sb(0\_) := (-1)^{m0} b^{(m0)}(0)$, *m0 is the multiplicity of* $u = 0$ *as a root of* $b(u)$, *and* $b^{(m0)}(0)$ *denotes the value at* $u = 0$ *of the m0-th derivative of* $b(u)$.

The following result, determines the number of real negative roots of a real polynomial.

Lemma 2. (Saadaoui & Ozguler, 2005) *A nonzero polynomial* $\Psi \in R[s]$, *such that* $\Psi(0) \neq 0$, *has r real negative roots without counting the multiplicities if and only if the signature of the polynomial* $\Psi(s^2) + \Psi'(s^2)$ *is 2r. All roots of* ψ *are real, negative, and distinct if and only if* $\Psi(s^2) + \Psi'(s^2) \in H$.

We now describe a slight extension of the constant stabilizing gain algorithm of (Saadaoui & Ozguler,2003). Given a plant

$$G_0(s) = \frac{p_0(s)}{q_0(s)} \tag{3}$$

where $p_0, q_0 \in R[s]$ are coprime with $m = \deg p_0$ less than or equal to $n = \deg q_0$, the set

$$\Phi_r(p_0, q_0) = \{\alpha \in R : \sigma[\varphi(s,\alpha)] = \sigma[q_0(s) + \alpha p_0(s)] = r\}$$

is the set of all real α such that $\phi(s,\alpha)$ has signature equal to *r*. Let (h,g) and (f,e) be the even-odd components of $q_0(s)$ and $p_0(s)$, respectively, so that

$$q_0(s) = h(s^2) + sg(s^2),$$

$$p_0(s) = f(s^2) + se(s^2).$$

Let (H,G) be the even-odd components of $q_0(s)p_0(-s)$. Also let $F(s^2) = p_0(s)p_0(-s)$. By a simple computation, it follows that (s^2 is replaced by u):

$$H(u) = h(u)f(u) - ug(u)e(u)$$

$$G(u) = g(u)f(u) - h(u)e(u) \tag{4}$$

$$F(u) = f^2(u) - ue^2(u)$$

With this setting, we have

$$[q_0(s) + \alpha p_0(s)]p_0(-s) = [H(s^2) + \alpha F(s^2)] + sG(s^2)$$

If $G \neq 0$ and if they exist, let the *real negative roots with odd multiplicities of* $G(u)$ be $\{v_1, v_2, \cdots v_l\}$ with the ordering $v_1 > v_2 > \cdots > v_l$, with $v_0 := 0$ and $v_{l+1} := -\infty$ for notational convenience.

The following algorithm determines whether $\Phi_r(p_0,q_0)$ is empty or not and outputs its elements when it is not empty (Saadaoui & Ozguler, 2005) :

Algorithm 1.

1. Consider all the sequences of signums

$$I = \begin{cases} \{i_0, i_1, \cdots, i_l\} & \textit{for odd } r-m, \\ \{i_0, i_1, \cdots, i_{l+1}\} & \textit{for even } r-m. \end{cases}$$

where

$$i_j \in \{-1,1\} \textit{ for } j = 0, 1, \cdots, l+1 .$$

2. Choose all the sequences that satisfy

$$r - \sigma(p_0) = \begin{cases} i_0 - 2i_1 + 2i_2 \cdots + 2(-1)^l i_l & \textit{for odd } r-m \\ i_0 - 2i_1 + 2i_2 \cdots + (-1)^{l+1} i_{l+1} & \textit{for even } r-m \end{cases}.$$

3. For each sequence of signums $I\{i_j\}$ that satisfy step 2, let

$$\alpha_{\max} = \max\left\{-\frac{H}{V}(v_j)\right\} \quad \forall v_j \; for which F(v_j) \neq 0 \textit{ and } i_j SF(v_j) = 1.$$

and

$$\alpha_{\min} = \min\left\{-\frac{H}{V}(v_j)\right\} \quad \forall v_j \; for which F(v_j) \neq 0 \textit{ and } i_j SF(v_j) = -1.$$

The set $\Phi_r(p_0,q_0)$ is non-empty if and only if for at least one signum sequence I satisfying step 2, $\alpha_{\max} < \alpha_{\min}$ holds.

4. $\Phi_r(p_0,q_0)$ is equal to the union of intervals $(\alpha_{\max}, \alpha_{\min})$ for each sequence of signums I that satisfy step 3.

The algorithm above is easily specialized to determine all stabilizing proportional controllers $C(s) = \alpha$ for the plant $G_0(s)$. This is achieved by replacing r in step 2 of the algorithm by n, the degree of $\phi(s,\alpha)$.

Remark 1. By Step 2 of Algorithm 1, a necessary condition for the existence of a $\alpha \in \Phi_r(p_0,q_0)$ is that the odd part of $[q_0(s) + \alpha p_0(s)]p_0(s)$ has at least $\bar{r} = \left\lfloor \frac{|r - \sigma(p_0)| - 1}{2} \right\rfloor$ real negative roots with odd multiplicities. When solving a constant stabilization problem, this lower bound is $\bar{r} = \left\lfloor \frac{r - \sigma(p_0) - 1}{2} \right\rfloor$.

2.1 Determining the admissible ranges of k_p using padé approximation

In this part, our aim is to find all admissible values of k_p. Replacing the time delay by a Padé approximation $e^{-Ls} \approx \frac{p(-s)}{p(s)}$ where $p(s) \in H$, we get the following closed-loop characteristic polynomial

$$\begin{aligned} \phi_0(s) &= sq(s)p(s) + (k_d s^2 + k_p s + k_i)p(-s) \\ &= q_0(s) + (k_d s^2 + k_p s + k_i)p_0(s) \end{aligned} \tag{5}$$

Where

$$q_0(s) = sq(s)p(s),$$

$$p_0(s) = p(-s).$$

Multiplying $\phi_0(s)$ by $p_0(-s)$, we obtain

$$\begin{aligned} \psi_0(s) &= \phi_0(s)p_0(-s) \\ &= (H(.) + s^2 k_d F(.) + k_i F(.)) + s(G(.) + k_p F(.)) \end{aligned} \tag{6}$$

where H, G and F are given by (4). Note that $p_0(-s) = p(s) \in H$, therefore by Remark 1 the odd part $G(u) + k_p F(u)$ of $\psi_0(s)$ must have all its roots real, negative, and distinct. At this step, two parameters k_d and k_i are eliminated and an auxiliary problem with only one parameter will be solved. Let

$$\varphi_1(s) = (G(s^2) + sG'(s^2)) + k_p(F(s^2) + F'(s^2))$$

using Lemma 2, finding values of k_p such that $G(u) + k_p F(u)$ has all its roots real, negative and distinct is equivalent to stabilizing the new constructed polynomial $\varphi_1(s)$. This new constructed problem can be solved using Algorithm 1 as only one parameter appears.

2.2 An alternative method for computing the admissible ranges of k_p

In this part an alternative method is used to get the admissible values of k_p without the need of using Padé approximation. Moreover this method allows the calculation of the admissible ranges of the parameters (k_p, k_d), where k_p denotes the proportional gain and k_d the derivative gain. The following result will be used in determining the admissible values of k_p (alternatively k_d).

Lemma 3: (Kharitonov et al., 2005) *Consider the quasi-polynomial,*

$$\Psi(s) = \sum_{i=0}^{n} \sum_{l=1}^{r} h_{il} s^{n-i} e^{\tau_l s}$$

such that $\tau_1 < \tau_2 < \cdots < \tau_r$ *, with main term* $h_{0r} \neq 0$ *, and* $\tau_1 + \tau_r > 0$ *. If* $\Psi(s)$ *is stable, then* $\Psi'(s)$ *is also a stable quasi-polynomial.*

The characteristic function of the closed loop system formed by the PID controller and the time delay system (2) is given by,

$$\Psi_1(s) = sQ(s) + (k_d s^2 + k_p s + k_i)e^{-Ls} \tag{7}$$

Since the term e^{-Ls} has no finite roots, the quasi-polynomial $\Psi_1(s)$ and $\Psi(s) = \Psi_1(s)e^{Ls}$ have the same roots, therefore stability of $\Psi(s)$ is equivalent to stability of $\Psi_1(s)$, see (Kharitonov et al., 2005). In the sequel, the quasi-polynomial,

$$\Psi(s) = sQ(s)e^{Ls} + (k_d s^2 + k_p s + k_i) \tag{8}$$

will be used to study stability of the closed loop system. Now, using Lemma 3, if $\Psi(s)$ is stable then $\Psi'(s)$ is also a stable quasi-polynomial, where $\Psi'(s)$ is given by,

$$\Psi'(s) = \left((Ls+1)Q(s) + sQ'(s)\right)e^{Ls} + 2k_d s + k_p \tag{9}$$

Note that only two parameters (k_p, k_d) appear in the expression of $\Psi'(s)$. By Lemma 3, stabilizing $\Psi'(s)$ is equivalent to calculating the admissible (k_p, k_d) values for the original problem.

The D-decomposition method (Hohenbicher & Ackermann, 2003; Gryazina & Polyak, 2006) is used to determine the stabilizing regions in the (k_p, k_d) plane. The D-decomposition method is based on the fact that roots of the quasi-polynomial (9) change continuously when the coefficients are changed continuously. Hence, a stable quasi-polynomial can become unstable if and only if at least one of its roots crosses the imaginary axis. Using this fact, the (k_p, k_d) plane can be partitioned into regions having the same number of roots of (9) in the left half plane. Stability can be checked by choosing a point inside a region and applying classical methods such as the Nyquist criterion.

Evaluating the characteristic function at the imaginary axis is equivalent to replacing s by $j\omega,\ \omega \geq 0$, in (9) we get,

$$\begin{aligned}\Psi'(jw) &= \left[R(w) - LwI(w) - wI'(w)\right] - \left[wLR(w) + wR'(w)\right]\sin(Lw) + k_p \\ &\left[(R(w) - LwI(w) - wI'(w))\sin(Lw) + (wLR(w) + wR'(w)\cos(Lw) + 2wk_d\right]\end{aligned} \tag{10}$$

where,

$$Q(j\omega) = R(\omega) + jI(\omega) \tag{11}$$

and,

$$Q'(j\omega) = R'(\omega) + jI'(\omega) \tag{12}$$

Applying the D-decomposition method implies equating the real and imaginary parts of (10) to zero. Two cases must be considered:

- **Case 1**: $\omega = 0$ which leads to the following equation,

$$k_p = -Q(0) \tag{13}$$

- **Case 2**: $\omega > 0$ in this case we get,

$$\begin{aligned} k_p = &-\left[R(w) - LwI(w) - wI'(w)\right]\cos(Lw) \\ &+\left[wLR(w) + wR'(w)\right]\sin(Lw) \end{aligned} \tag{14}$$

$$\begin{aligned} k_d = &\left[(R(\omega) - L\omega I(\omega) - \omega I'(\omega))\sin(L\omega)\right. \\ &\left.+ (\omega LR(\omega) + \omega R'(\omega))\cos(L\omega)\right](2\omega)^{-1} \end{aligned} \tag{15}$$

By sweeping over values of $\omega > 0$, the (k_p, k_d) plane can be partitioned into root invariant regions. Using (13), (14) and (15) stabilizing regions in the (k_p, k_d) plane can be determined.

3. PID controller design

In this section, first the set of all stabilizing PID controllers are determined, this forms the set of admissible solutions for optimization. Next, GAs are used to optimize these PID parameters.

3.1 Stabilizing regions

The admissible (k_p, k_d) values are calculated in section 2, now, we go back to the original problem by considering (7). Our method consists of fixing one parameter k_p or k_d and determining the stabilizing regions in the plane of the remaining two parameters. By sweeping over the admissible values of the first parameter (k_p or k_d) the complete set of the stabilizing regions is found. Once again the D-decomposition method is used.

For $\omega = 0$, we get the following equation,

$$k_i = 0 \tag{16}$$

For $\omega > 0$, two cases will be investigated. First, let us fix k_p and find the stabilizing regions in the (k_d, k_i) plane. Using (7), replacing s by $j\omega$ and equating the real and imaginary parts to zero we get,

$$\begin{bmatrix} \cos(L\omega) & -\omega^2\cos(L\omega) \\ \sin(L\omega) & -\omega^2\sin(L\omega) \end{bmatrix} \begin{bmatrix} k_i \\ k_d \end{bmatrix} = \begin{bmatrix} k_p\omega\sin(L\omega) + \omega I(\omega) \\ -k_p\omega\cos(L\omega) + \omega R(\omega) \end{bmatrix} \tag{17}$$

Note that the matrix at the left-hand of (17) is singular. The singular frequencies (Hohenbicher & Ackermann, 2003) are determined as the solutions of equation (18),

$$k_p + I(\omega)\sin(L\omega) - R(\omega)\cos(L\omega) = 0 \tag{18}$$

and are denoted by $\omega_i, i = 1,2,\cdots$. For each singular frequency ω_i, an equation of a straight line in the (k_d, k_i) plane is defined by,

$$k_i = \omega_i^2 k_d + \omega_i R(\omega_i)\sin(L\omega_i) + \omega_i I(\omega_i)\cos(L\omega_i) \tag{19}$$

which partition the plane into root-invariant regions among which the stabilizing regions, if any, have to be determined. Alternatively, one may fix k_d first and repeat the above procedure. In this case we have,

$$\begin{bmatrix} \cos(L\omega) & -\omega\sin(L\omega) \\ \sin(L\omega) & \omega\cos(L\omega) \end{bmatrix} \begin{bmatrix} k_i \\ k_p \end{bmatrix} = \begin{bmatrix} k_d\omega^2\cos(L\omega) + \omega I(\omega) \\ k_d\omega^2\sin(L\omega) - \omega R(\omega) \end{bmatrix} \tag{20}$$

solving this system we get,

$$k_i(\omega) = \omega^2 k_d + \omega I(\omega)\cos(L\omega) - \omega R(\omega)\sin(L\omega) \tag{21}$$

$$k_p(\omega) = -I(\omega)\sin(L\omega) - R(\omega)\cos(L\omega) \tag{22}$$

For $\omega > 0$, the above two equations partition the (k_p, k_i) plane into root-invariant regions. Hence stabilizing regions, if any, can be found. It is interesting to note here that depending on which parameter we fix first, either k_p or k_d, the obtained stabilizing region is different. This can be explained by the fact that in the first case the matrix is always singular and in the second case it is always non-singular.

3.2 PID controller design using GAs

Determining the total set of stabilizing PID controllers is a first step in the design process. Once this set of stabilizing PID controllers is found, it is natural to search within this set, controllers that meet extra performance specifications. Four performance measures will be considered:

- Maximum percent overshoot (OS).
- Settling time (ST).
- Rise time (RT).
- Integral square error (ISE) $ISE = \int_0^{t_f} |e(t)|^2\, dt$, where e(t) is the error at time t.

Optimization is done using genetic algorithms. The first step in this design procedure, which consists of determining the total set of stabilizing PID controllers, is very important. It enhances the application of the genetic algorithm by fixing the search space, unlike other works on optimizing PID controllers using GAs (Chen et al., 2009; Chen & Chang, 2006; Jan et al., 2008) where the ranges of (k_p, k_i, k_d) are set arbitrary. Moreover, it improves the optimization time and increases the chances of obtaining the global optimum. Optimization of the PID controllers is done by minimizing each of the four cost functions: maximum percent overshoot, settling time, rise time and integral square error. GAs tries to find the global optimum by evaluating many points of the solution space in each generation. There are mainly three operations: reproduction, crossover and mutation, used to form new generations until a termination condition is reached (Goldberg, 1989; Haupt & Haupt, 2004; Melanie, 1998). In what follows, we briefly describe the parameters and tools used in the implementation of the genetic algorithm:

1. Since stability is an essential property for any control system, the search space consists of the stabilizing values of the PID controller which are determined in section 3.1. Population size is chosen to be 150 individuals per population. In the beginning the initial population is randomly chosen within the stabilizing values of the PID controller. These individuals are candidate solutions to the problem. The number of generations is chosen as 100.
2. Evaluation of a generation is done by calculating the cost function for each individual. Since minimization problems are considered, the fitness function will be the inverse of the cost function.
3. Genetic operators: reproduction, crossover and mutation are used to form new generations. Individuals are chosen for reproduction according to their fitness value. Fittest ones have larger probability of selection. The Roulette Wheel selection method is used.
4. In order to improve next generations, crossover which is a process of combining parts of the parent individuals to produce new offsprings is done. Crossover value used is 0.8.

To avoid local minimum and explore new parts of the search space, mutation process is applied. It consists of randomly modifying individuals in the generation.

1. Mutation probability is always set to a small value so that the search algorithm is not turned to a random search algorithm. The value chosen in our case is 0.01.
2. Steps 2-5 are repeated until a termination condition is reached. The maximum generation termination condition is adopted.

4. Examples

In this section three illustrative examples are given.

Example 1:

Consider stabilizing the second order integrating system,

$$G(s) = \frac{11.32}{s^2 + 11.32s} e^{-s} \tag{23}$$

by a PID controller. This transfer function represents a system composed of master-slave parts. The slave part is a model of a mobile robot which can move in one direction (Seuret et al., 2006). This robot is controlled through a communication network which introduces delays in the control loop. See (Seuret et al., 2006) for more details about this system. First, the admissible values of (k_p, k_d) are calculated, see Fig. 1. The first approach proposed consists of fixing k_p and using (16), (18) and (19) to determine stabilizing regions in the plane of the remaining two parameters.

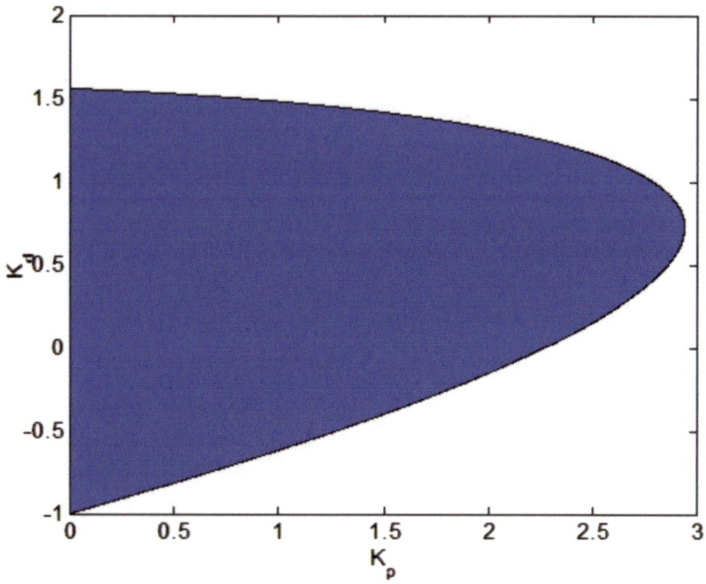

Fig. 1. Admissible values of (k_p, k_d).

By sweeping over the admissible values of k_p, the complete stabilizing regions of (k_p, k_i, k_d) values are found as shown in Fig.2.

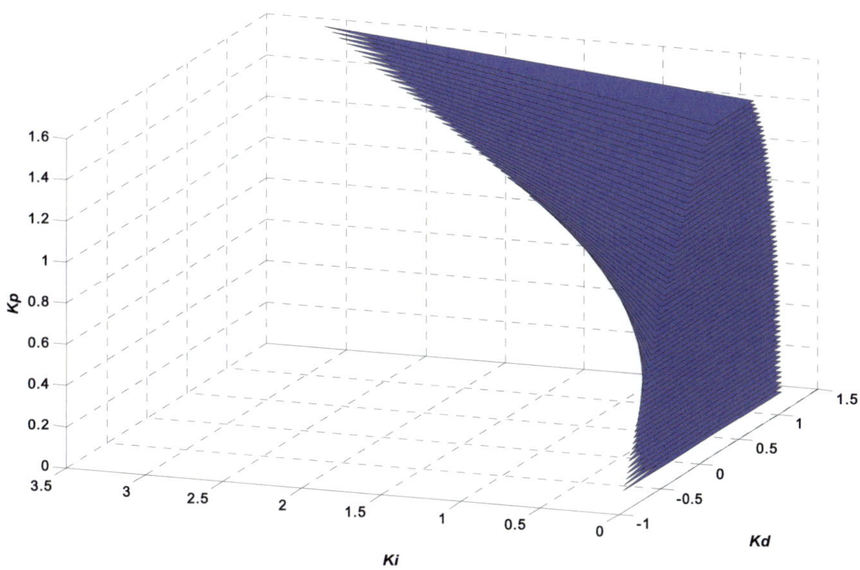

Fig. 2. Complete set of stabilizing PID controllers for example1.

Alternatively, we can fix k_d and use (16), (21) and (22) to find the stabilizing regions of (k_p, k_i, k_d) values as shown in Fig.3.

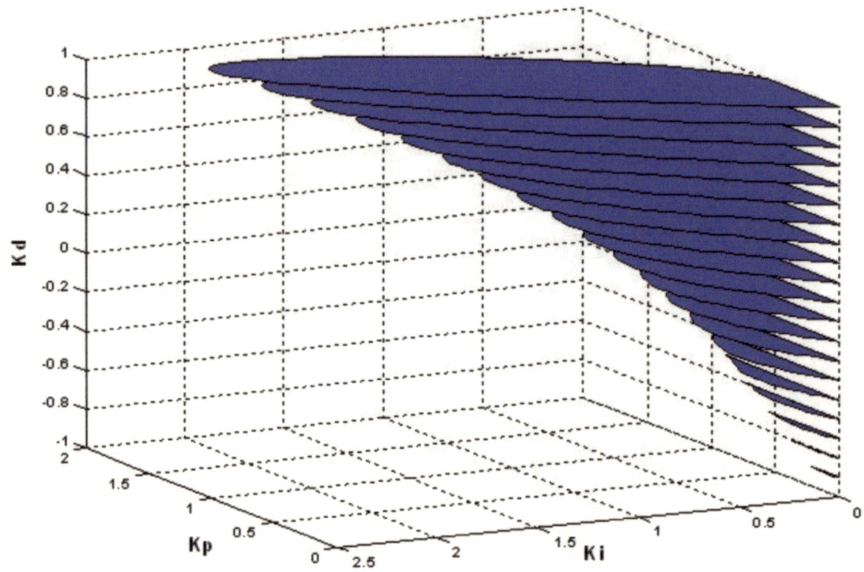

Fig. 3. Complete set of stabilizing PID controllers for example1.

Fixing $k_d = 0$ is equivalent to using a PI controller. For $k_d = 0$, we obtain the stabilizing region of (k_p, k_i) values as given in Fig. 4.

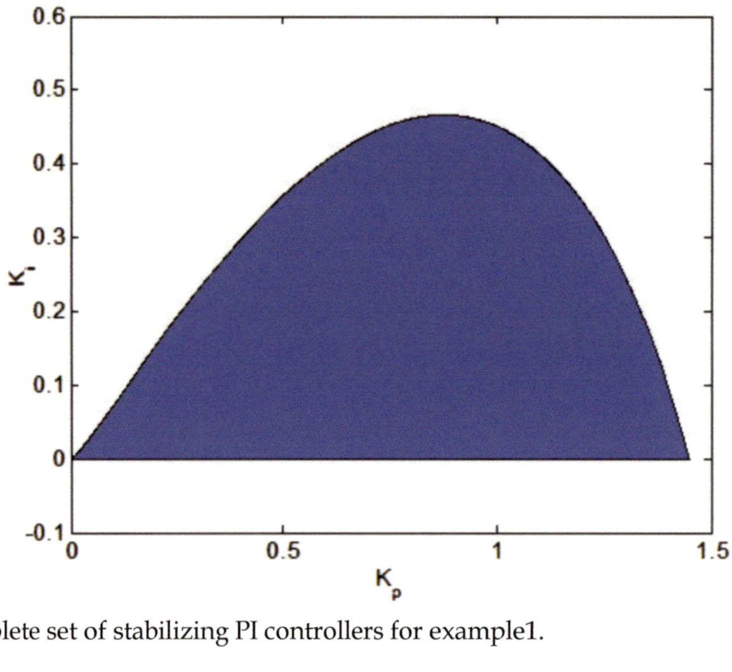

Fig. 4. Complete set of stabilizing PI controllers for example1.

Using this stabilizing region of (k_p, k_i) values, we apply the genetic algorithm described in section 3.2, to minimize each of the performance indices: maximum percent overshoot (Opt1), settling time (Opt2), rise time (Opt3) and integral square error (opt4). Although a simple controller, PI controller, is used for an integrating system with delay, the results obtained are satisfactory as shown by the step response of the closed loop system in Fig. 5.

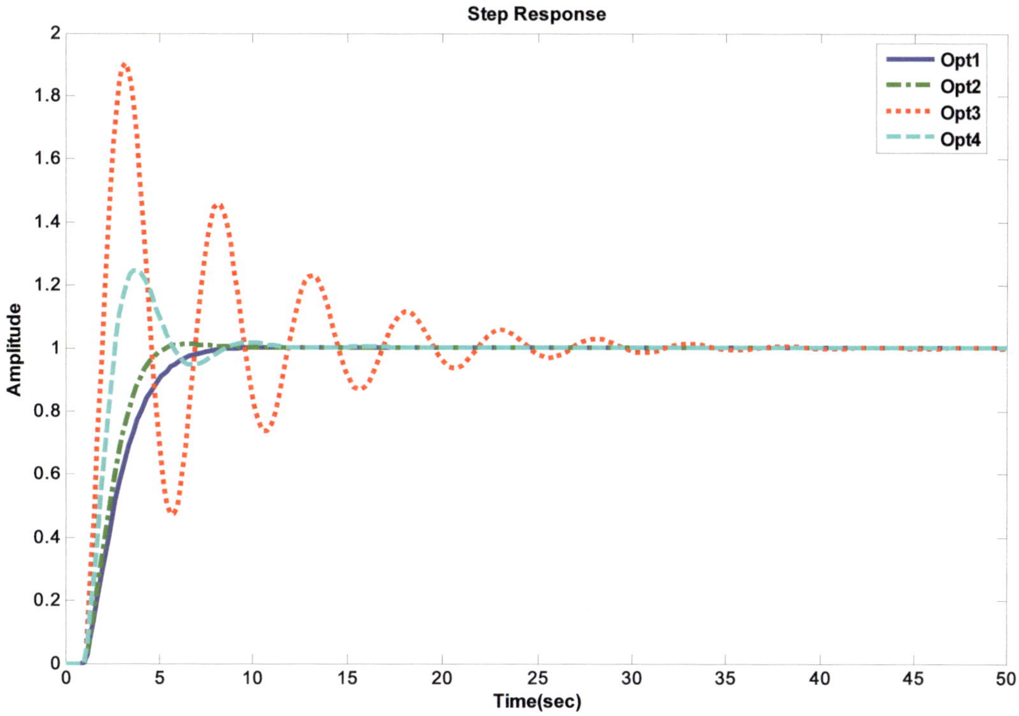

Fig. 5. Step responses of the closed loop system.

Example 2:

Consider stabilizing the first order system given in (Silva et al., 2005),

$$G(s) = \frac{1}{3s+1}e^{-2.8s} \tag{24}$$

by a PID controller. For comparison reasons, we apply three classical methods. These methods are: Ziegler-Nichols method (ZN), Chien Hornes & Reswich method (CHR) and finally Cohen-Coon method (CC). After obtaining the stabilizing regions of (k_p, k_i, k_d) values, we apply the genetic algorithm to minimize the four performance indices described in section 3.2. Step responses of the closed loop system obtained with classical methods are given in Fig 6.

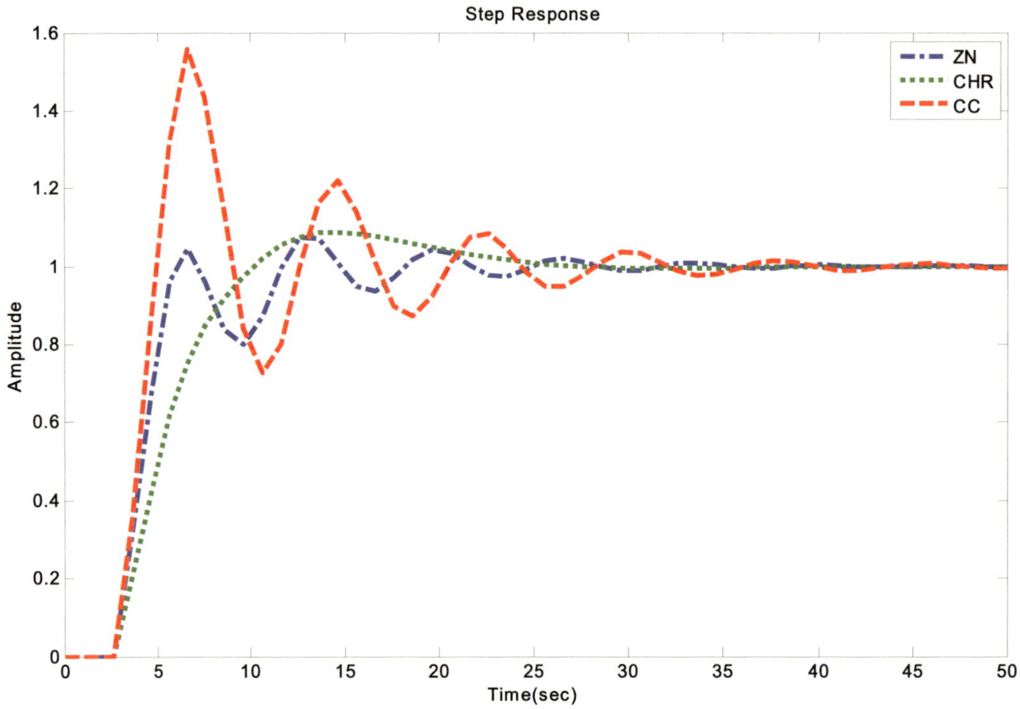

Fig. 6. Step responses of the closed loop system for example 2.

Determining the set of stabilizing controllers and using GAs to optimize these stabilizing values, gives the step responses shown in Fig 7.

Table 1 summarizes the different results obtained using our method and classical methods. For each performance criteria we can see that our method gives better results. However it should be stressed here that minimizing one performance criteria can deteriorate other performance indices, for example we minimize overshoot at the expense of a larger settling time as shown in the results below. Optimizing the integral square error (opt4) gives better results as an overall performance. Another advantage of our method is to answer the question: what is the best performance we can get with such a controller? This is possible as optimization is done over all the admissible stabilizing values of the controller.

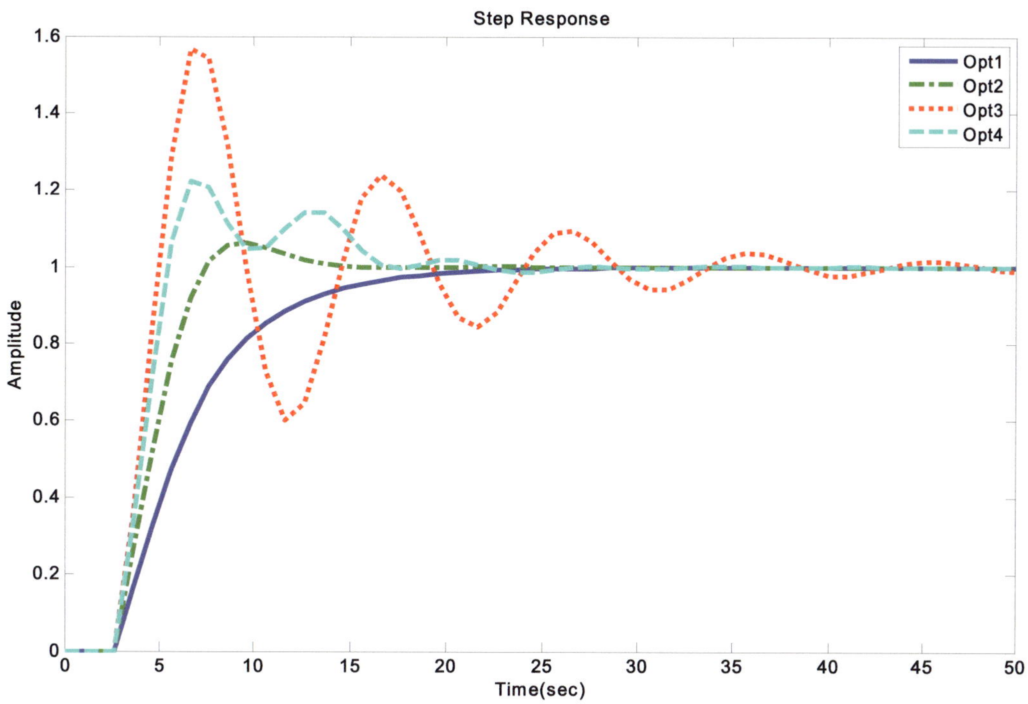

Fig. 7. Step responses of the closed loop system for example 2.

Method	(K_p,K_i,K_d)	OS (%)	ST (s)	RT (s)	ISE
ZN	(1.28, 0.22, 1.8)	7.9	24.9	5.9	3.17
CHR	(0.64, 0.21, 0.9)	9.1	24	8	3.75
CC	(1.69, 0.39, 1.49)	51	32.9	5.8	3.33
Opt 1	(0.52, 0.14, 0.28)	0	19	9	4.77
Opt 2	(0.84, 0.22, 0.39)	4.9	13	7	3.73
Opt 3	(1.69, 0.35, 0.55)	58	48	6	4.32
Opt 4	(1.22, 0.35, 1.67)	21.7	21.9	5.9	2.95

Table 1. Results obtained by different methods.

Example 3:

Consider stabilizing the second order plant given in (Wang, 2007a) by

$$G(s) = \frac{k}{(1+T_1 s)(1+T_2 s)} e^{-Ls},$$

by a PID controller.

Let $k=1$ and $L=0.2$ applying the procedure developed in this work, we get (the four cases given in (Wang, 2007a) will be investigated):

- Case 1: For $T_1=0.2$ and $T_2=0.5$ the admissible range of k_p values is obtained as $-1<k_p<5.56$ with a Padé approximation of order 3.
- Case 2: For $T_1=-1$ and $T_2=-0.5$ the admissible range of k_p values is obtained as $-1<k_p<0.58$ with a Padé approximation of order 2.
- Case 3: For $T_1=1$ and $T_2=-0.5$ the admissible range of k_p values is obtained as $-5.21<k_p<-1$ with a Padé approximation of order 3.
- Case 4: For $T_1=-1$ and $T_2=0.5$ the admissible range of k_p values is obtained as $-9.79<k_p<-1$ with a Padé approximation of order 3.

Although a necessary condition is used to determine the admissible ranges of k_p, in the four cases above the intervals obtained are the same as the ones given in (Wang, 2007a).

5. Conclusion

In this work, a new method is given for calculating all stabilizing values of a PID controller which is applied to a special class of time delay systems. The proposed approach is based on determining first the admissible ranges of one of the controller's parameters. This step is solved using two different approaches. Next, the D-decomposition method is applied to obtain the stabilizing regions in the controller's parameter space. The genetic algorithm optimization method is then applied to find among those stabilizing controllers those that satisfy further performance specifications. Four time domain measures, which are maximum percent overshoot, settling time, rise time and integral square error, were minimized. Application of the proposed method to uncertain time delay system is under investigation.

6. References

Chen, C. K.; Koo, H. H.; Yan, J. J. & Liao, T. L. (2009). GA-based PID active queue management control design for a class of TCP communication networks, *Expert Systems with Applications*, Vol. 36, pp. 1903-1913.

Chen, C. K. & Chang, S. H. (2006). Genetic algorithms based optimization design of a PID controller for an active magnetic bearing. *International Journal of Computer Science and Network*, Vol. 6, pp. 95-99.

Datta, A.; Ho, M. T. & Bhattacharyya, S. P. (2000). *Structure and synthesis of PID controller,* Springer-Verlag, ISBN 1-85233-614-5, London, Great Britain.

Goldberg, D. E. (1989). *Genetic algorithms in search, optimization and machine learnin,* Addison-Wesley, ISBN 0201157675.

Gryaznia, E. N. & Polyak, B. T. (2006). Stability regions in the parameter space: D-decomposition revisited. *Automatica*, Vol. 42, pp. 13-26.

Haupt, R. L. & Haupt, S. E. (2004) *Practical genetic algorithms,* John Wiley & Sons, ISBN 0471188735.

Hohenbicher, N. & Ackermann, J. (2003). Computing stable regions in parameter spaces for a class of quasi-polynomials, Proc. 4th IFAC Workshop on Time Delay Systems TDS'03, Recquencourt, France.

Jan, R. M.; Tseng, C. S. & Lin, R. J. (2008). Robust PID control design for permanent magnet synchronous motor: A genetic approach. *Electric Power Systems Research*, Vol. 78, pp. 1161-1168.

Melanie, M. (1998) *An introduction to genetic algorithms,* MIT Press, ISBN 0262631857.

Normey-Rico, J. E. & Camacho, E. F. (2007). *Control of dead time process.* Springer, London.

Kharitonov, V. L.; Niculescu, S.; Moreno, J. & Michiels, W. (2005). Static output stabilization: Necessary conditions for multiple delay controllers. *IEEE Transactions on Automatic Control*, Vol. 50, pp. 82-86.

Saadaoui, K. & Ozguler, A. B. (2003). On the set of all stabilizing first order controllers, Proc. American Control Conference, Denver, CO, USA, 2003.

Saadaoui, K. & Ozguler, A. B. (2005). A new method for the computation of all stabilizing controllers of a given order. *International Journal of Control,* Vol. 78, pp. 14-28.

Silva, G. J.; Datta, A. & Bhattacharyya, S. P. (2001). PI stabilization of first order systems with time delay. *Automatica*, Vol. 37, pp. 2225-2031.

Silva, G. J.; Datta, A. & Bhattacharyya, S. P. (2002). New results on the synthesis of PID controllers. *IEEE Transactions on Automatic Control*, Vol. 47, 2002, pp. 241-252.

Silva, G. J.; Datta, A. & Bhattacharyya, S. P. (2005) *PID controllers for time delay systems,* Birkhauser, ISBN 978-0817642662, Boston USA.

Seuret, A.; Michaut, F.; Richard, J. P. & Divoux, T. (2006). Network control using GPS synchronization, Proc. American Control Conference, Minneapolis, Minnesota, USA.

Wang, D. J. (2007). Further results on the synthesis of PID controllers. *IEEE Transactions on Automatic Control,* Vol. 52, pp. 1127-1132.

Wang, D. J. (2007). Stabilizing regions of PID controllers for n-th order all pole plants with dead time. *IET Control Theory & Applications,* Vol. 1, pp. 1068-1074.

Zhong, Q. C. (2006). *Robust control of time delay systems,* Springer-Verlag, ISBN 978-1846282645.

8

Practical Control Method for Two-Mass Rotary Point-To-Point Positioning Systems

Fitri Yakub, Rini Akmeliawati and Aminudin Abu
Malaysia-Japan International Institute of Technology (MJIIT), Universiti Teknologi Malaysia International Campus (UTM IC)
Malaysia

1. Introduction

Motion control systems play an important role in industrial engineering applications such as advanced manufacturing systems, semiconductor manufacturing system, computer numerical control (CNC) machining and robot systems. In general, positioning system can be classified into two types, namely point-to-point (PTP) positioning systems and continuous path (CP) control system (Crowder R.M, 1998). PTP positioning systems, either of one-mass or multi-mass systems, is used to move an object from one point to another point either in angular or linear position. For example, in application with one-mass system, such as CNC machines, PTP positioning is used to accurately locate the spindle at one or more specific locations to perform operations, such as drilling, reaming, boring, tapping, and punching. In multi-mass systems application, such as in spot-welding robot, which has a long arm for linear system or long shaft in rotary system, PTP positioning is used to locate the manipulator from one location to another.

PTP positioning system requires high accuracy with a high speed, fast response with no or small overshoot and to be robust to parameter variations and uncertainties. Therefore, the most important requirements in PTP positioning systems are the final accuracy and transition time whereas the transient path is considered as the second important. However, it is not easy to achieve high precision performances because of non-linearities and uncertainties exist in the motion control systems. One significant nonlinearity is friction that causes steady state error and/or limit cycles near the reference position (Armstrong et al., 1994). Another source of nonlinearity in motion control system is saturation of the actuator and/or electronic power amplifier. Saturation causes slow motion and may effect the stability of the performances (Slotine & Li, 1991). In PTP applications, the system performance is expected to be the same or as close as its performance when the system is in normal condition. Thus, robustness is also an important requirement in order to maintain the stability of the positioning systems.

In order to satisfy the design requirements, a good controller is required. Many types of controllers have been proposed and evaluated for positioning systems. The use of proportional-integral-derivative (PID) controllers are the most popular controller used in industrial control systems including motion control systems due to their simplicity and also satisfactory performances (Yonezawa et al., 1990). However, it is difficult to achieve a fast response with no or small overshoot simultaneously.

In practical applications, an engineer does not need deep knowledge or be an expert in control systems theory while designing controllers. Thus, easiness of controller design process, simplicity of the controller structure and no requirement of exact object model and its parameters are very important and preferable in real applications. To achieve these, nominal characteristic trajectory following (NCTF) controller for one-mass rotary systems had been proposed as a practical controller for PTP positioning systems in (Wahyudi, 2002). The controller design procedure is simple and easily implemented since it is only based on a simple open-loop experiment. In addition, an exact object model and its parameters does not required while designing the controller. Thus, this controller is easy to design, adjustable and understands.

The NCTF controller had been proposed as a practical controller for PTP positioning systems. However, the NCTF controller is designed based on one-mass rotary positioning systems. Positioning system was considered as a one-mass system when a rigid coupling with high stiffness is used. The existing NCTF controller does not work for two-mass rotary system because of the vibration happened due to mechanical resonance of the plant such as flexible coupling or a long shaft with low stiffness are used. This vibration gives the unstable performance response of the plant. Therefore, enhancement and improvement design of NCT and a compensator are required to make the NCTF controller suitable for two-mass rotary positioning systems, which have long shaft between the actuator and load. This chapter is an attempt to address the problem of NCTF for two-mass system or multi-mass systems.

2. Basic concept of NCTF controller

The structure of the NCTF control system is shown in Figure 1 (Mohd Fitri Mohd Yakub et al., 2010) consists of:

a. A nominal characteristic trajectory (NCT) that is constructed based on measured θ_l, and $\dot{\theta}_l$, which were obtained by a simple open-loop experiment. Thus, the NCT provides information of characteristics of the system, which can be used to design a compensator.
b. A compensator which is used to force the object motion to reach the NCT as fast as possible, control the object motion to follow the NCT, and end it at the origin of the phase-plane $(e = 0, \dot{e} = 0)$ as shown in Figure 2.

Therefore, the controller is called as the NCTF controller.

As shown in Figure 1, the controller output is signal u. This signal is used to drive the object. The input to the controller are error, e, and object motion, $\dot{\theta}_l$. In principle, the controller compares the object motion input, $\dot{\theta}_l$ with the error-rate, $\dot{e}$, provided by predetermined NCT, at certain error. The difference between the actual error-rate of the object and that of the NCT is denoted as signal u_p, which is the output of the NCT. If the object motion perfectly follows the NCT, the value of signal u_p is zero. Thus, no action is performed by the compensator. When the signal u_p is not zero, the compensator is used to drive the value of signal u_p to zero.

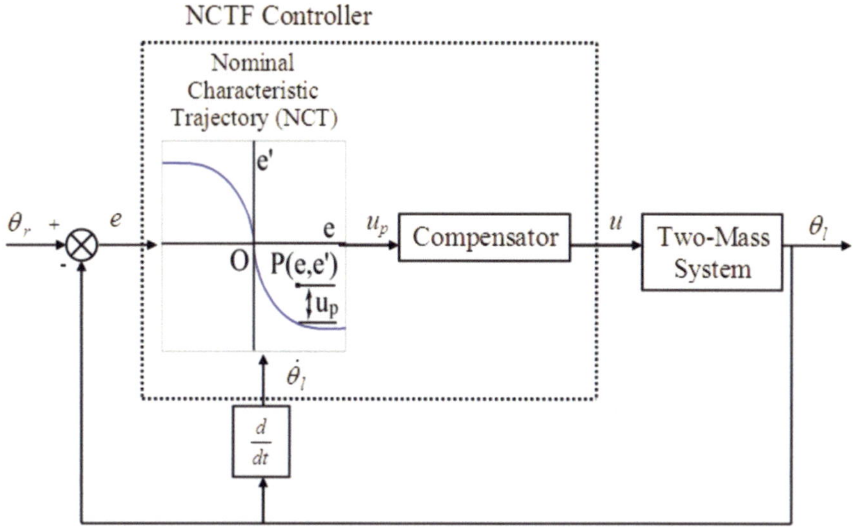

Fig. 1. Structure of NCTF control system

In Figure 2, the object motion is divided into two phases, the reaching phase and the following phase. During the reaching phase, the compensator forces the object motion to reach the NCT as fast as possible. Then, in the following phase, the compensator controls the object motion to follow the NCT and end at the origin. The object motion stops at the origin, which represents the end of the positioning motion. Thus, the NCT governs the positioning response performance.

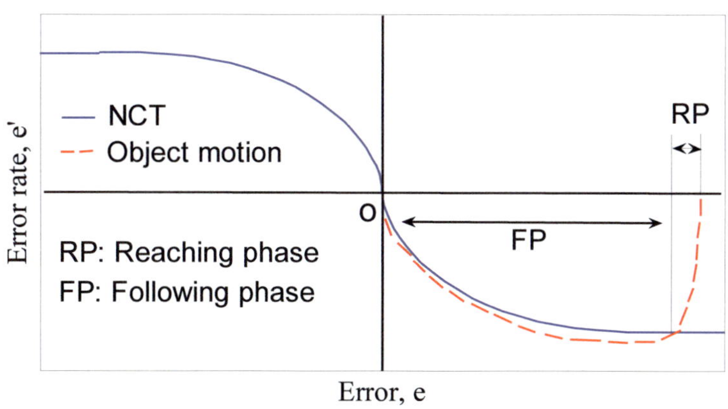

Fig. 2. NCT and object motion

The electric motor is assumed as the actuator in this discussion. To drive the object to reach the NCT, the actuator needs to reach its maximum velocity. The characteristic of the actuator when it stops from its maximum velocity influences the final accuracy of the PTP positioning operation. Thus, this characteristic is required to design the controller. In order to obtain the required characteristics, conducting a simple open-loop experiment is a simple and practical way.

In summary, the structure of the NCTF control system for PTP positioning systems shown in Figure 1 only works under the following two conditions (Wahyudi et al., 2003):

i. A DC or an AC servomotor is used as an actuator of the object.
ii. The reference input, θ_r is constant and $\dot{\theta}_r = 0$.

3. Practical controller design

The NCTF controller consists of NCT, which is constructed based on a simple open-loop experiment of the object, and PI compensator, which is designed based on the obtained NCT. Therefore, the design of NCTF controller can be described by the following steps:

i. The object is driven with an open loop stepwise input and its load displacement and load velocity responses are measured.
ii. Construct the NCT by using the object responses obtained during the deceleration process. Since the NCT is constructed based on the actual responses of the object, it contains nonlinear characteristics such as friction and saturation. The NCTF controller is expected to avoid impertinent behaviour by using the NCT.
iii. Design the compensator based on the open-loop responses and the NCT information. In order to consider the real characteristic of the mechanism in designing compensator, a practical stability analysis is determined.

The NCT includes information of the actual object parameters. Therefore, the compensator can be designed by using only the NCT information. Due to the fact that the NCT and the compensator are constructed from a simple open-loop experiment of the object, the exact model including the friction characteristic and the conscious identification task of the object parameters are not required to design the NCTF controller. The controller adjustment is easy and the aims of its control parameters are simple and clear.

3.1 NCT constructions

As mentioned earlier, in order to construct the NCT, a simple open-loop experiment has to be conducted. In the experiment, an actuator of the object is driven with a stepwise input and, load displacement and load velocity responses of the object are measured. Figure 3 shows the stepwise input, load velocity and load displacement responses of the object. In this case, the object vibrates due to its mechanical resonance. In order to eliminate the influence of the vibration on the NCT, the object response must be averaged.

In Figure 4(a), moving average filter is used to get the averaged response because of its simplicity (Oppenheim & Schafer, 1999). The moving average filter operates by averaging a number of points from the object response to produces each point in the averaged response. Mathematically, it can be expressed as follows:

$$\theta_{av}(i)=\frac{1}{M}\sum_{j=-(M-1)/2}^{(M-1)/2}\theta(i+j) \quad (1)$$

$$\dot{\theta}_{av}(i)=\frac{1}{M}\sum_{j=-(M-1)/2}^{(M-1)/2}\dot{\theta}(i+j) \quad (2)$$

where θ represents the object displacement in rad, θ' the object velocity in rad/s, θ_{av} the averaged object displacement, θ'_{av} the averaged object velocity and M is the number of data points of the object responses used in the averaging process. The averaged velocity and displacement responses are used to determine the NCT. Since the main problem of the PTP motion control is to stop an object at a certain position, a deceleration process (curve in area A of Figure 4) is used. The phase plane of the NCT has a horizontal axis of error, e, and a vertical axis of error-rate, $\dot{e}$, as shown in Figure 4(b). The horizontal axis shows the error from the desired position. Therefore, the values are taken from the displacement data of the open-loop responses. The error is the displacement data when it started entering deceleration process subtracted by the maximum displacement. The error-rate is the velocity data within the deceleration process. In Figure 4(a), h is the maximum velocity which depends on the input step height, and is gradually decreasing until zero. Thus, the pair data of error and error-rate construct the NCT, i.e. the phase-plane diagram of e and $\dot{e}$.

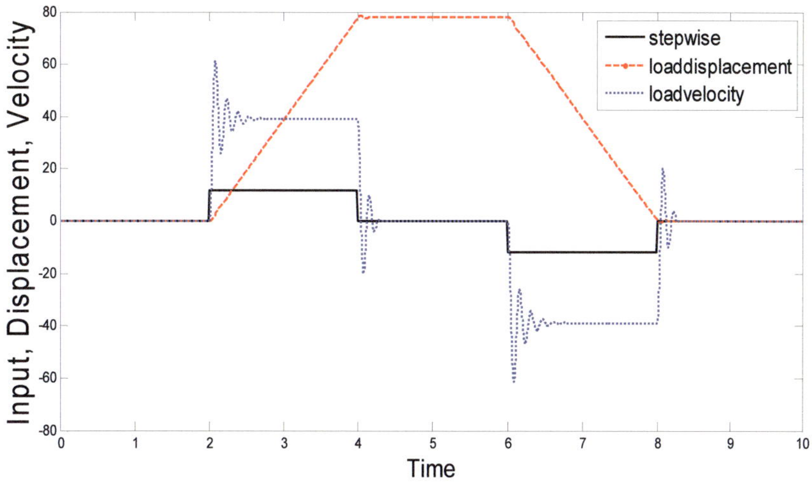

Fig. 3. Input and actual object response

From the response, the maximum velocity of the actuator is obtained. Important characteristics of the actuator can be identified from the data within the deceleration range. In the deceleration process, the actuator moves freely from its maximum velocity until it stops.

In the deceleration process, non-linearity characteristics such as friction and saturation of the object have taken its effect. Therefore, the data can be translated into the form that represents the system characteristics including friction and saturation. From the curve in area *A* and *h* in Figure 4(a), the NCT in Figure 4(b) is determined.

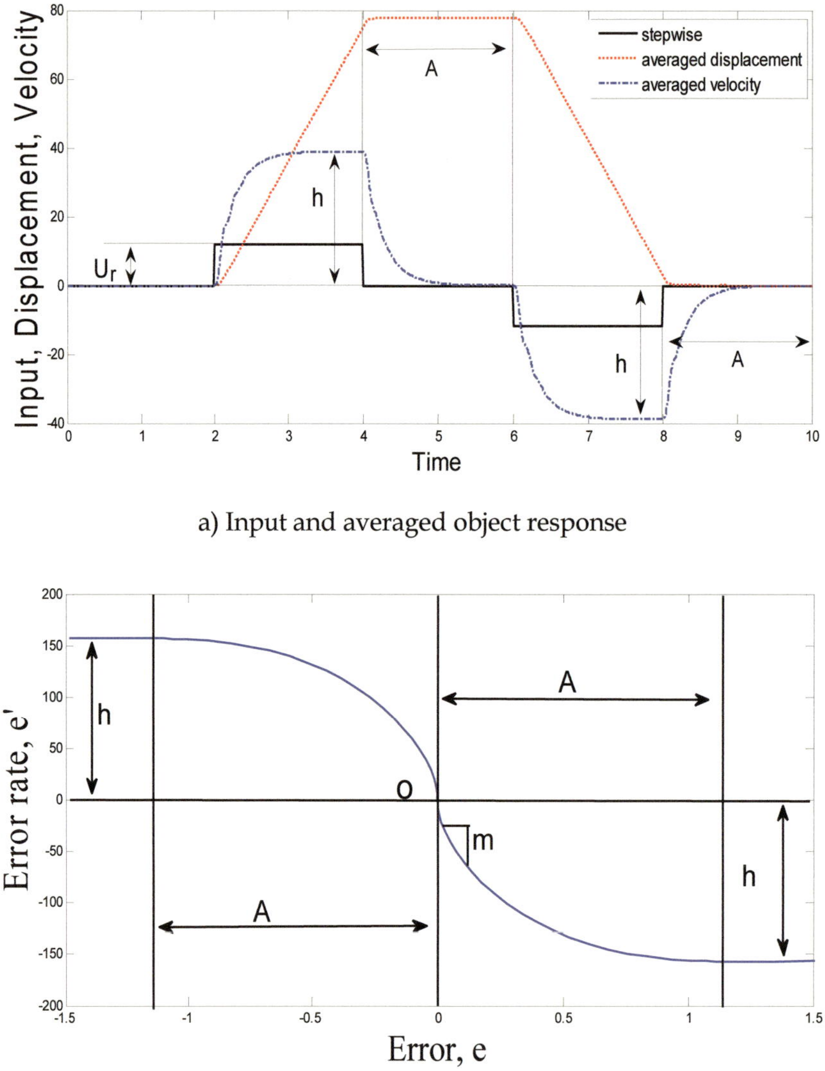

a) Input and averaged object response

b) Nominal characteristic trajectory

Fig. 4. Construction of the NCT.

There are two important parameters in the NCT as shown in Figure 4(b), the maximum error indicated by *h* and the inclination of the NCT near the origin indicated by *m*. This parameter is related to the dynamics of the object. Therefore, these parameters can be used to design the compensator.

3.2 Simplified object parameters

An exact modeling, including friction and conscious identification processes, is not required in the NCTF controller design. The compensator is derived from the parameter m and h of the NCT. Since the DC motor is used as the actuator, the simplified object can be presented as a following fourth-order system (Dorf, & Bishop, 2008):

$$G_o(s) = \frac{\theta_l(s)}{U(s)} = K\frac{\alpha_2}{s(s+\alpha_2)}\frac{\omega_f^2}{s^2+2\zeta_f\omega_f s+\omega_f^2} \tag{1}$$

where $\theta_l\ (s)$ represents the displacement of the object, $U(s)$, the input to the actuator and K, ζ, a_2 and ω_f are simplified object parameters. The NCT is determined based on the averaged object response which does not include the vibration. So, it can be assumed that the averaged object response is a response to the stepwise inputs of the averaged object model as follows:

$$\frac{\theta_l(s)}{U(s)} = K\frac{\alpha_2}{s(s+\alpha_2)} \tag{2}$$

where $\theta_l(s)$ represents the displacement of the object, $U(s)$ is an input to the actuator, and, K and a_2 are simplified object parameters. This simplified model is reasonable since it is assumed that a DC or AC servomotor is used to drive the positioning systems. The simplified object model of Eq. (2) can be described in the state space representation as follow:

$$\frac{d}{dt}\begin{bmatrix}\theta_l\\ \dot{\theta}_l\end{bmatrix} = \begin{bmatrix}0 & 1\\ 0 & -\alpha_2\end{bmatrix}\begin{bmatrix}\theta_l\\ \dot{\theta}_l\end{bmatrix} + \begin{bmatrix}0\\ \alpha_2 K\end{bmatrix}u \tag{3}$$

Characteristic of the positioning systems near the reference position is very important because system stability and positioning accuracy depend on it characteristic. The inclination m of the NCT near the origin is related to the simplified object parameter α_2. The NCT in Figure 4(b) shows,

$$\frac{d\dot{e}}{de} = m \tag{4}$$

The NCT inclination near the origin m is constructed when the input to the object is zero $(u = 0)$. By letting $u = 0$ in Eq. (3),

$$\frac{d\dot{\theta}_l}{d\theta_l} = -\alpha_2 \tag{5}$$

As considering $e = \theta_r - \theta$ and θ_r is constant in the case of PTP positioning systems. Then, Eq. (5) is translated to the following form,

$$\frac{d\dot{e}}{de} = -\alpha_2 \tag{6}$$

Hence, from Eqs. (4) and (6), we obtain,

$$\alpha_2 = -m \tag{7}$$

The value of the maximum error rate h of the NCT is related to the steady-state velocity due to input actuator u_p. Based on the final value theorem, the steady-state velocity is Ku_r. Then,

$$h = -Ku_r \tag{8}$$

Thus, K can be expressed as,

$$K = \frac{h}{u_r} \tag{9}$$

The friction characteristic influences the NCT inclination near the origin m and the maximum error rate of h of the NCT. Therefore, the simplified object model includes the effect of friction when they are determined with Eqs. (8) and (9).

3.3 Compensator design

The following proportional integral (PI) and notch filter (NF) compensator is proposed for two-mass systems:

$$G_c(s) = \frac{(K_p s + K_i)}{s}\left(\frac{K_{dc}(s^2 + 2\zeta_f \omega_f s + \omega_f^2)}{(s^2 + 2\zeta_o \omega_o s + \omega_o^2)}\right) \tag{10}$$

where K_p is the proportional gain, K_i is the integral gain, K_{dc} is the filter gain, ζ_f and ω_f represent zeros of NF, while ζ_o and ω_o represent poles of NF. The PI compensator is adopted for its simplicity to force the object motion to reach the NCT as fast as possible and control the object motion to follow the NCT and stop at the origin.

Figure 5 shows the block diagram of the continuous close-loop NCTF control system with the simplified object near the NCT origin where the NCT is linear and has an inclination a_2=-m. The signal u_p near the NCT origin in Figure 5 can be expressed as the following equation:

$$u_p = \dot{e} + \alpha_2 e = \alpha_2 e - \dot{\theta}_l \tag{11}$$

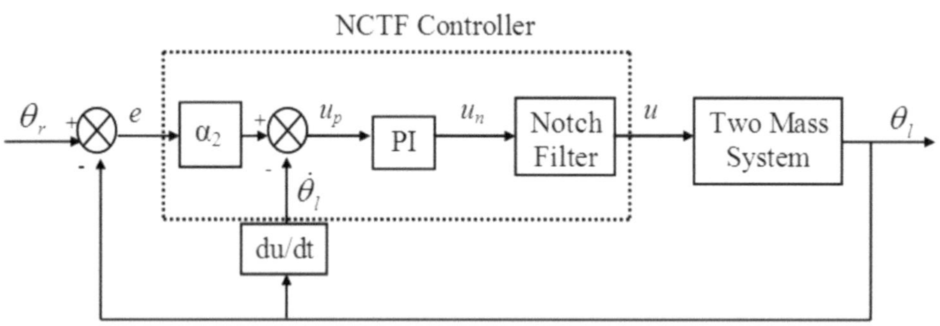

Fig. 5. Simplified NCTF control system

The PI compensator is designed based on the simplified object parameters, which are also related to the NCT, i.e h and m. Due to its simplicity, the following PI compensator is adopted.

$$G_c(s) = K_p + \frac{K_i}{s} \tag{12}$$

According to the signal u_p given by Eq. (11) and the simplified object model in Eq. (2), the NCTF control system in Figure 1 can be represented by block diagram as shown in Figure 6. Using the PI compensator parameters K_p and K_i, and the simplified object model in Eq. (2), the transfer function near the origin of the closed-loop system for the NCTF controller in Figure 1 is,

$$\frac{\theta_l(s)}{\theta_r(s)} = \frac{\alpha_2}{s+\alpha_2} G(s) \tag{13}$$

where,

$$G(s) = \frac{\alpha_2 K K_p s + \alpha_2 K K_i}{s^2 + \alpha_2 K K_p s + \alpha_2 K K_i} \tag{14}$$

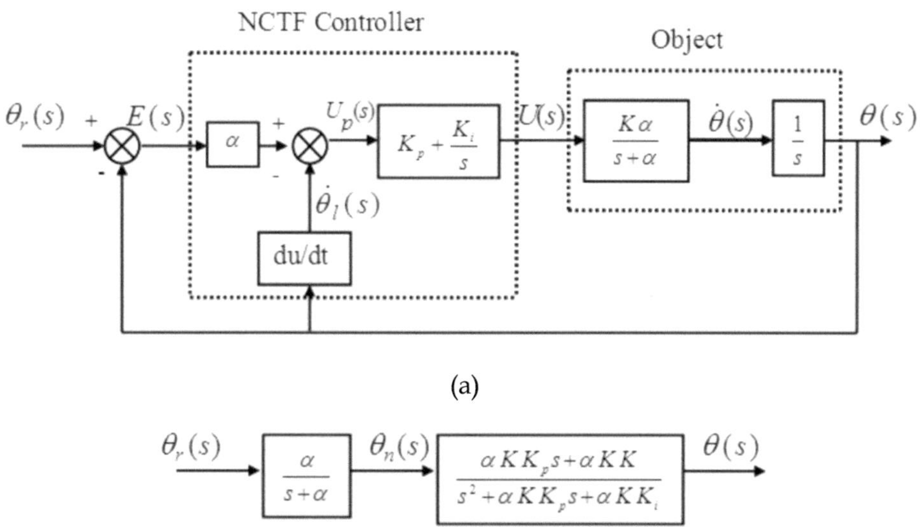

(b)

Fig. 6. NCTF control system near the origin: (a) Linearized block diagram, (b) Linearized closed-loop transfer function

G(s) given by Eq. (13) can be rewritten in the following form;

$$G(s) = \frac{2\zeta\omega_n s + \omega_n^2}{s^2 + 2\zeta\omega_n s + \omega_n^2} \tag{15}$$

where,

$$\alpha_2 K K_p = 2\zeta\omega_n \tag{16}$$

$$\alpha_2 K K_i = \omega_n^2 \tag{17}$$

In the case of $G(s) = 1$, Eq. (13) become;

$$\frac{\theta_l(s)}{\theta_r(s)} = \frac{\alpha_2}{s+\alpha_2} \tag{18}$$

In PTP positioning systems, Eq. (18) can be rewritten in time domain as follow:

$$\dot{e} + \alpha_2 e = 0 \tag{19}$$

From Eq. (11), it can be said that Eq. (19) represents the condition for $u_p = 0$. Since u_p indicates the difference between the object motion and the NCT, it can be concluded that the object motion follows the NCT perfectly if $G(s) = 1$. Therefore, the PI compensator parameters should be designed based on ω_n and ζ so that $G(s) = 1$.

The PI compensator parameters, K_p and K_i can be expressed as a function of the natural frequency and the damping ratio as follows:

$$K_p = \frac{2\zeta\omega_n}{\alpha_2 K} = \frac{2\zeta\omega_n u_r}{mh} \tag{20}$$

$$K_i = \frac{\omega_n^2}{\alpha_2 K} = \frac{\omega_n^2 u_r}{mh} \tag{21}$$

A higher ω_n and a larger ζ are preferable in the compensator design. However, while choosing ζ and ω_n, the designer must consider the stability of the control system. In continuous system, a linear stability limit can be calculated independently of the actual mechanism characteristic. However, the stability limit is too limited because of neglected Coulomb friction is known to increase the stability of the system, allowing for the use of higher gains than those predicted by a linear analysis (Townsend WT & Kenneth Salisbury J, 1987).

The higher gains are expected to produce a higher positioning performance. Thus the practical stability limit is necessary for selecting the higher gain in a design procedure. The selection of ω_n and ζ are chosen to have 40% of the values of ζ_{prac}, so that the margin safety of design is 60% (Guilherme Jorge Maeda & Kaiji Sato, 2007). Figure 7 shows the various margin safeties based on practical stability limit, ζ_{prac}. During the design parameter selection, the designer may be tempted to use large values of ω_n and ζ in order to improve the performance. However, excessively large values of ω_n will cause the controller to behave as a pure integral controller, which may lead to instability. Therefore, the choice of ω_n should start with small values and progress to larger one and not the other hand.

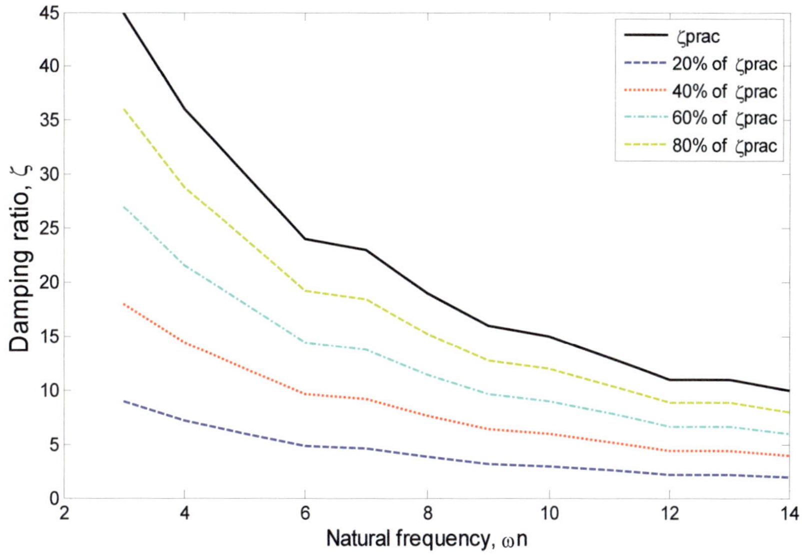

Fig. 7. Practical stability limit respecting a various margin of safety

Figure 8 shows the positioning performance response due to 0.5 deg step input. It is shown that the best performance was in between 60% to 80% margin of safety.

In many systems, the mechanical couplings between the motor, load, and sensor are not perfectly rigid, but instead act like springs. Here, the motor response may overshoot or even oscillate at the resonance frequency resulting in longer settling time. The most effective way to deal with this torsional resonance is by using an anti resonance notch filter.

According to standard frequency analysis, resonance is characterized by a pair of poles in the complex frequency plane. The imaginary component indicates the resonant frequency,

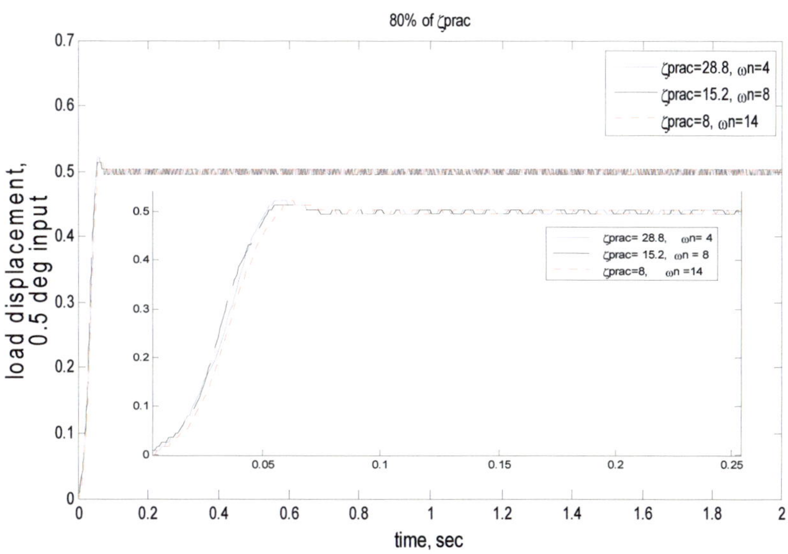

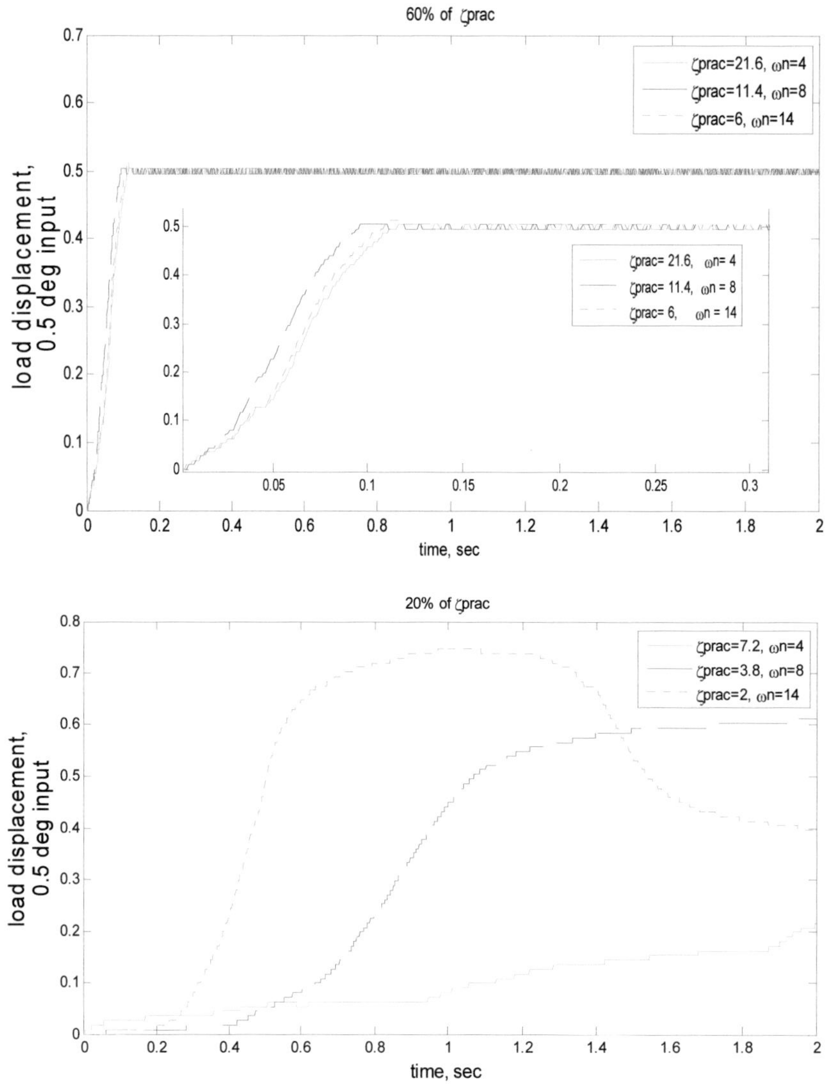

Fig. 8. Response of the compensator due to 0.5 deg step input

while the real component determines the damping level. The larger the magnitude of the real part, the greater the damping (William East & Brian Lantz, 2005).

A NF consists of a pair of complex zeros and a pair of complex poles. The purpose of the complex zeros defined by ω_f and ζ_f is to cancel the resonance poles of the system. The complex poles defined by ω_o and ζ_o, on the other hand, create an additional resonance and to improve the stability, by increasing gain margin of the plant. If the magnitude of the real value of the poles is large enough, it will result in a well-damped response. The ratio between ζ_f and ζ_o will determine how deep the notch in order to eliminate the resonant frequency of the plant. Parameter K_{dc} will be affected in steady state condition when the transfer function of the NF becomes one.

Figure 9(a) shows where the poles and zeros of the system without the controller are located on the s-plane, which gives unstable responses. In Figure 9(b), NF is added to the system, the poles marked *A* are the ones due to the mechanical resonance. These are cancelled by the complex zeros marked by *B*. Although it is assumed that the NF completely cancels the resonance poles, perfect cancellation is not required. As long as the notches zeros are close enough to the original poles, they can adequately reduce their effect, thereby improving system response.

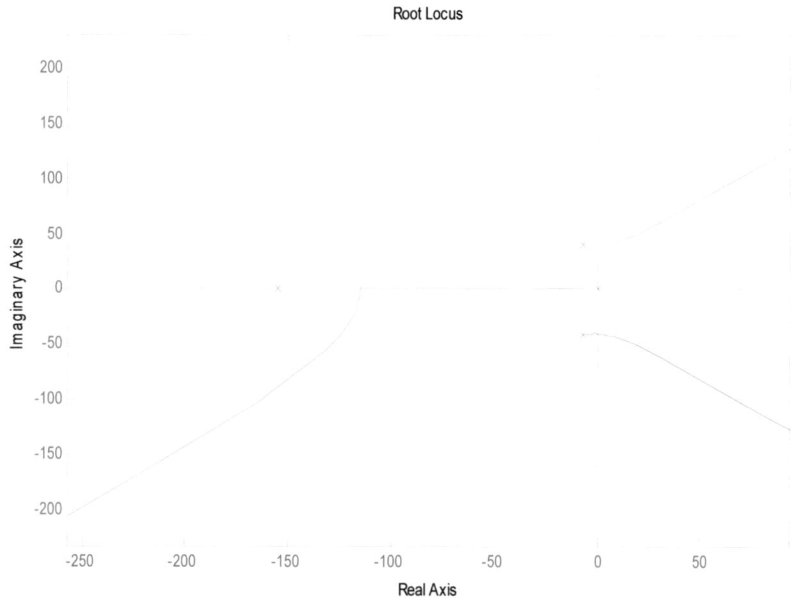

a) Root locus of the system

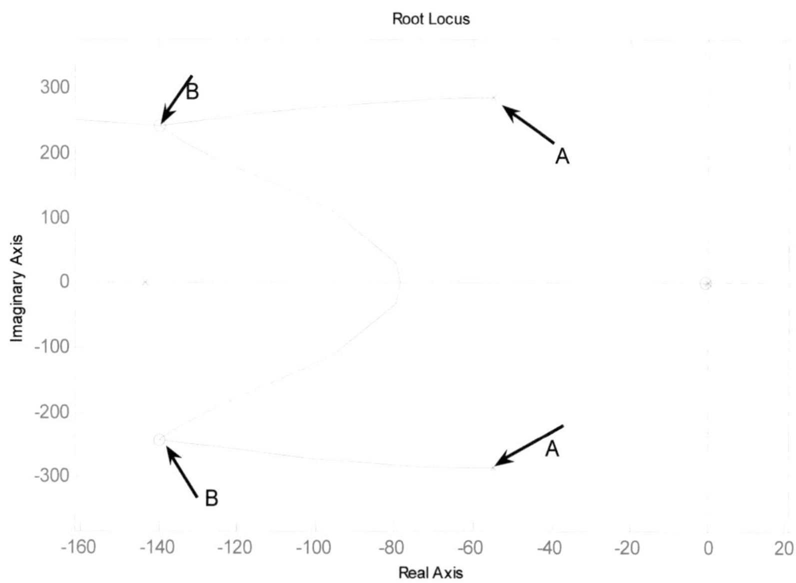

b) Poles and Zeros cancellation of the notch filter

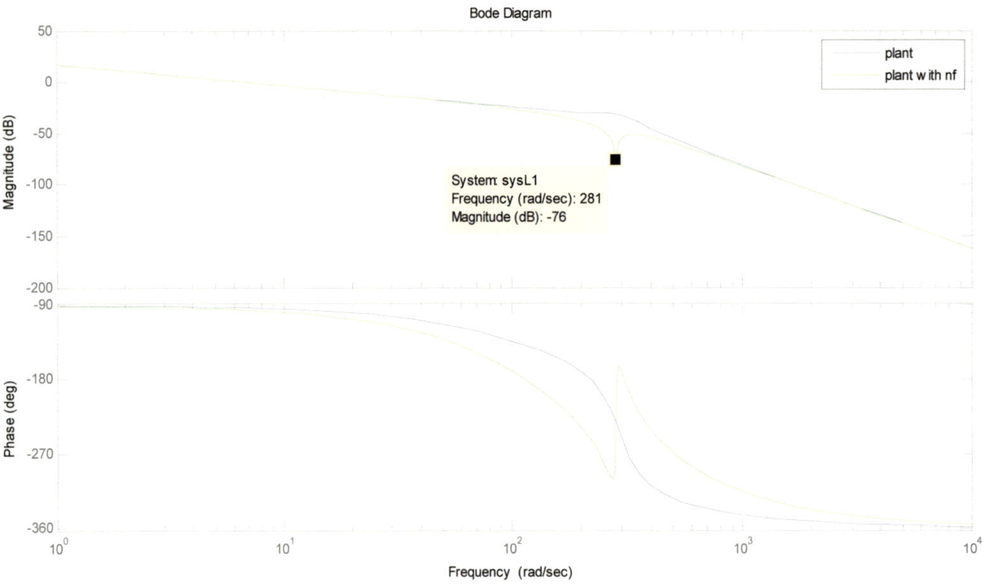

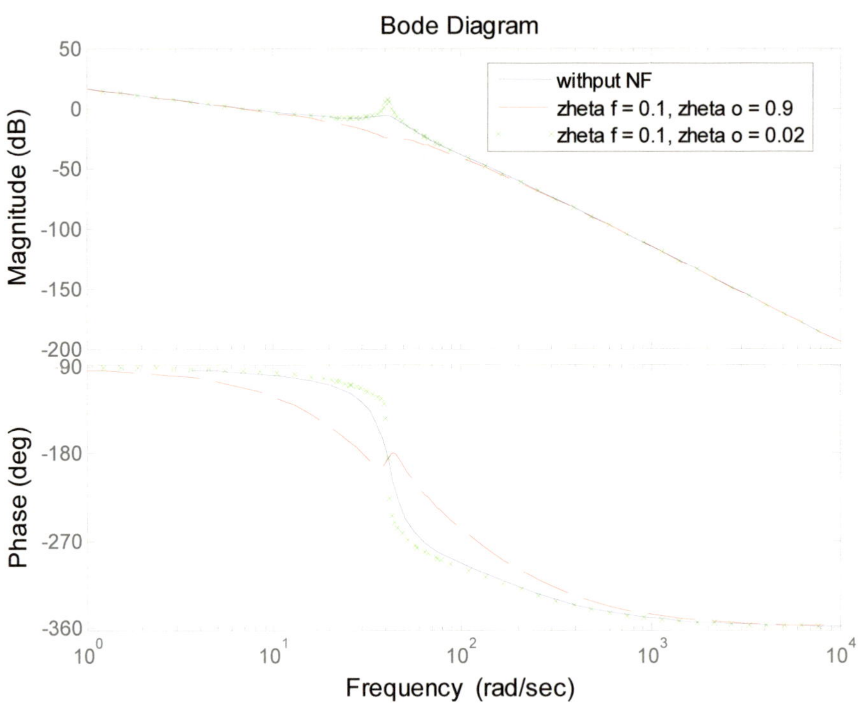

c) Bode diagram of the compensated system with difference ratio of ζ_f and ζ_o

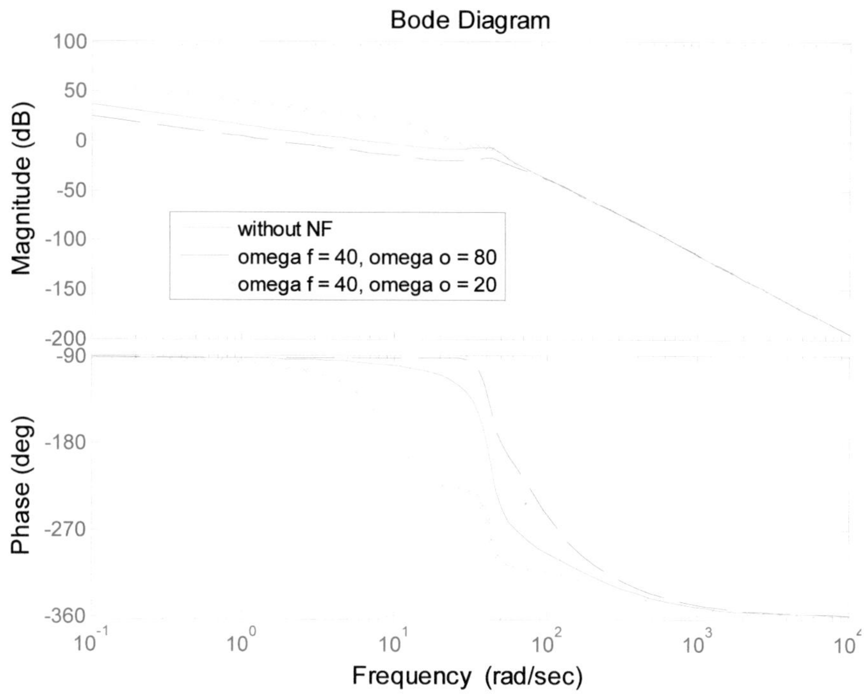

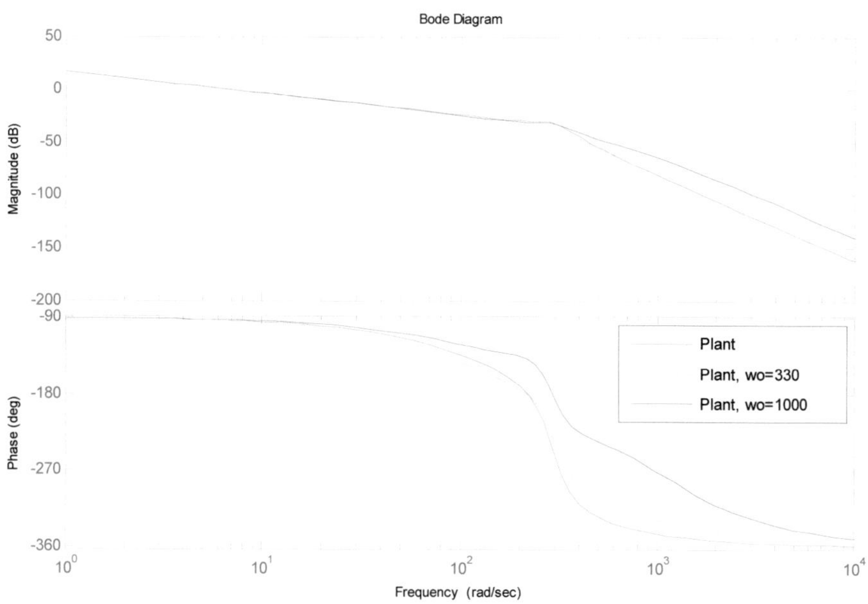

d) Bode diagram of the compensated system with difference ratio of ω_f and ω_o.

Fig. 9. System response

4. Modeling and experimental setup of the systems

Figure 10 shows the simplified diagram of a two-mass rotary positioning system. It consists of mechanical and electromechanical components. The electromechanical (electrical and mechanical) component is the DC motor, which performs the conversion of electrical energy to mechanical energy, and the rest are mechanical components. Two masses, having the moments of inertia, J_m and J_l, are coupled by low stiffness shaft which has the torsion stiffness, K_s and a damping. For the case that the system can be accurately modeled without considering the major nonlinear effects by the speed dependent friction, dead time and time delay, a linear model for two-mass mechanical system can be obtain using the conventional torque balance rule (Robert L.Woods & Kent L.Lawrence, 1997):

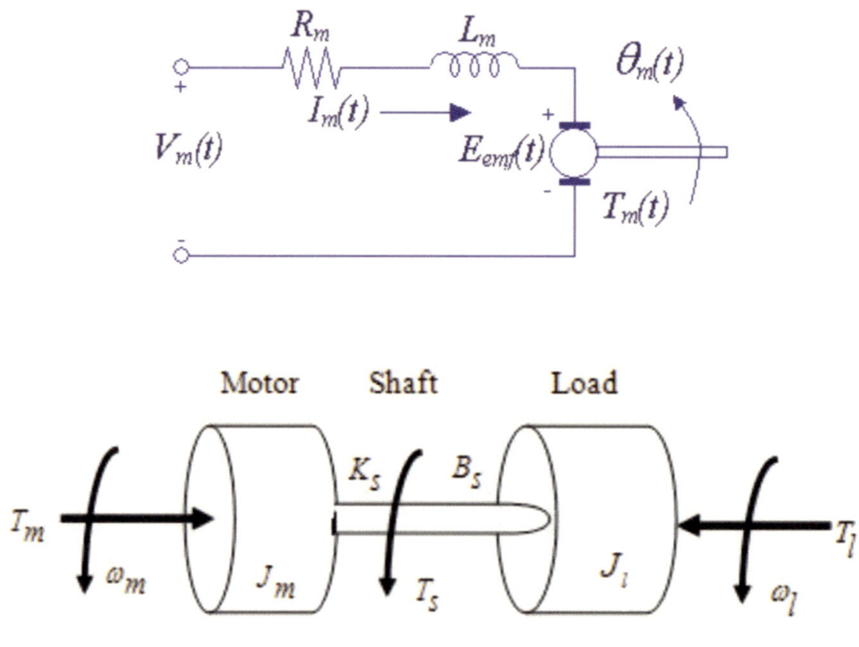

Fig. 10. Schematic diagram of two-mass system

The electrical part of the DC motor is derived by using Kirchoff Voltage Law (KCL),

$$V_m(t) - E_{emf}(t) = L_m \frac{di_m(t)}{dt} + R_m i_m(t) \tag{22}$$

Next, modeling on the mechanical parts of the system is done by applying Newton's second law of motion to the motor shaft,

$$J_l\left(\frac{d\omega_l}{dt}\right) = T_s - T_l - B_l\omega_l \tag{23}$$

$$T_s = K_s(\theta_m - \theta_l) + B_s(\omega_m - \omega_l) \tag{24}$$

with,

$$\frac{d\theta_m}{dt} = \omega_m, \frac{d\theta_l}{dt} = \omega_l \quad (25)$$

where V_m is input voltage, E_{emf} is electromagnetic field, L_m is motor inductance, R_m is motor resistance and i_m is current. J_m and J_l (kgm^2) are the motor and load moments of inertia, w_m and w_l (rad/s) are the motor and load angular speed, T_m and T_l (Nm) are the motor and load disturbance torque, T_s (Nm) is the transmitted shaft torque, B_m and B_l (Nms/rad) are the viscous motor and load frictions, B_s (Nms/rad) is the inner damping coefficient of the shaft and K_s (Nm/rad) is the shaft constant. SI units are applicable for all notations.

The improved NCTF controller is applied to an experimental two-mass rotary positioning systems as shown in Figure 11. It is a two-mass rotary PTP positioning system, which is used in various sequences in industrial applications. It consists of a direct current (DC) motor with encoder, a driver, a low stiffness shaft, load inertia and a rotary encoder. The reading of the encoder is used as a feedback to the controller. The motor is driven by control signals, which are sent to a DC motor driver powered by a power supply. This basic configuration is considered as a normal condition of the system. To measure a load position as a feedback signal of the system, a rotary encoder with 10000 pulses per revolution (ppr) made by Nemicon is used. The object consists of a direct current motor, a driver, inertia load motor, flexible shaft, coupling and inertia load mass.

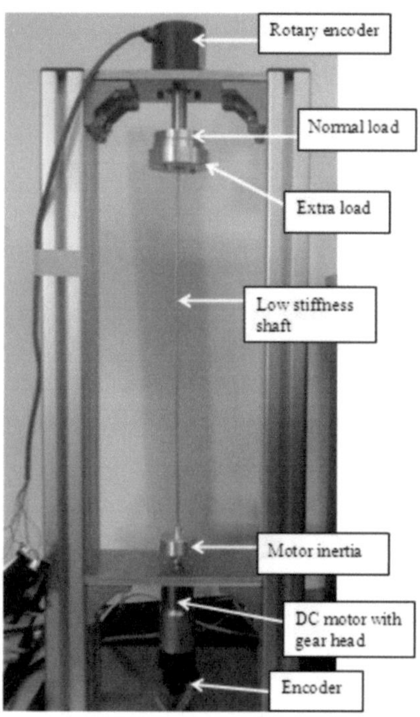

Fig. 11. Experimental two-mass rotary positioning systems

When a load mass has a heavy movable mass, the flexible shaft between the motor mass and load mass work as a spring elements which lead to a vibration. Figure 12 shows step responses to 0.5 deg step input to the experimental rotary system in a normal object condition without the controller. The object will vibrate due to a mechanical resonance and its vibrating frequency is 40 Hz.

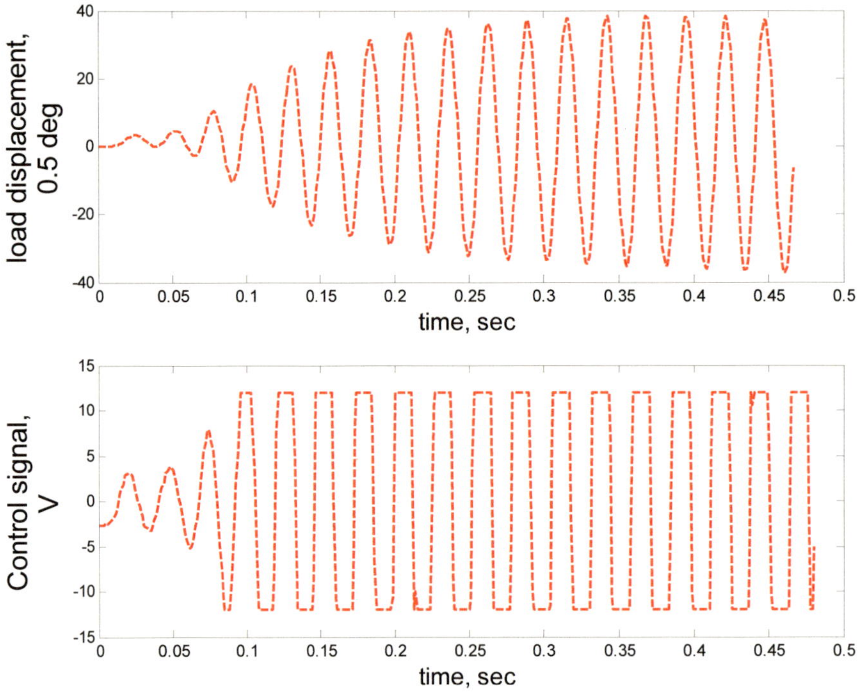

Fig. 12. Vibration responses due to 0.5 deg step response

The detailed block diagram of the model of the object used only for making simulations is shown in Figure 13. In the detailed model of the object, friction and saturation are taken into consideration (De Wit C et al, 1993). The significance of this research lies in the fact that a simple and practical controller can be designed for high precision positioning systems. By improving the NCTF controller, the controller will be more reliable and practical for

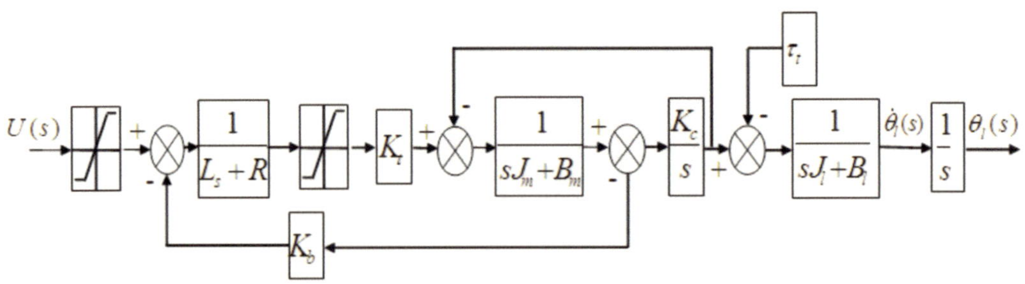

Fig. 13. Dynamic model of the object

realizing high precision positioning systems for two-mass positioning systems compared with conventional PID in term of controller performances. In this section, the continuous time, a linearized mathematical model of the lab scale two-mass rotary system is developed. In order to identify and estimate the parameters, which constitute the two-mass rotary system transfer function, system identification is utilized. Table 1 shows the parameter values of the model for each mechanism.

Parameter	Value	Unit
Motor inertia, J_m	17.16e-6	Kgm^2
Inertia load, J_l	24.17e-6	Kgm^2
Stiffness, K_c	0.039	Nm/rad
Motor resistance, R	5.5	Ω
Motor inductance, L	0.85e-3	H
Torque constant of the motor, K_t	0.041	Nm/A
Motor voltage constant, K_b	0.041	Vs/rad
Frictional torque, τ_t	0.0027	Nm
Motor viscous friction, B_m	8.35e-6	Nms/rad
Load viscous friction, B_l	8.35e-6	Nms/rad

Table 1. Model parameters

5. Experimental results and discussions

Conventional PID controllers were designed based on a Ziegler Nichols (Z-N) and Tyres Luyben (T-L) closed-loop method, using proportional control only. The proportional gain is increased until a sustained oscillation output occur which giving the sustained oscillation, K_u, and the oscillation period, T_u are recorded. The tuning parameter can be found in Table 2 (Astrom K. & Hagglund T, 1995).

Controller	K_p	T_i	T_d
Ziegler Nichlos	$K_u/1.7$	$T_u/2$	$T_u/8$
Tyres Luyben	$K_u/2.2$	T_ux2.2	$T_u/6.3$

Table 2. Controller tuning rule parameters

The significance of this research lies in the fact that a simple and easy controller can be designed for high precision positioning system which is very practical. According to Figure 4, the inclination, m and maximum error rate, h of the NCT are 81.169 and 61.6 respectively. Based on the practical stability limit from Figure 7 and its responses on Figure 8, design parameters for ζ and ω_n are chosen as 9.5 and 10.5 in order to evaluate the performance of NCTF controller. The object will vibrate due to a mechanical resonance and its vibrating frequency is 40 Hz.

From Figure 9, parameters for nominator of NF (ζ_f and ω_f) are selected as 0.7 and 40 Hz while denominator (ζ_o and ω_o) of NF is selected as 0.9 and 100 Hz. Selection of NF parameters are also based on Ruith-Huwirt stability criterion. In order to obtain an always stable continuous closed-loop system, the following constraint needs to be satisfied.

$$2\,\zeta_o\,\omega_o \geq -\alpha_2 \tag{26}$$

In order to evaluate the effectiveness of improved NCTF controller designed for a two-mass system, the controller is compared with PID controller, which are tuned using Ziegler-Nichols and Tyres-Luyben methods. The PI compensator parameters are calculated from the simplified object parameters (K and a_2) and the design parameters (ω_n and ζ). Table 3 shows the parameters of the compensator of the improved NCTF controller and PID controllers. In order to evaluate the robustness of the improved NCTF control system, the experiments were conducted in two conditions: with normal load and increasing the load inertia as shown in Figure. 14. Normal object is the two-mass experimental rotary positioning with nominal object parameter. Table 4 shows different object parameters for nominal and increased load inertia.

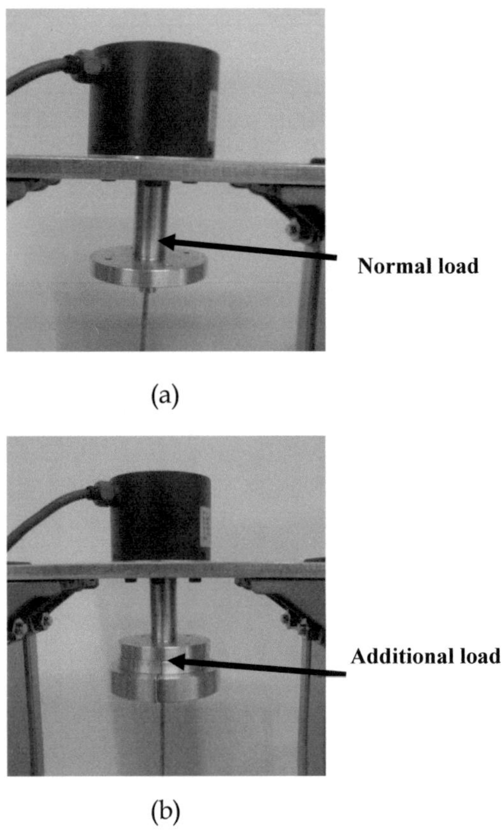

Fig. 14. Nominal and increased object inertia: (a) Nominal object, (b) Increased object inertia

Controller	K_p	K_i	K_d	ς_f	ω_f	ς_o	ω_o
NCTF	4.79e-1	2.65e-1	-	0.7	40	0.9	60
Z-N PID	78.696	4918.5	0.314	-	-	-	-
T-L PID	59.618	846.85	0.303	-	-	-	-

Table 3. Controller parameters

Object	Inertia
Normal load	$J_l = 14.17 \times 10^{-6}$ kgm^2
Increased inertia load	$2 \times J_l$ $5 \times J_l$ $10 \times J_l$

Table 4. Object parameter comparison

Controller			Overshoot (%)	Settling Time (sec)	Ess (deg)
J_l	0.5 deg	Z-N	22.4	0.2	0.108
		T-L	17	0.35	0.108
		NCTF	0	0.083	0.036
	1 deg	Z-N	106.1	0.05	0.072
		T-L	8.9	0.055	0.072
		NCTF	1.7	0.03	0.036
	5 deg	Z-N		unstable	
		T-L		unstable	
		NCTF	6.92	0.043	0.036
$2 \times J_l$	0.5 deg	Z-N	67.47	0.08	0.94
		T-L	18.7	0.25	0.036
		NCTF	0	0.032	0.036
	1 deg	Z-N		unstable	
		T-L	33.2	0.36	0.72
		NCTF	8.9	0.044	0.036
	5 deg	Z-N		unstable	
		T-L		unstable	
		NCTF	11.78	0.052	0.036
$5 \times J_l$	0.5 deg	Z-N	81.8	0.082	0.072
		T-L	13.4	0.25	0.072
		NCTF	0.15	0.072	0.036
	1 deg	Z-N		unstable	
		T-L	39.5	0.3	0.072
		NCTF	47.6	0.09	0.036
	5 deg	Z-N		unstable	
		T-L		unstable	
		NCTF	20.42	0.85	0.036
$10 \times J_l$	0.5 deg Input	Z-N	94.4	0.075	0.072
		T-L	0	0.31	0.072
		NCTF	20.6	0.07	0.036
	1 deg Input	Z-N		unstable	
		T-L	47.6	0.35	0.072
		NCTF	54.8	0.09	0.036
	5 deg input	Z-N		unstable	
		T-L		unstable	
		NCTF	36.44	0.14	0.036

Table 5. Experimental positioning performance comparison

Figure 15 shows the improved NCTF controller for a two-mass rotary positioning system in nominal object and increased object inertia. Figure 16 shows step responses to 0.5, 1 and 5 deg step input when the improved NCTF controller are used to control a nominal object. The positioning performance is evaluated based on percentage overshoot, settling time and positioning accuracy. Figure 17 shows step responses to 0.5, 1 and 5 deg step input to control the increased object with load inertia of twice the nominal object (2 x J_l). Figure 18 and Figure 19 show step responses to 0.5, 1 and 5 deg step input to control the increased object inertia with five and ten times of the nominal object one, i.e 5 x J_l and 10 x J_l. Table 5 presents the experimental positioning performance. An average of 20 similar experiments was conducted.

With nominal object, the improved NCTF controller gives the smallest percentage of overshoot, fastest settling time and a better positioning accuracy than both PID controllers. Unlike the improved NCTF controller for the two-mass rotary system, the PID controllers results in a vibrating response which is 40 Hz. This vibration caused by mechanical resonance of the system. For increased object inertia, the improved NCTF controller demonstrates good positioning accuracy and stable responses. On the other hand, PID based on Ziegler-Nichols rule resulted in unstable response, faster than Tyres-Luyben method for increased object inertia. Therefore, improved NCTF controller is much more robust to inertia variations compared with the two PID controllers. Moreover, for the experimental results, the improved NCTF controller resulted in positioning accuracy near the sensor resolution.

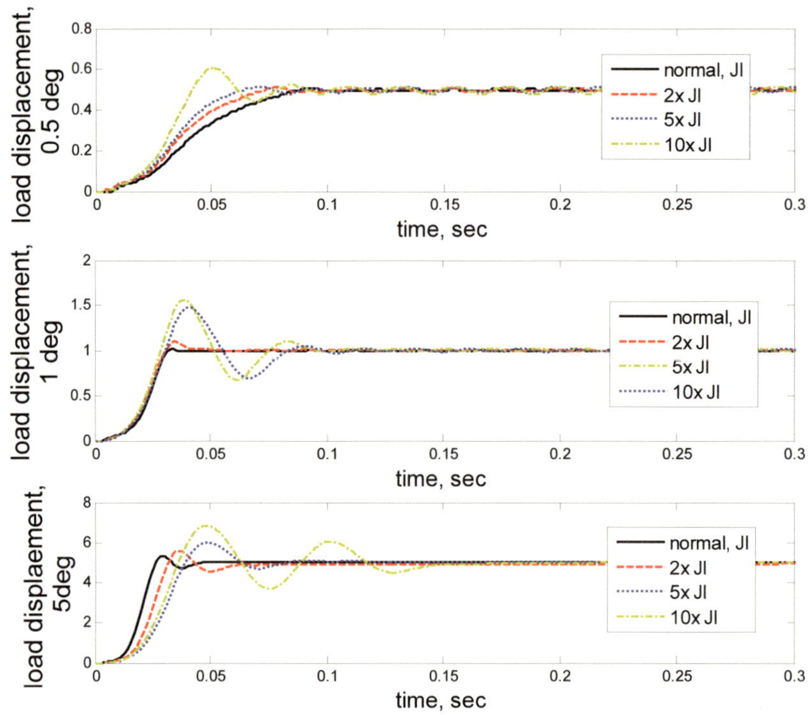

Fig. 15. Step response of improved NCTF controller

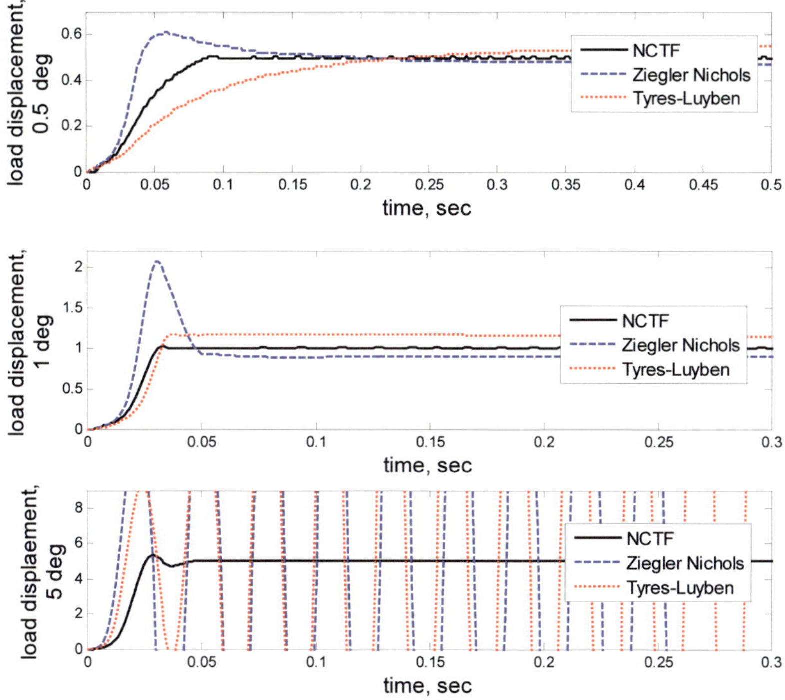

Fig. 16. Step response comparison, nominal object condition

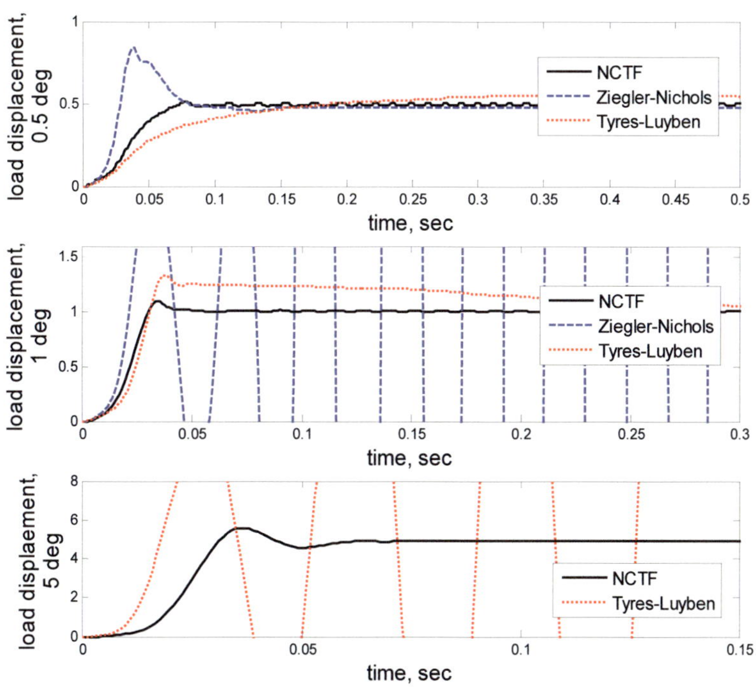

Fig. 17. Step response comparison, increased object inertia (2 x J_l)

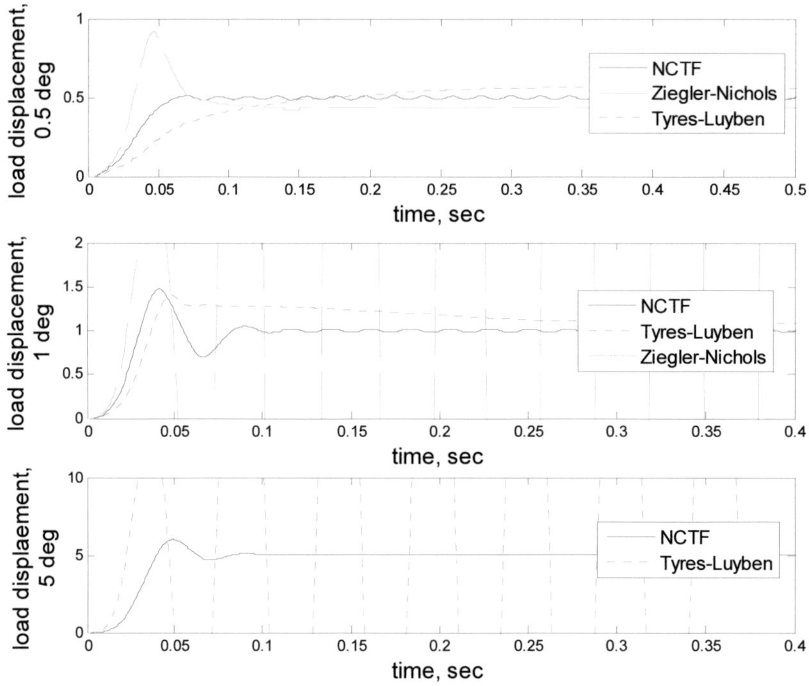

Fig. 18. Step response comparison, increased object inertia (5 x J_l)

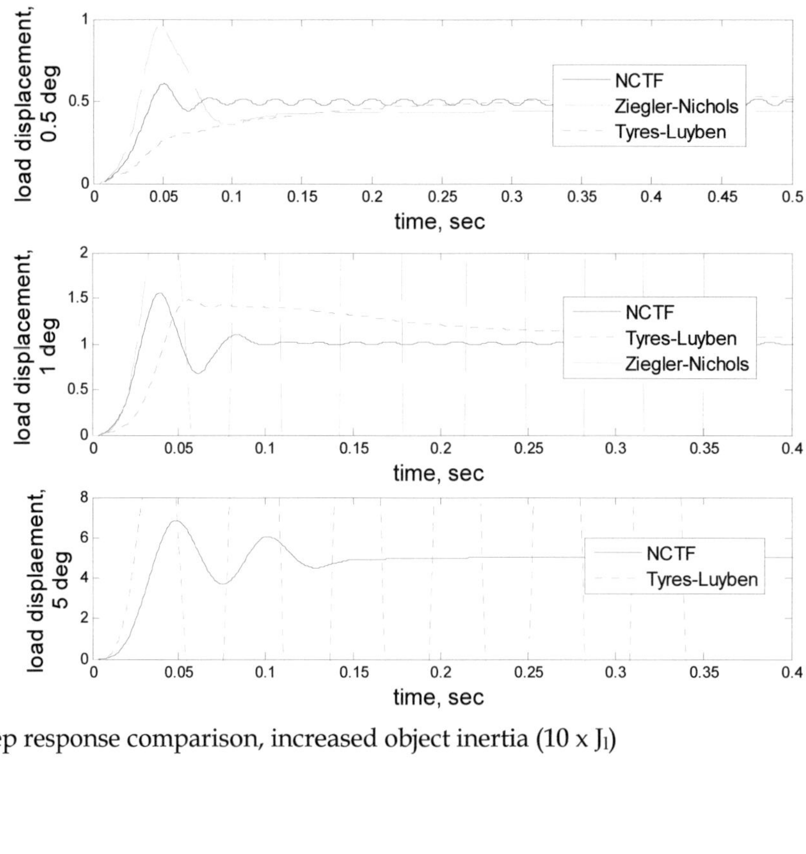

Fig. 19. Step response comparison, increased object inertia (10 x J_l)

6. Conclusion

This paper has discussed and described the improved practical design of the NCTF controller for two-mass rotary positioning systems. First, the NCT determination was discussed. In order to eliminate the influence of the vibration on the NCT, the vibrating responses are averaged using simple moving averaged filter. Then, the PI with NF compensator is designed based on the NCT information and the object responses. The PI is adopted for its simplicity to force the object motion to reach the NCT as fast as possible and control the object motion to follow the NCT and stop at the origin. The NF is adopted to eliminate the vibration frequency caused by mechanical resonance of the system.

Through experiments using two-mass rotary experimental positioning systems, the effectiveness and robustness of the improved NCTF controller was evaluated. The experimental results confirmed that the improved NCTF controller for a two-mass system has better performance in term of percentage of overshoot, settling time and positioning accuracy than PID controllers. The improved NCTF controller for two-mass system resulted in an insignificant vibration. Moreover, experimental resulted prove that the improved NCTF controller was much more robust to inertia variations than PID controllers designed based on Ziegler-Nichols and Tyres-Luyben methods. Therefore, the improved NCTF controller for two-mass rotary positioning systems that includes the NF will have a better performance than the PID controllers with NF. Applications of the NCTF concept to multi-mass positioning systems can be done for further study.

7. Acknowledgment

This research is supported by Ministry of Higher Education Malaysia under vot 78606 and Malaysia-Japan International Institute of Technology (MJIIT), Universiti Teknologi Malaysia International Campus (UTM IC), Kuala Lumpur, Malaysia.

8. References

Armstrong-Helouvry B., Dupont P, and De Wit C. (1994). A survey of models, analysis tools and compensation method for the control of machines with friction. *Automatica*, Vol. 30, No. 7, pp. 1083-1138.

Astrom K. & Hagglund T, PID controllers: Theory, design and tuning. *Instrument Society of America*, 1995.

Crowder R.M. (1998). *Electric drives and theier controls*, Oxford: Oxford University Press

De Wit C., Olsson H., Astrom K.J & Lischinssky, Dynamic friction models and control design, *Proceedings American Control Conference,* San Francisco, USA, 1993, pp. 1920-1926.

Dorf, R.C., & Bishop, R.H. (2008). *Modern Control System*. Pearson Prentice Hall.

Guilherme Jorge Maeda & Kaiji Sato, Practical control method for ultra-precision positioning using a ballscrew mechanism. *Precision Engineering Journal,* Vol. 32(2008)309-318, November 2007.

Mohd Fitri Mohd Yakub, Wahyudi Martono and Rini Akmeliawati, "Performance Evaluation of Improved Practical Control Method for Two-Mass PTP Positioning Systems", *Proceedings of the 2010 IEEE Symposium on Industrial Electronics & Applications (ISIEA2010)* 3–5 October 2010, Penang, pp. 550-555.

Oppenheim A.V. & Schafer R.W, Discrete Time Signal Processing. *Englewood Cliffs*, Prentice Hall, 1999.

Robert L.Woods and Kent L.Lawrence, Modelling and Simulation of Dynamic Systems, Prentice Hall Inc, 1997.

Slotine, J. J. & Li, W. P. (1991). Applied Nonlinear Control, Prentice-Hall, Inc., New Jersey

Townsend WT and Kenneth Salisbury J, The effect of coulomb friction and stiction on force control, *IEEE Int Conf Robot Automat* 1987;4:883-9.

Wahyudi. (2002). *New practical control of PTP positioning systems*. Ph.D. Thesis. Tokyo Institute of Technology.

Wahyudi, Sato, K., & Shimokohbe, A. (2003) Characteristics of practical control for point-to-point positioning systems: effect of design parameters and actuator saturation. *Precision Engineering*, 27(2), 157-169.

William East & Brian Lantz, Notch Filter Design, August 29, 2005.

Yonezawa H., Hirata H. and Sasai H. (1990). Positioning table with high accuracy and speed. *Annal CIRP*. Vol. 39, pp. 433-436.

Part 3

Robust PID Controller Design

Performance Robustness Criterion of PID Controllers

Donghai Li[1], Mingda Li[2], Feng Xu[1],
Min Zhang[2], Weijie Wang[3] and Jing Wang[2]
[1]*State Key Lab of Power Systems Dept of Thermal Engineering, Tsinghua University, Beijing*
[2]*Metallurgical Engineering Research Institute, University of Science and Technology Beijing, Beijing*
[3]*Beijing ABB Bailey Engineering Company, Beijing*
P. R. China

1. Introduction

PID is one of the earliest and most popular controllers. The improved PID and classical PID have been applied in various kinds of industry control fields, as its tuning methods are developing. After the PID controller was first proposed by Norm Minorsky in 1922, the various PID tuning methods were developing and the advanced and intelligent controls were proposed. In the past few decades, Z-N method which is for first-order-plus-time-delay model was proposed by Ziegler and Nichols (Ziegler & Nivhols, 1943), CHR method about generalized passive systems was proposed by Chien, Hrones and Reswick (Chien et al., 1952), and so many tuning methods were developed such as pole assignment and zero-pole elimination method by Wittenmark and Astrom, internal model control (IMC) by Chien (Chien & Fruehauf, 1990). The gain and phase margin (GPM) method was proposed by Åström and Hägglund (Åström & Hägglund, 1984), the tuning formulae were simplified by W K Ho (Ho et al., 1995).

In classical feedback control system design, the PID controller was designed according to precise model. But the actual industrial models has some features as follows:

1. The system is time variant and uncertain because of the complex dynamic of industrial equipment.
2. The process is inevitably affected by environment and the uncertainty is introduced.
3. The dynamic will drift during operation.
4. The error exists with the dynamic parameter measurement and identification.

So there are two inevitable problems in control system designing. One is how to design robust PID controller to make the closed-loop system stable when the parameters are uncertain in a certain range. The other is the performance robustness which must be considered seriously when designing PID controllers. The performance robustness is that when the parameters of model change in a certain interval, the dynamic performances of system are still in desired range.

This chapter discusses the new idea mentioned previous - Performance Robustness. Based on the famous Monte-Carlo method, the performance robustness criterion is proposed. The performance robustness criterion could give us a new view to study the important issue that how the PID controller performs while the parameters of model are uncertain. Not only the stability, but also the time-domain specifications such as overshoot and adjusting time, and the frequency-domain specifications such as gain margin and phase margin can be obviously clear on the specification figures.

The structure of this chapter is as follows. A brief history of Monte-Carlo method is given in section 2. The origin, development and latest research of Monte-Carlo method are introduced. The performance robustness criterion is discussed in detail. This section also contains several formulas to explain the proposed criterion. In section 3, the performance robustness criterion is applied on typical PID control systems comparison, the detailed comparisons between DDE method and IMC method, and between DDE method and GPM method. Finally, section 4 gives out a conclusion.

2. Monte-Carlo method in performance robustness criterion

2.1 A brief history of Monte-Carlo method

Monte-Carlo method is also called random sampling technology or statistical testing method. In 1946, a physicist named Von Neumann simulated neutron chain reaction on computer by random sampling method called Monte-Carlo method. This method is based on the probability statistics theory and the random sampling technology. With the further development of computer, the vast random sampling test became viable. So it was consciously, widely and systematic used in mathematical and physical problems. The Monte-Carlo method is also a new important branch of computational mathematics.

In the late 20th century, Monte-Carlo method is closely linked the computational physics, computational statistical probability, interface science of computer science and statistics, and other boundary discipline. In addition, the Monte-Carlo method also plays a role for the development of computer science. In order to show the new performance evaluation method of mainframe which has multi-program, variable word length, random access and time-shared system, the performance of developed computer was simulated and analysed on the other computer. The relationship could be clear via the study on different target.

Large numbers of practical problems on nuclear science, vacuum technology, geological science, medical statistics, stochastic service system, system simulation and reliability were solved by Monte-Carlo method, and the theory and application results have gained. It was used in simulation of continuous media heat transfer and flow (Cui et al., 2000), fluid theory and petroleum exploration and development (Lu & Li, 1999). Monte-Carlo method was combined with heat network method to solve the temperature field of spacecraft, and the steady-state temperature field of satellite platform thermal design was calculated and analysed (Sun et al., 2001). In chemical industry, Yuan calculated the stability of heat exchanger with Monte-Carlo method, and it was used in selection and design (Yuan, 1999).

In power system, Monte-Carlo method was applied in reliability assessment of generation and transmission system, the software was design and the application was successful (Ding & Zhang, 2000).

2.2 Performance robustness criterion based on Monte-Carlo method

Consider the SISO system as follows:

$$G(s) = \frac{N(s)}{D(s)} e^{-Ls} \tag{1}$$

In this system, N(s) and D(s) are coprime polynomials, and D(s)'s order is greater than or equal N(s)'s order, L is rational number greater than or equal to zero. The controlled model is some uncertain, and the parameters of N(s) and D(s) are variable in bounded region. So, the model is a group of transfer function denoted by {G(s)}. The control system is shown in figure 1.

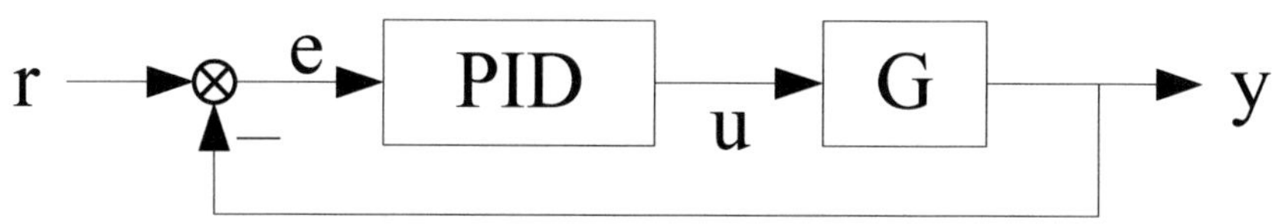

Fig. 1. Control system structure

The controller is PID controller:

$$u(s) = K_p(1 + \frac{1}{T_i s} + T_d s)e(s) \tag{2}$$

or

$$u(s) = (K_p + \frac{K_i}{s} + K_d s)e(s)$$

The parameters K_p, K_i, K_d are positive number, and all of the PID controllers compose a controller group denoted by {PID}.

The PID tuning methods are used on the nominal controlled models, and the closed-loop systems are obtained. The overshoot $\sigma\%$ and adjustment time T_s are considered as dynamic performance index. Because the controlled models are a group of transfer function, the dynamic performance index is a collection, denoted by:

$$\{\sigma\%, T_S\} \tag{3}$$

Obviously, it is a collection of two-dimension vector an area in plane plot. The distance between this area and origin reflects the quality of control system, and the size of this area shows the dispersion of performance index, that is the performance robustness of control system.

The comparison study on PID tuning methods should follow the steps below:

1. Confirm the controlled model transfer function and parameter variety interval, and the transfer function group is obtained.
2. Confirm the compared PID tuning methods, and choose the appropriate experiment times N to ensure the dispersion of performance index invariable when the N is larger.
3. Tuning PID controller for the nominal model.
4. In every experiment, a specific model is selected from the transfer function group by a rule (random in this paper). With the PID controller obtained in step three, the step response of closed-loop PID control system is tested, and the overshoot and adjustment time could be measured.
5. Repeat the step 4 N times, and plot the performance index on coordinate diagram. So, the N points compose an area on the coordinate diagram.
6. Repeat the step 3-5 by different tuning methods.
7. Compare the performance index of different tuning methods.

In next section, performance robustness is applied on PID control system comparison.

3. Performance robustness comparisons

3.1 Performance robustness comparison of typical PID control systems

In this section, we consider four typical models as follows:

1. First-order-plus-time-delay model (FOPTD)

$$G(s)=\frac{k}{1+sT}e^{-sL} \quad \text{k, T, L>0.} \tag{4}$$

2. Second-order-plus-time-delay model (SOPTD)

$$G(s)=\frac{k}{(1+sT_1)(1+sT_2)}e^{-sL} \quad k, T_1, T_2, L>0 \tag{5}$$

or

$$G(s)=\frac{\omega_n{}^2}{s^2+2\xi\omega_n s+\omega_n{}^2}e^{-sL} \quad \omega_n>0\,,1>\xi>0\,,L>0.$$

3. High-order model

$$G(s)=\frac{k}{(1+sT)^n} \quad \text{k, T>0, } n\geq 3 \text{ and } n\in N\,. \tag{6}$$

4. Non-minimum model

$$G(s)=\frac{k(-s+a)}{(1+sT_1)(1+sT_2)} \quad k, T_1, T_2, a>0. \tag{7}$$

The classical PID tuning methods are showed in table 1.

Tuning methods	K_p	T_i	T_d
Z-N	1.2T/kL	2L	L/2
CHR	0.6T/kL	T	L/2
Cohen-Coon	$\frac{1.35T}{kL}\left(1+\frac{0.18L}{T}\right)$	$\frac{0.5L+2.5T}{0.61L+T}L$	$\frac{0.37T}{0.19L+T}L$
IMC	$\frac{0.5L+T}{k(L+T_f)}$	T+L/2	$\frac{LT}{L+2T}$
IST²E	$\frac{0.968}{k}\left(\frac{L}{T}\right)^{-0.904}$	$\frac{T}{0.977-0.253(L/T)}$	$0.316T\left(\frac{L}{T}\right)^{0.892}$
GPM	$\frac{W_pT}{A_mk}$	$\left(2W_p-\frac{4W_p^2L}{\pi}+\frac{1}{T}\right)^{-1}$	
	$W_p=\frac{A_m\Phi_m+0.5\pi A_m(A_m-1)}{(A_m^2-1)L}$		

Table 1. Formulas of classical PID tuning method

If the tuning object is zero overshoot, the selection of IMC method free parameter T_f will only correlate to delay-time L. We fit the approximate relation between L and T_f.

$$\begin{cases} T_f = p_1L^3 + p_2L^2 + p_3L + p_4 & L \le 100 \\ T_f = L/2 & L > 100 \end{cases} \tag{8}$$

where

$p_1=-1.7385\times10^{-5}$, $p_2=3.0807\times10^{-3}$, $p_3=0.3376$, $p_4=5.6400$.

The different transfer function models can be simplified and transferred to FOPTD model(Xue, 2000).

Suppose the FOPTD (4).

Calculate the first and second derivative and then we obtain

$$\frac{G_1'(s)}{G_1(s)} = -L - \frac{T}{1+Ts} \tag{9}$$

and

$$\frac{G_1''(s)}{G_1(s)} - \left(\frac{G_1'(s)}{G_1(s)}\right)^2 = \frac{T^2}{(1+Ts)^2}. \tag{10}$$

when s=0,

$$T_{ar} = -\frac{G_1'(0)}{G_1(0)} = L + T \tag{11}$$

and

$$T^2 = \frac{G_1''(0)}{G_1(0)} - T_{ar}^2 \tag{12}$$

We can get L and T from equation above, and the system gain can be obtained directly by k=G(0).

So, in actual application, if we have the transfer functions, the more accurate FOPTD equivalent models will be get.

For example, the transfer function is

$$G(s) = \frac{1}{(20s+1)^3} . \tag{13}$$

The approximate FOPTD model is

$$G_1(s) = \frac{1}{34.64s+1} e^{-25.36s} . \tag{14}$$

The step response is shown in figure 2.

For FOPTD model (4), the L/T is very important. So, there are three cases to be discussed L<T, L≈T and L>T. The parameters and simulation results are shown in table 2, 3, figure 3, 4 and 5.

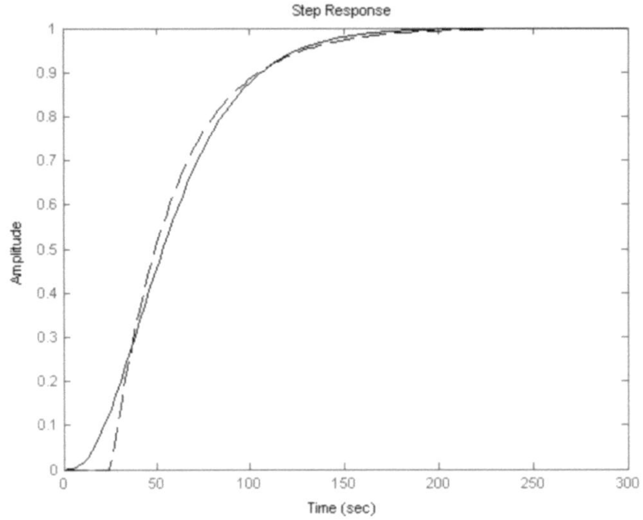

Fig. 2. Step response comparison (the solid line is original system and the dotted line is approximate system)

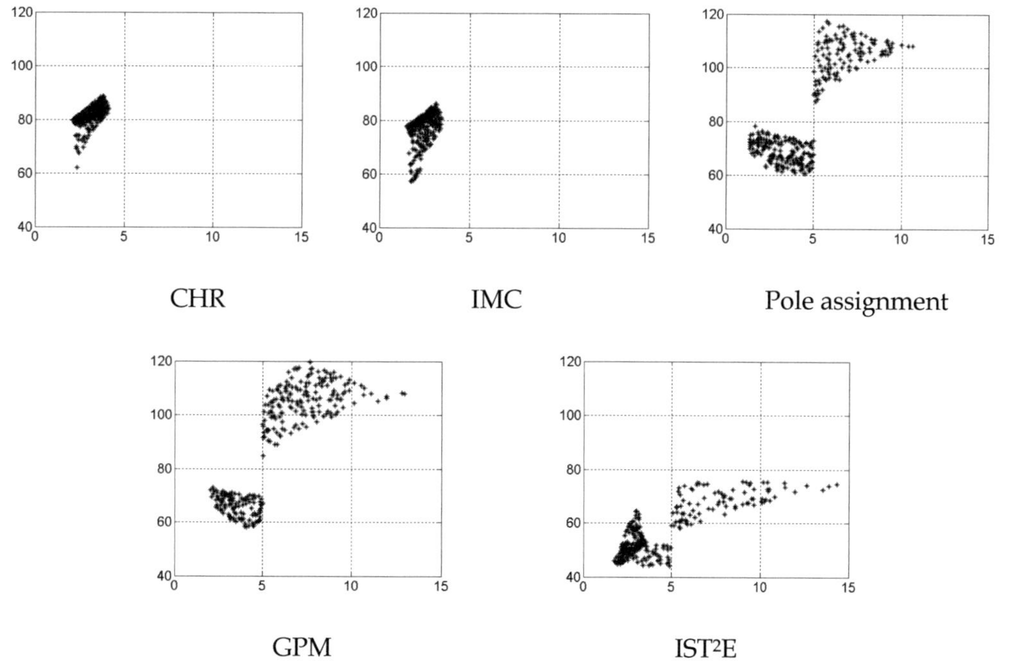

Fig. 3. Simulation results of FOPTD model when L<T (the abscissa represents overshoot and the ordinate represents adjustment time)

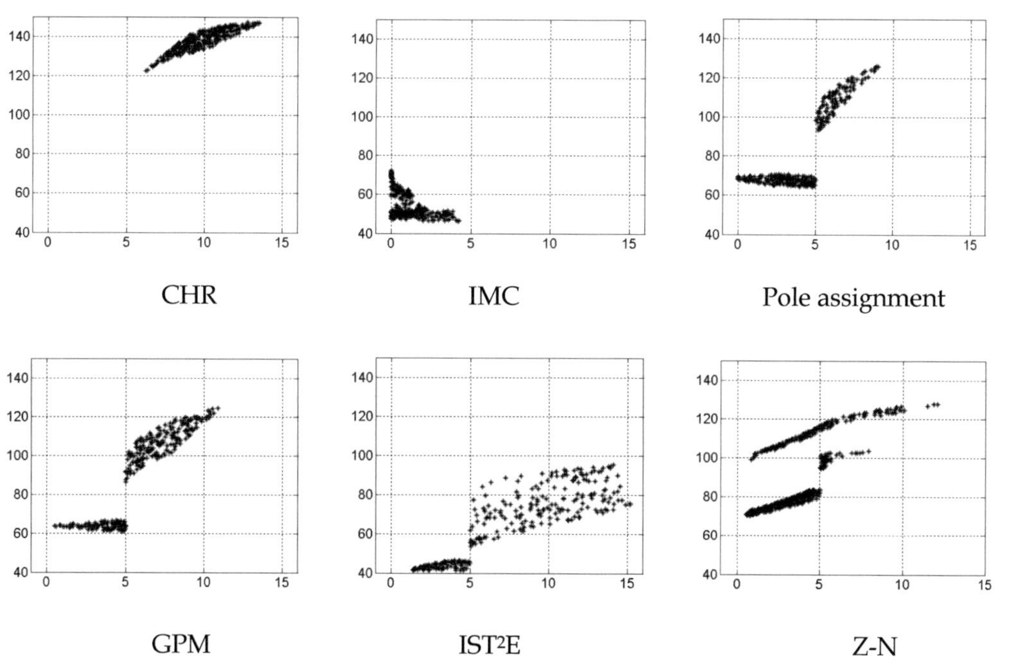

Fig. 4. Simulation results of FOPTD model when L≈T (the abscissa represents overshoot and the ordinate represents adjustment time)

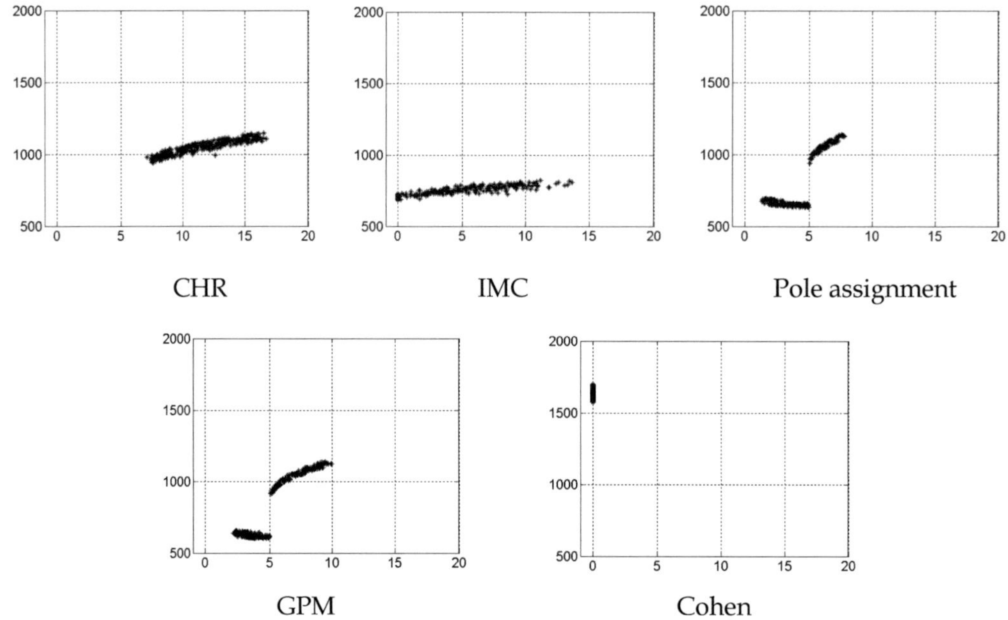

Fig. 5. Simulation results of FOPTD model when L>T (the abscissa represents overshoot and the ordinate represents adjustment time)

In order to compare different method visualized, the figures which have too long adjustment time or too large overshoot are not included in figure 3, 4, 5, 7 and 8.

	L	T	k
L<T	[18,22]	[180,220]	1
L≈T	[18,22]	[18,22]	1
L>T	[180,220]	[18,22]	1

Table 2. Parameters of FOPTD model

		CHR	IMC	Pole assignment	GPM	IST²E	Cohen	Z-N
L<T	Overshoot (%)	2.08~4.08 (3.10)	1.49~3.41 (2.48)	1.37~10.6 (4.54)	2.04~12.9 (5.86)	1.75~14.3 (4.42)	64.4~122 (91.2)	49.6~102 (74.3)
	Adjustment time	62.2~88.6 (81.6)	57.4~86.1 (77.2)	60.7~117 (83.2)	58.1~120 (89.7)	44.1~75.5 (56.3)	113~477 (181)	105~214 (140)
L≈T	Overshoot (%)	6.22~13.5 (9.83)	0~4.27 (1.13)	0~9.03 (4.20)	0.50~10.9 (5.93)	1.40~15.2 (7.48)	21.6~52.5 (36.8)	0.57~12.0 (3.58)
	Adjustment time	122~147 (138)	46.3~72.3 (54.8)	64.3~126 (83.4)	61.2~125 (91.6)	41.3~95.4 (64.3)	74.8~225 (136)	70.5~128 (90.1)
L>T	Overshoot (%)	7.11~16.6 (11.6)	0~13.6 (4.01)	1.36~7.79 (4.34)	2.20~9.94 (5.65)	Not stable	0	0
	Adjustment time	940~1147 (1051)	681~821 (743)	635~1137 (815)	602~1140 (855)	Not stable	1571~1701 (1642)	>6000

Table 3. Performance index of FOPTD model

For SOPID model (5), we choose $T_1, T_2 \in [16,24]$ and $L \in [80,100]$. The nominal parameters are T_1=T_2=20, L=90. The simulation results are shown in table 4 and figure 6.

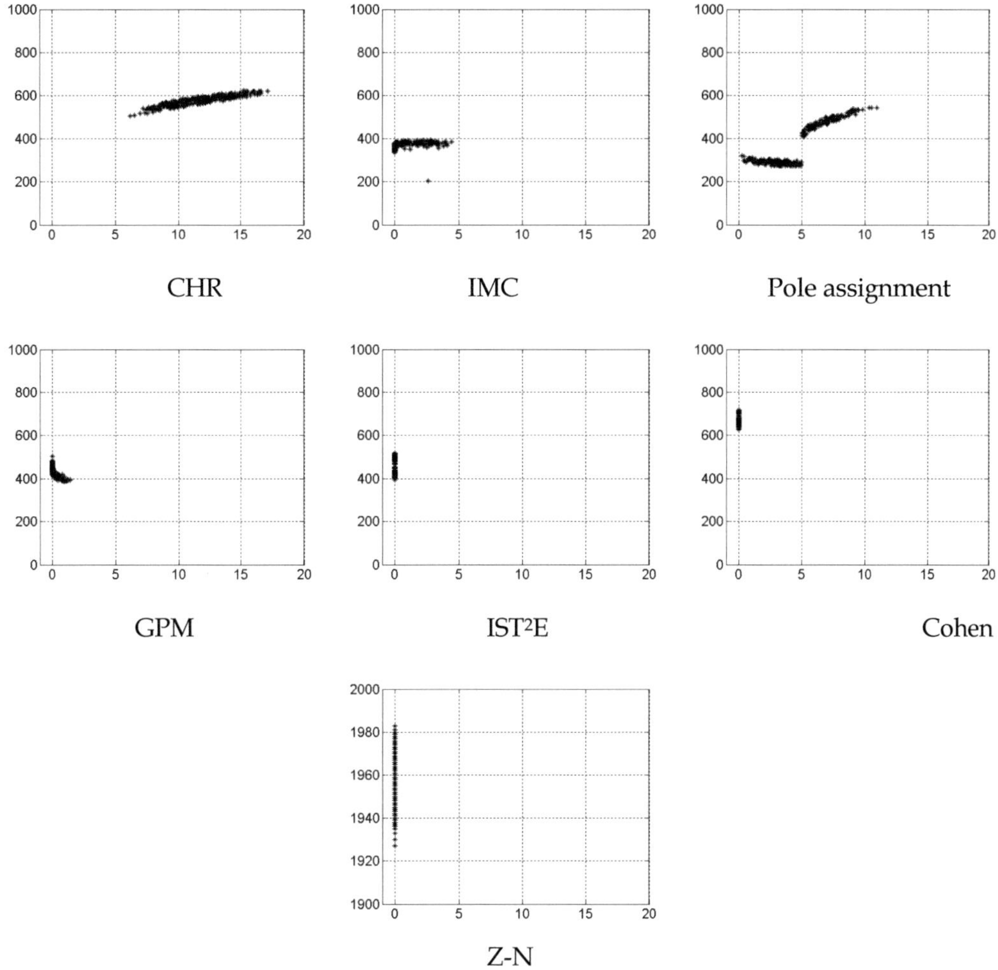

Fig. 6. Simulation results of SOPTD model (the abscissa represents overshoot and the ordinate represents adjustment time)

For High-order model (6), we choose $T \in [16,24]$ and $k \in [0.8,1.2]$. The nominal parameters are T=20, k=1 and n=3. The simulation results are shown in table 5 and figure 7.

	CHR	IMC	Pole assignment	GPM	IST^2E	Cohen	Z-N
Overshoot (%)	6.20~17.1 (11.7)	0~4.44 (0.65)	0.25~11.0 (4.66)	0~1.41 (0.17)	0	0	0
Adjustment time	504~623 (577)	202~394 (365)	272~543 (370)	389~505 (434)	394~518 (436)	627~719 (665)	1927~1983 (1956)

Table 4. Performance index of SOPTD model

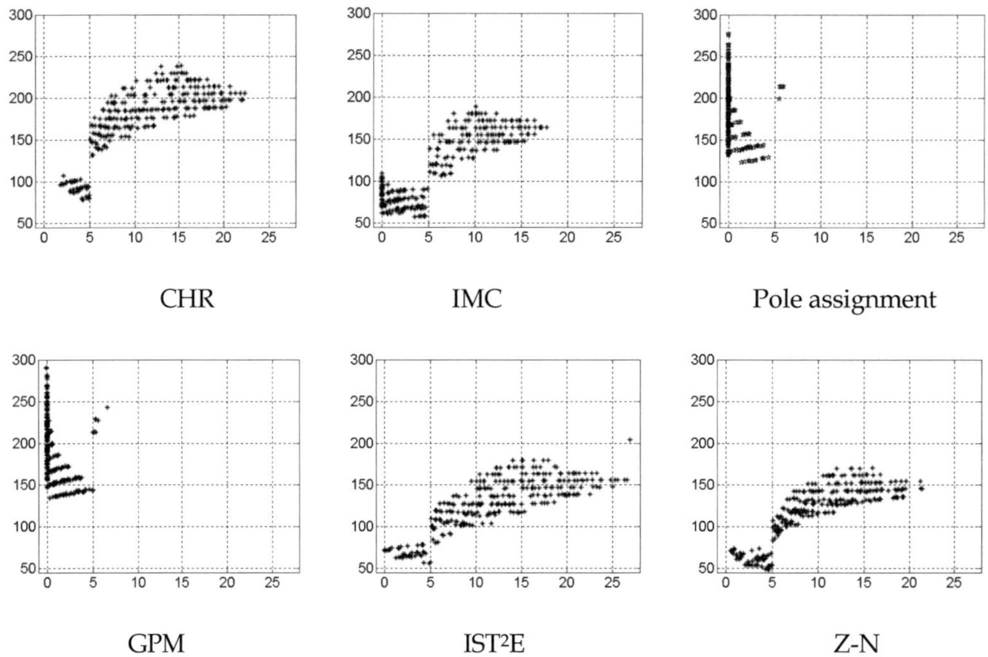

Fig. 7. Simulation results of High-order model (the abscissa represents overshoot and the ordinate represents adjustment time)

	CHR	IMC	Pole assignment	GPM	IST²E	Cohen	Z-N
Overshoot (%)	1.79~22.3 (11.1)	0~17.8 (6.46)	0~5.93 (0.420)	0~6.59 (0.820)	0~26.9 (12.3)	5.49~35.3 (20.0)	0.493~21.4 (9.79)
Adjustment time	77.9~238 (174)	57.5~188 (118)	122~277 (187)	134~290 (189)	56.0~204 (128)	69.1~185 (113)	48.4~170 (117)

Table 5. Performance index of High-order model

For Non-minimum model (7), we choose $T_1 \in [4.5,5.5]$, $T_2 \in [0.36,0.44]$, $a \in [1,1.5]$ and $k \in [3.2,4.8]$. The nominal parameters are T_1=5,T_2=0.4, a=1.25 and k=4. The simulation results are shown in table 6 and figure 8.

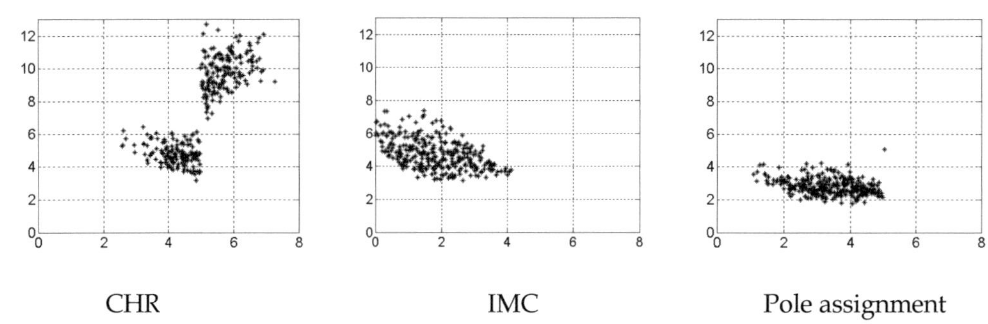

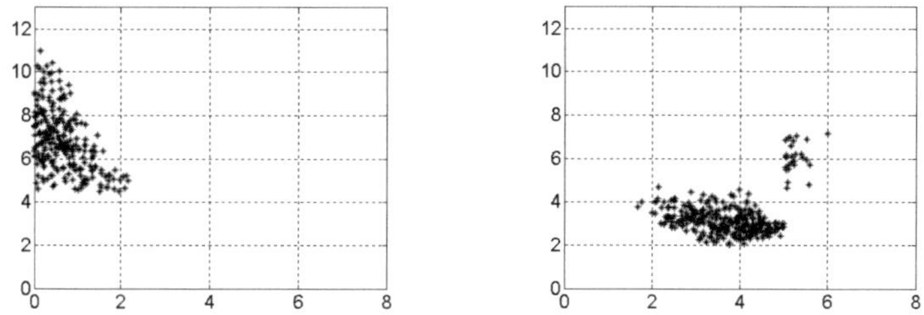

Fig. 8. Simulation results of Non-minimum model (the abscissa represents overshoot and the ordinate represents adjustment time)

	CHR	IMC	Pole assignment	GPM	IST2E	Cohen	Z-N
Overshoot(%)	2.56~7.27 (5.00)	0~4.13 (1.79)	1.10~5.04 (3.38)	0~2.16 (0.554)	1.68~5.99 (3.75)	Not stable	Not stable
Adjustment time	3.18~12.7 (7.37)	3.14~7.38 (4.78)	1.76~5.06 (2.83)	4.49~11.1 (6.95)	2.04~7.16 (3.39)	Not stable	Not stable

Table 6. Performance index of Non-minimum model

From the simulation results above, it is clear that the GPM method and IMC method are superior to other compared tuning methods.

3.2 Performance robustness comparison of DDE and IMC

The desired dynamic equation method (DDE) is proposed for unknown models. This two-degree-of-freedom (2-DOF) controller designing can meet desired setting time, and has physical meaning parameters (Wang et al., 2008).

In this section, we consider 15 transfer function models as follows.

$$G_1(s) = \frac{1}{(s+1)(0.2s+1)} \tag{15}$$

$$G_2(s) = \frac{(-0.03s+1)(0.08s+1)}{(2s+1)(s+1)(0.4s+1)(0.2s+1)(0.05s+1)^3} \tag{16}$$

$$G_3(s) = \frac{2(15s+1)}{(20s+1)(s+1)(0.1s+1)^2} \tag{17}$$

$$G_4(s) = \frac{1}{(s+1)^4} \tag{18}$$

$$G_5(s) = \frac{1}{(s+1)(0.2s+1)(0.04s+1)(0.0008s+1)} \tag{19}$$

$$G_6(s) = \frac{(0.17s+1)^2}{s(s+1)^2(0.028s+1)} \tag{20}$$

$$G_7(s) = \frac{-2s+1}{(s+1)^3} \tag{21}$$

$$G_8(s) = \frac{1}{s(s+1)^2} \tag{22}$$

$$G_9(s) = \frac{e^{-s}}{(s+1)^2} \tag{23}$$

$$G_{10}(s) = \frac{1}{(20s+1)(2s+1)^2} e^{-s} \tag{24}$$

$$G_{11}(s) = \frac{-s+1}{(6s+1)(2s+1)^2} e^{-s} \tag{25}$$

$$G_{12}(s) = \frac{(6s+1)(3s+1)}{(10s+1)(8s+1)(s+1)} e^{-0.3s} \tag{26}$$

$$G_{13}(s) = \frac{2s+1}{(10s+1)(0.5s+1)} e^{-s} \tag{27}$$

$$G_{14}(s) = \frac{-s+1}{s} \tag{28}$$

$$G_{15}(s) = \frac{-s+1}{s+1} \tag{29}$$

Case	DDE-PID settings					Approximation				IMC settings
	h_0	h_1	l	t_{sd}	$\{K_P, K_I, K_D, b\}$	k	θ	τ_1	τ_2	$\{K_C, \tau_I, \tau_D\}$
G_1(PI)	2.35	-	3	2	{4.12,7.83,3.33}	1	0.1	1.1	-	{5.5,0.8}
G_2(PI)	0.45	-	13	10	{0.80,0.35,0.77}	1	1.47	2.5	-	{0.85,2.5}
G_2PID	0.61	1.6	8	9	{2.02,0.86,1.44,1.94}	1	0.77	2	1.2	{1.30,2,1.2}
G_3(PI)	2	-	7	2.5	{1.71,2086,0.43}	1.5	0.15	1.05	-	{2.33,1.05}
G_3PID	16	8	4	3	{24,40,4.5,20}	1.5	0.05	1	0.15	{6.67,0.4,0.15}
G_4(PI)	0.45	-	16	12	{0.65,0.28,0.63}	1	2.5	1.5	-	{0.3,1.5}
G_4PID	0.59	1.5	12	15	{1.33,0.49,0.96,1.28}	1	1.5	1.5	1	{0.5,1.5,1}
G_5(PI)	2	-	5	3	{2.4,4,2}	1	0.148	1.1	-	{3.72,1.1}
G_5PID	16	8	1	3	{96,160,18,80}	1	0.028	1.0	0.22	{17.9,0.22,0.22}
G_6(PI)	0.14	-	31	29	{0.33,0.045,0.32}	1	1.69		-	{0.296,13.5}
G_6PID	0.85	1.9	2	13	{9.66,4.26,5.92,9.23}	1	0.358		1.33	{1.40,2.86,1.33}
G_7(PI)	0.53	-	30	16	{0.35,0.18,0.33}	1	3.5	1.5	-	{0.214,1.5}
G_7PID	0.60	1.5	31	11	{0.57,0.23,0.38,0.55}	1	2.5	1.5	1	{0.3,1.5,1}
G_8(PI)	0.12	-	35	33	{0.30,0.036.0.29}	1	1.5		-	{0.33,12}

Case	DDE-PID settings					Approximation				IMC settings
	h_0	h_1	l	t_{sd}	$\{K_P,K_I,K_D,b\}$	k	θ	τ_1	τ_2	$\{K_C,\tau_I,\tau_D\}$
G_8PID	0.32	1.1	8	15	{1.46,0.40,1.39,1.42}	1	0.5		1.5	{1.5,4,1.5}
G_9(PI)	0.63	-	15	10	{0.71,0.42,0.67}	1	1.5	1.5	-	{0.5,1.5}
G_9PID	1	2	18	12	{1.17,0.56,0.67,1.11}	1	1	1	1	{0.5,1,1}
G_{10}(PI)	0.11	-	3	38	{3.37,0.35,3.33}	1	2	21	-	{2.25,16}
G_{10}PID	0.03	0.4	1	11	{3.67,0.33,10.4,3.64}	1	1	20	2	{10,8,2}
G_{11}(PI)	0.14	-	9	28	{1.13,0.16,1.11}	1	5	7	-	{0.7,7}
G_{11}PID	0.07	0.5	3	15	{1.80,0.24,3.51,1.78}	1	3	6	3	{1,6,3}
G_{12}(PI)	1.33	-	1.1	3	{10.3,12.1, 9.09}	0.23	0.3	1	-	{7.41,1}
G_{13}(PI)	0.5		3	8	{3.50,1.67,3.33}	0.65	1.25	4.5	-	{2.88,4.50}
G_{14}(PI)	0.4	-	15	10	{0.69,0.27,0.67}	1	1		-	{0.5,8}
G_{15}(PI)	0.8	-	18	5	{0.64,0.76,0.59}	1	1	1	-	{0.5,1}

Table 7. Controller parameters

The DDE and IMC method are used on them to compare the performance robustness. The controller parameters are shown in table 7. ±10% parameter perturbation is taken for performance robustness experiment with 300 times.

In order to compare the two methods easily, we divide them into four types shown in table 8.

No.	Type	Model
1	Normal model	G_1、G_9、G_{12}、G_{13}
2	High-order model	G_3、G_4、G_5、G_{10}
3	Non-minimum model	G_2、G_7、G_{11}、G_{15}
4	Model with integral	G_6、G_8、G_{14}

Table 8. Four types of models

The Normal model is simple and easy to control. The simulation results are shown in table 9 and 10.

Model	Controller	Step response	Performance robustness	
			DDE	IMC
G_1	PI	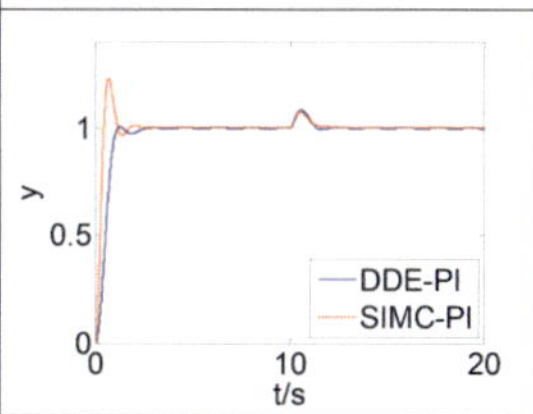	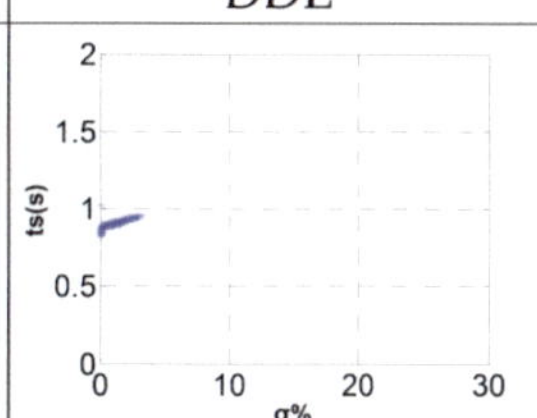	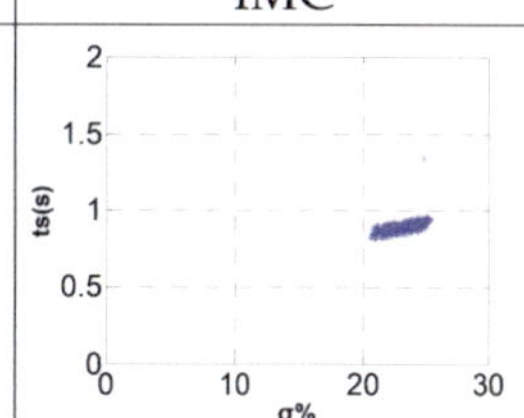
G_9	PI	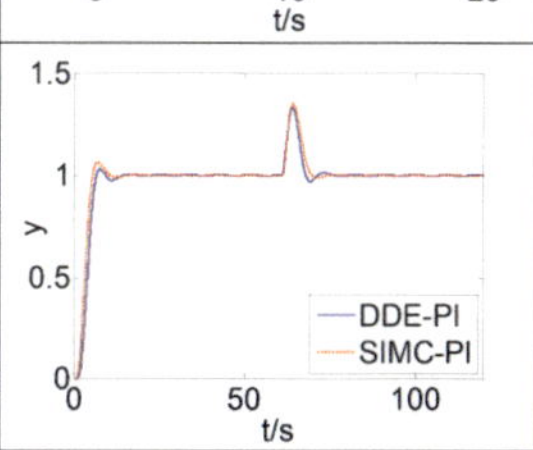	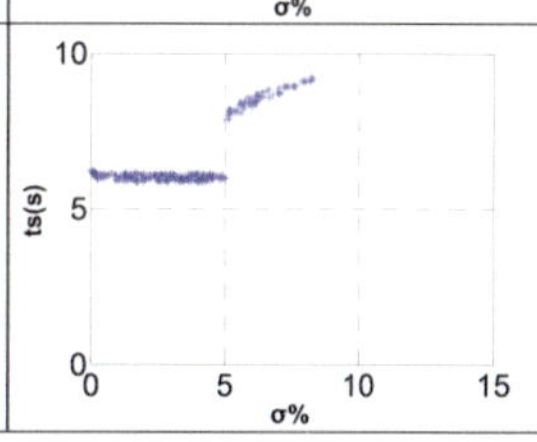	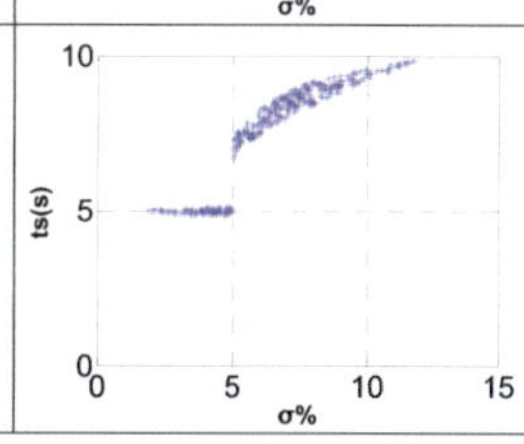

Model	Controller	Step response	Performance robustness	
			DDE	IMC
	PID	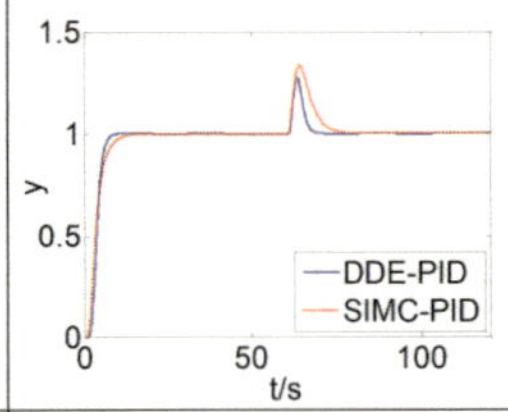	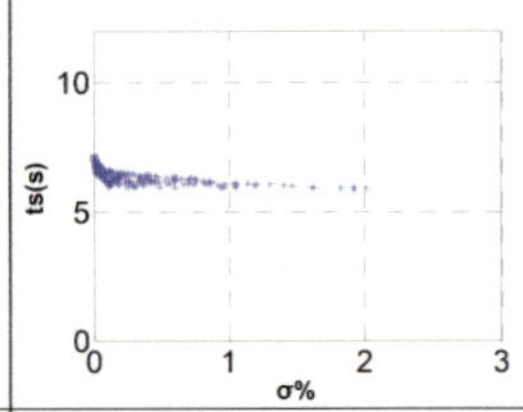	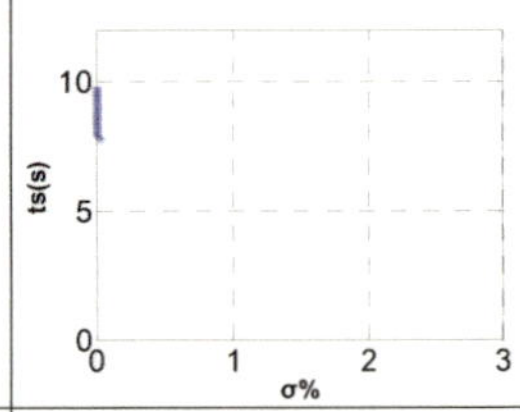
G_{12}	PI	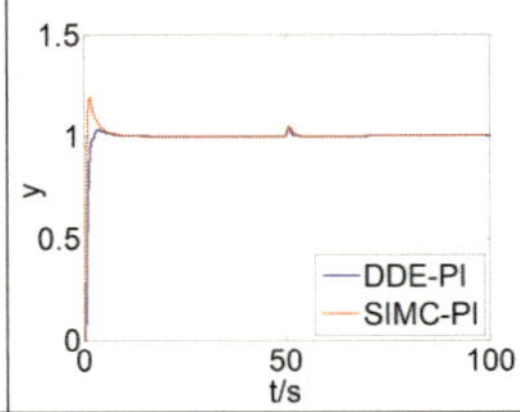	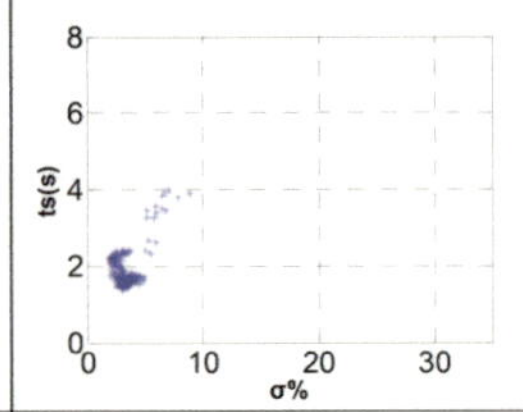	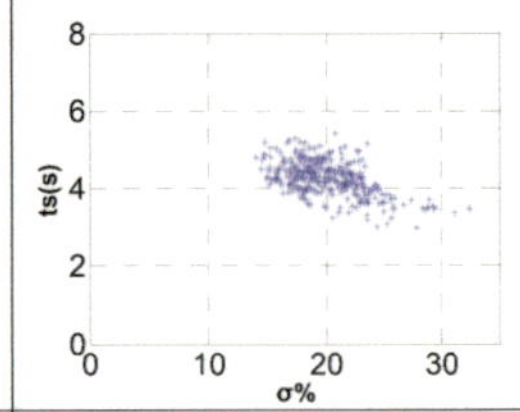
G_{13}	PI	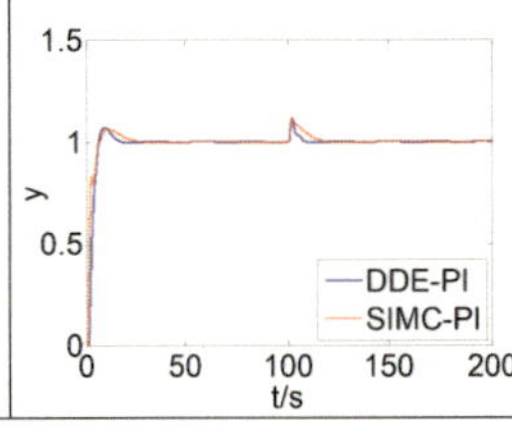	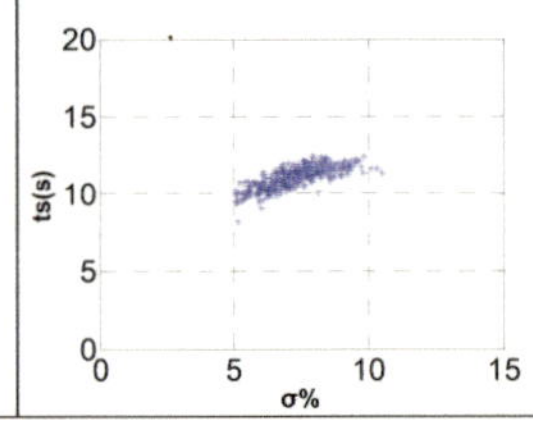	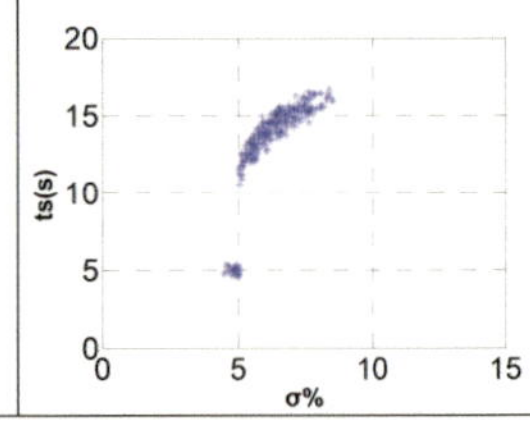

Table 9. Simulation results of Normal model

Model	Method	Overshoot(%)			Adjustment time(s)		
		Scope	Mean	Variance	Scope	Mean	Variance
G_1	DDE-PI	0~3.19	0.76	0.75	0.83~1.03	0.89	0.001
	IMC-PI	20.0~24.1	21.5	2.36	0.82~1.34	0.90	0.002
G_9	DDE-PI	0 ~8.74	3.15	5.07	5.88~9.13	6.68	1.19
	IMC-PI	1.25~11.4	6.50	5.10	4.88~9.84	7.54	2.75
	DDE-PID	0~2.00	0.240	0.13	5.87~7.34	6.61	1.15
	IMC-PID	0~0.011	0	0	4.88~9.84	7.54	2.87
G_{12}	DDE-PI	1.59~7.51	3.12	0.57	1.42~3.98	1.80	0.102
	IMC-PI	14.0~32.4	20.2	10.23	2.98~5.41	4.25	0.209
G_{13}	DDE-PI	5.03~10.5	7.22	1.36	8.19~12.5	11.1	0.53
	IMC-PI	4.43~8.50	6.27	0.74	4.70~16.6	13.7	5.81

Table 10. Performance index of Normal model

For Normal model, the control effects of two tuning method are similar. Because the IMC method is based on FOPTD model and SOPTD model, the approximation error can be ignored and the DDE method is effective.

Most of High-order model is series connection of inertial element in industry field (Quevedo, 2000). But, the simple PID is hard to control them because of the delay cascaded by inertial elements. The simulation results are shown in table 11 and 12.

Model	Controller	Step response	Performance robustness	
			DDE	IMC
G_3	PI	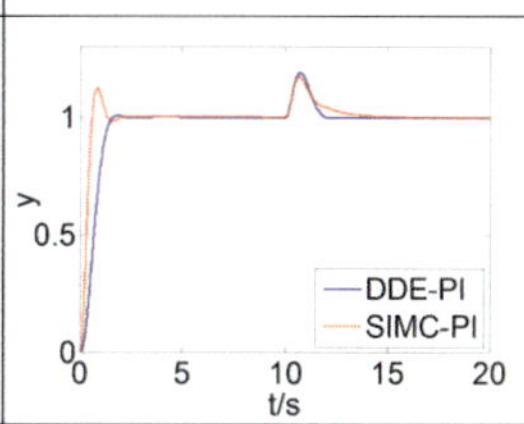	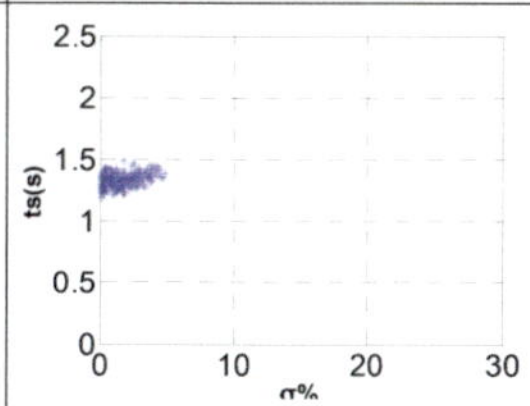	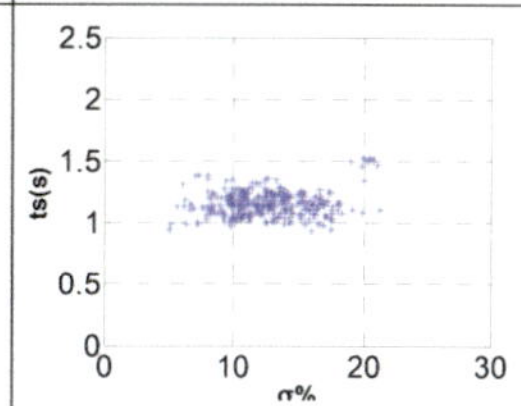
	PID	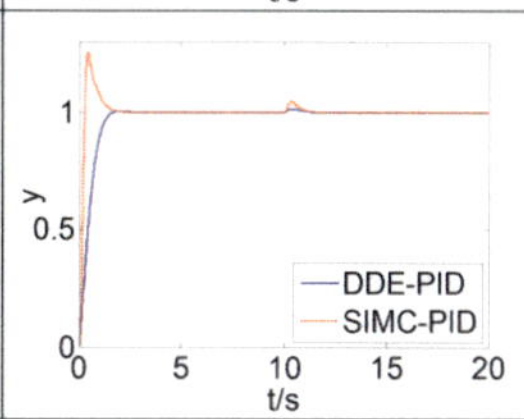	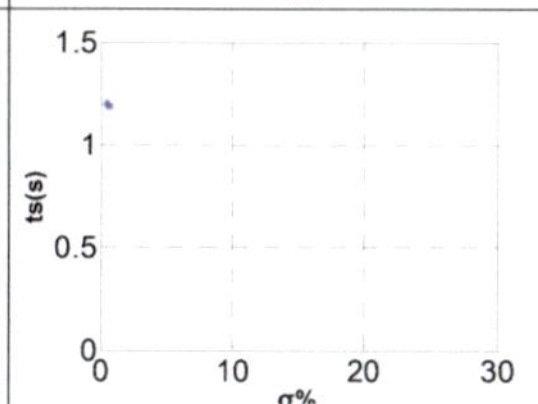	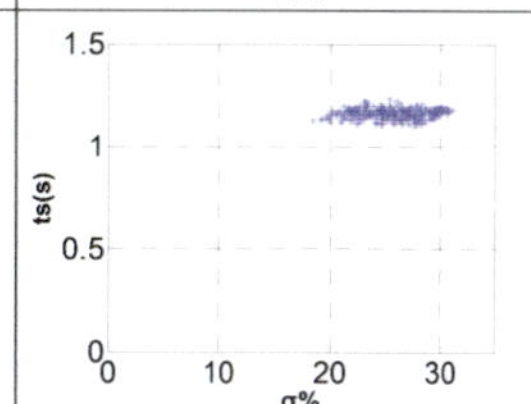
G_4	PI	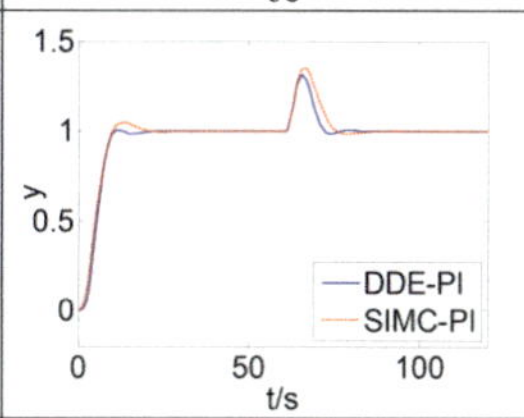	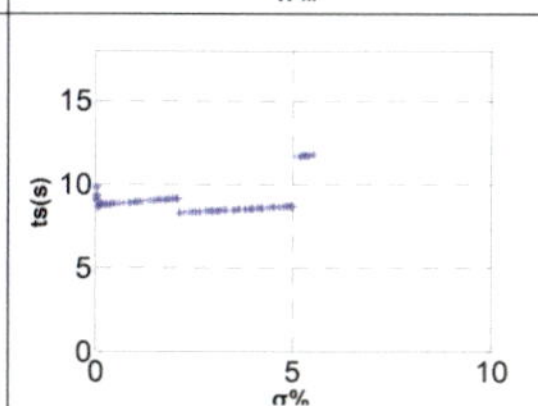	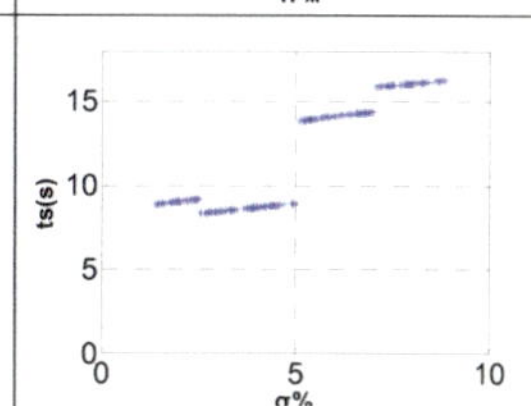
	PID	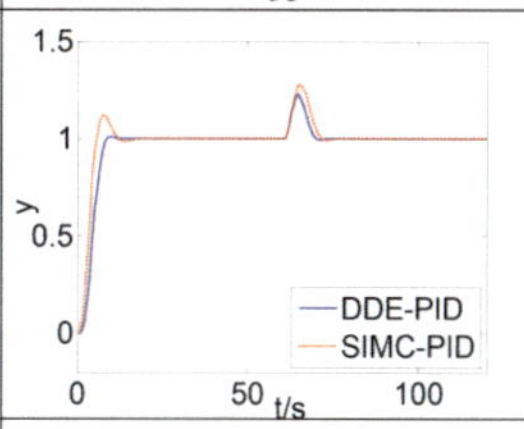	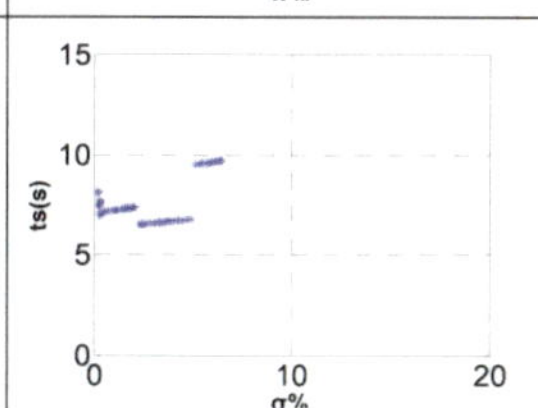	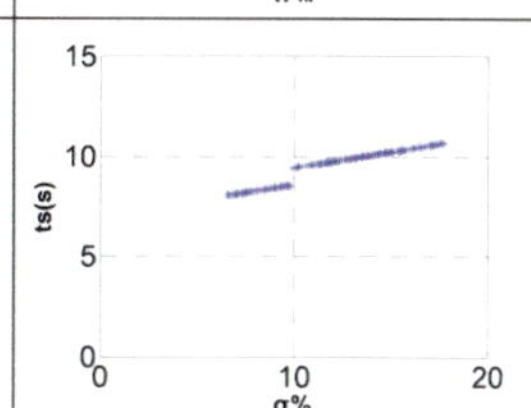
G_5	PI	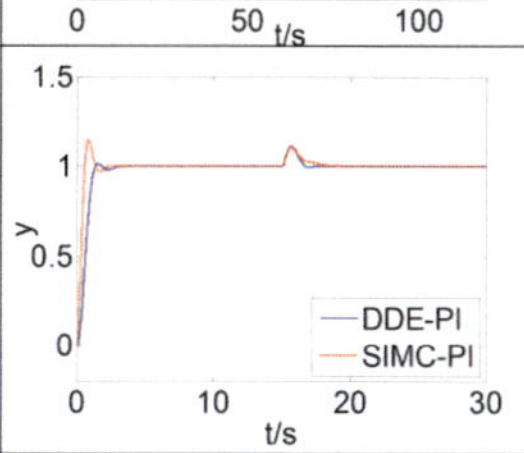	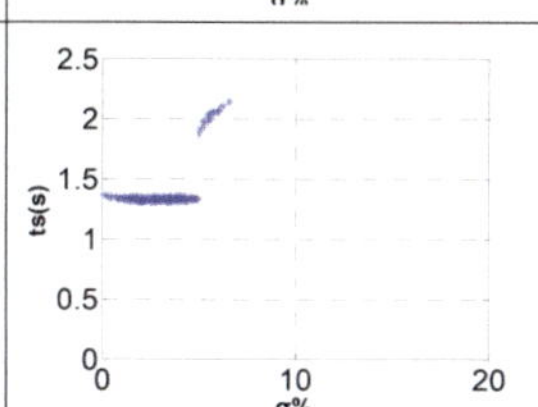	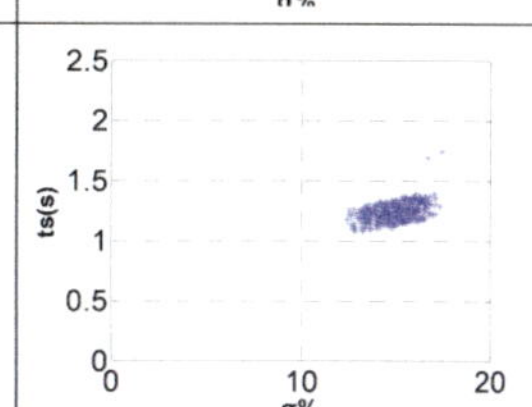
	PID	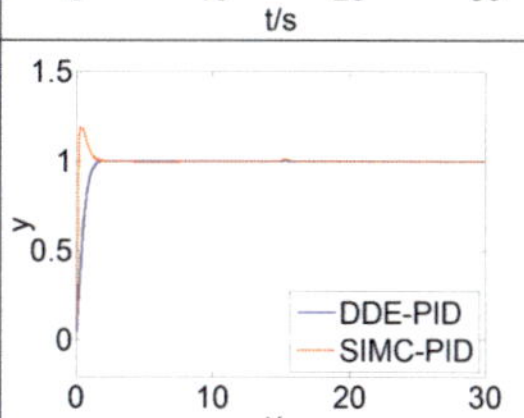	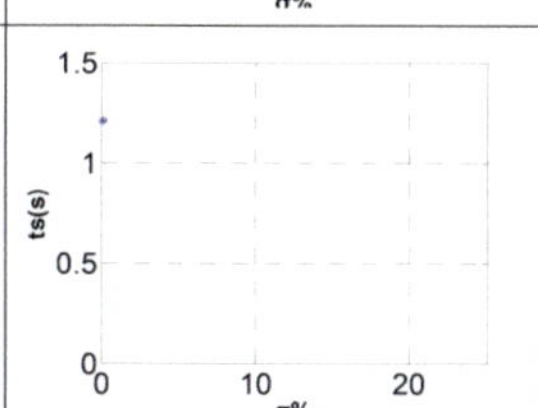	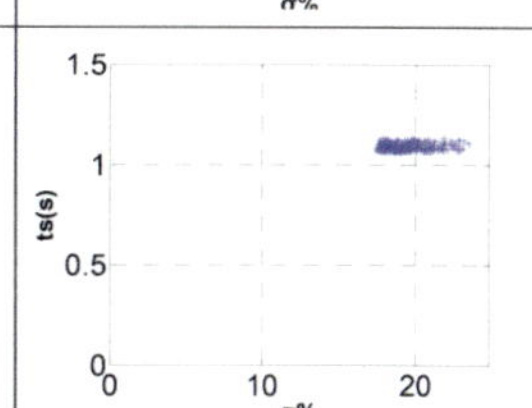
G_{10}	PI	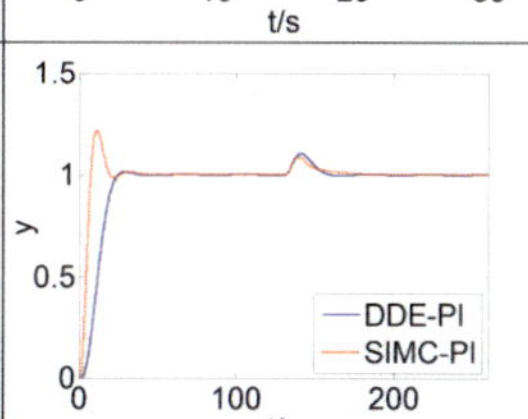	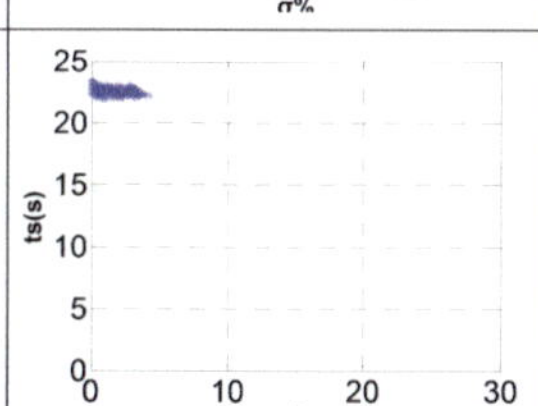	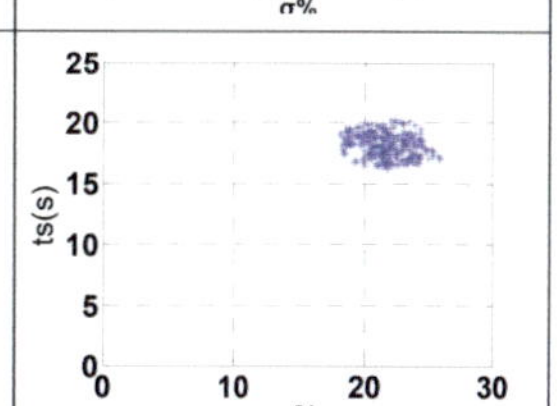

Model	Controller	Step response	Performance robustness	
			DDE	IMC
	PID	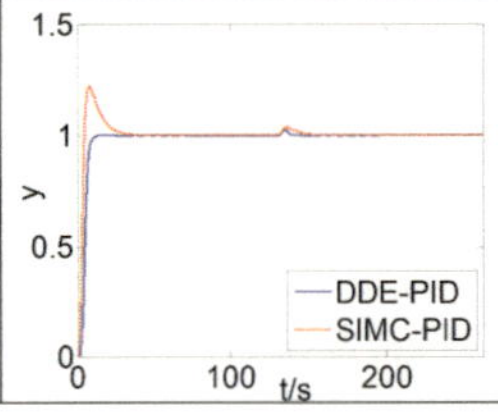	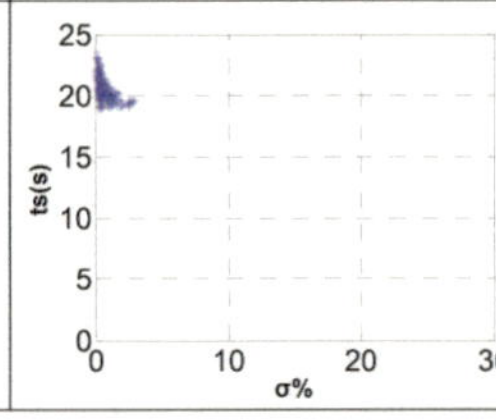	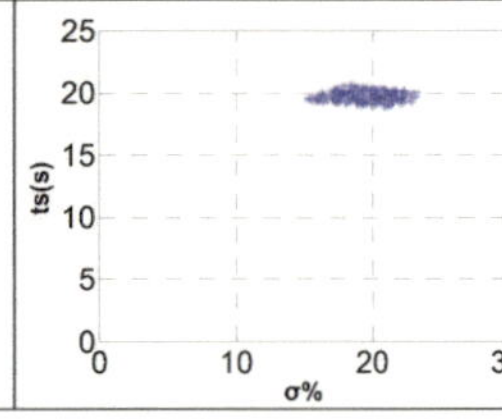

Table 11. Simulation results of High-order model

Model	Method	Overshoot(%)			Adjustment time(s)		
		Scope	Mean	Variance	Scope	Mean	Variance
G_3	DDE-PI	0.09~5.54	1.24	1.47	1.18~1.53	1.34	0.004
	IMC-PI	5.87~21.5	12.5	9.91	0.94~1.53	1.14	0.011
	DDE-PID	0.43~0.72	0.53	0.004	1.19~1.21	1.20	0
	IMC-PID	18.3~31.5	25.5	8.82	1.09~1.23	1.16	0
G_4	DDE-PI	0~6.92	2.18	4.97	7.96~11.8	9.15	1.81
	IMC-PI	1.35~8.83	4.89	4.71	8.38~16.3	11.9	10.2
	DDE-PID	0.12~6.35	2.15	4.49	6.51~9.72	7.53	0.935
	IMC-PID	6.48~17.5	11.9	10.4	8.04~10.7	9.45	0.764
G_5	DDE-PI	0.08~6.51	3.17	1.92	1.31~2.14	1.40	0.04
	IMC-PI	12.2~17.2	14.7	1.50	1.09~1.73	1.25	0.007
	DDE-PID	0.47~0.73	0.61	0.004	1.23~1.23	1.23	0
	IMC-PID	17.4~23.5	19.3	1.92	1.07~1.14	1.09	0
G_{10}	DDE-PI	0~4.03	1.42	1.39	22.0~22.6	22.5	0.084
	IMC-PI	17.6~26.1	21.5	3.15	16.3~20.2	18.3	1.00
	DDE-PID	0.014~1.46	0.287	0.05	9.63~10.8	10.1	0.055
	IMC-PID	15.1~23.7	19.5	3.73	18.9~19.7	19.7	0.168

Table 12. Performance index of High-order model

It is clear that DDE method is as fast as IMC method on High-order model, but the overshoot is almost zero. DDE method also has good performance robustness especially on G_3 and G_5.

The Non-minimum model has the zeros and poles on right half complex plane or time delay. The simulation results are shown in table 13 and 14.

Model	Controller	Step response	Performance robustness	
			DDE	IMC
G_2	PI	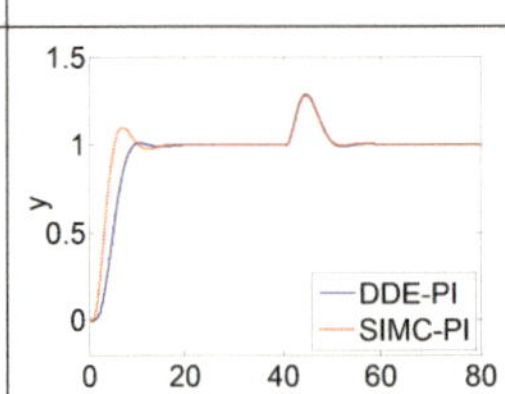	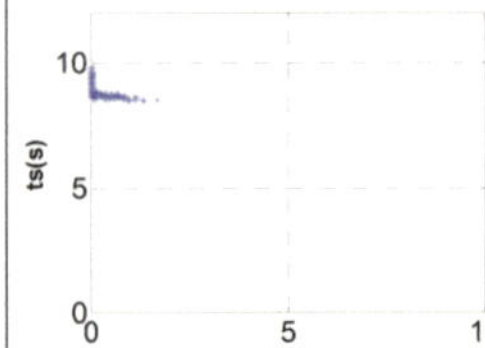	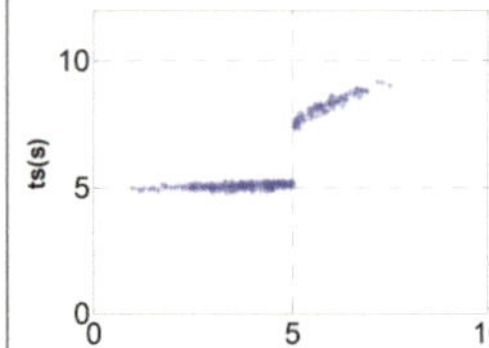

Model	Controller	Step response	Performance robustness	
			DDE	IMC
	PID	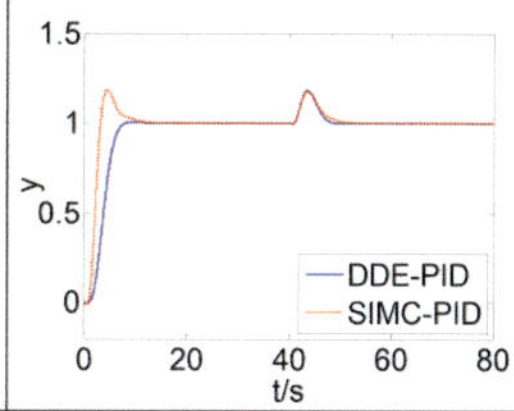	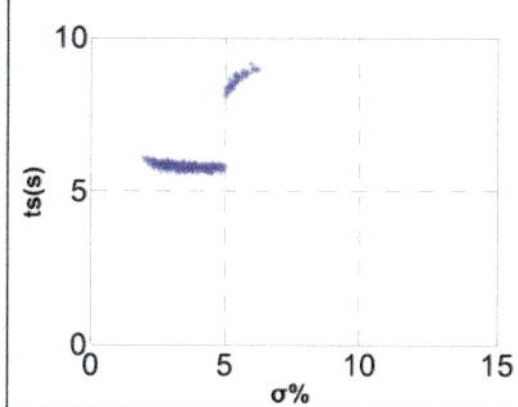	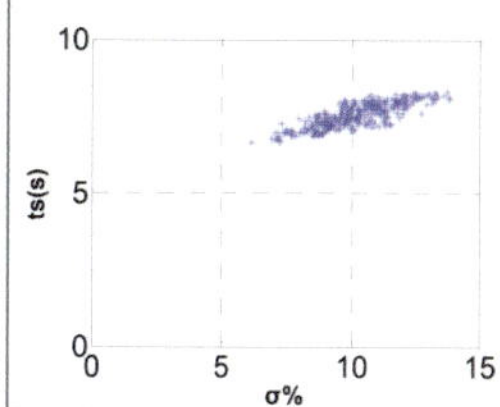
G_7	PI	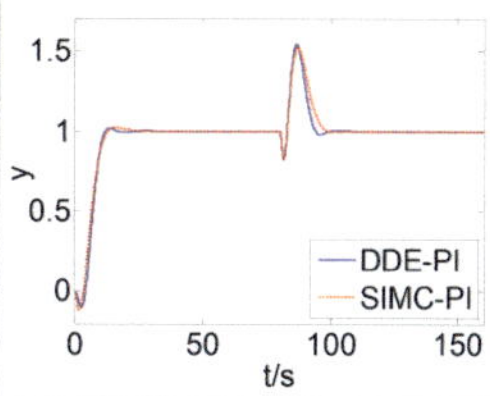	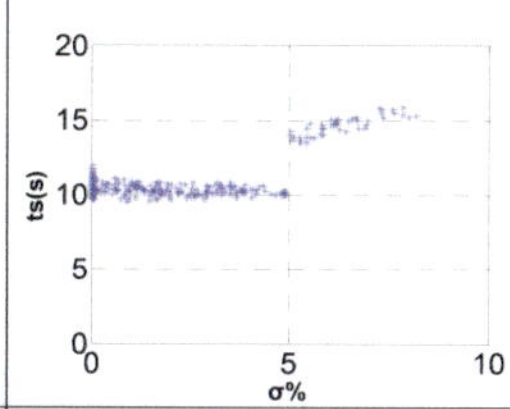	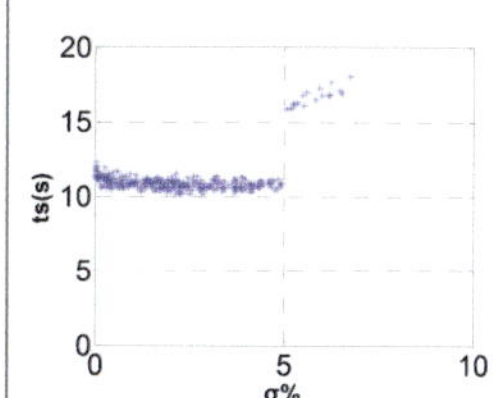
	PID	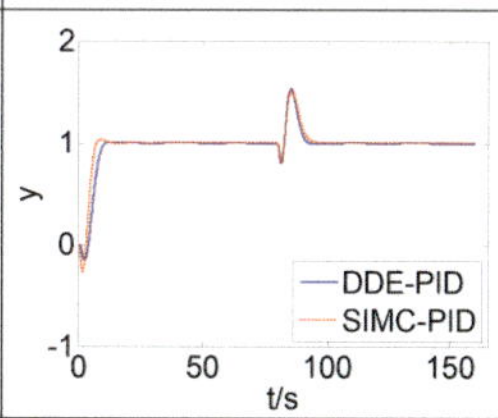	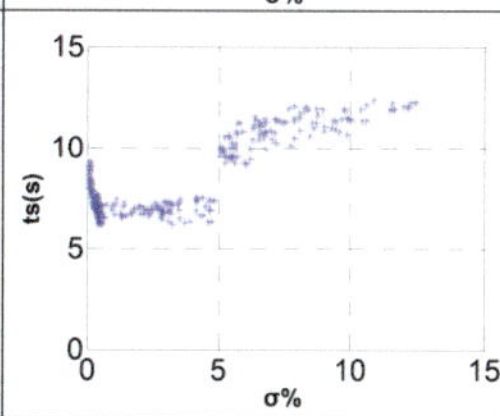	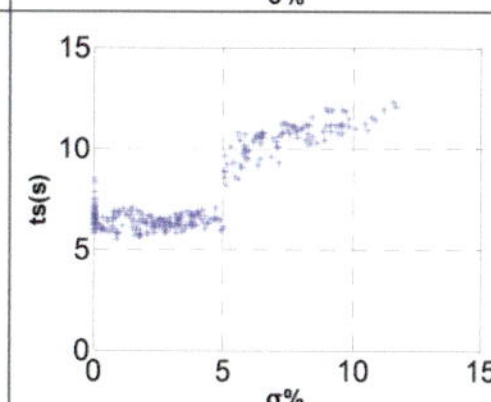
G_{11}	PI	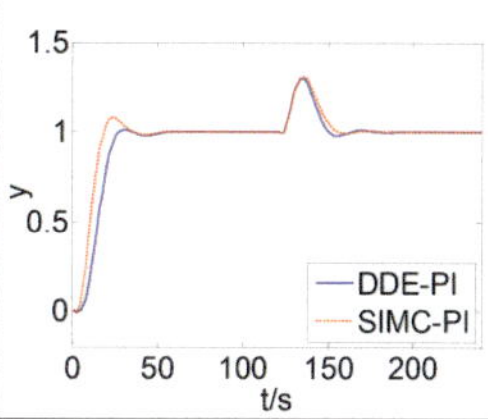	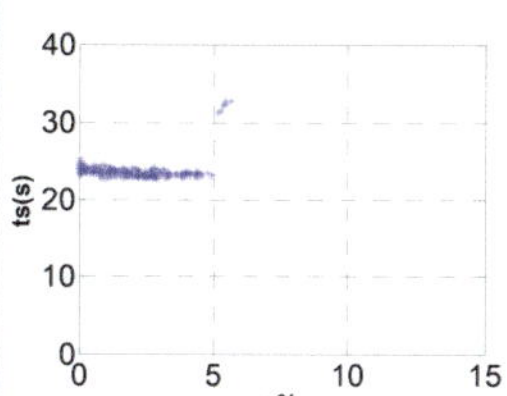	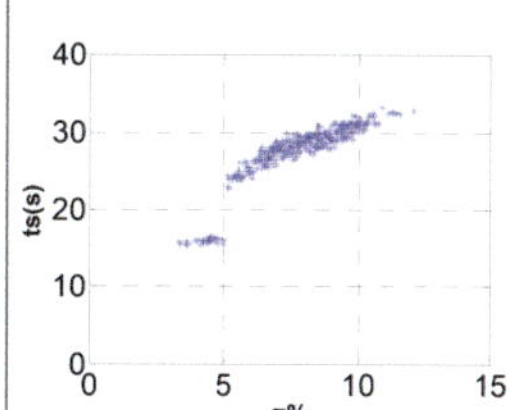
	PID	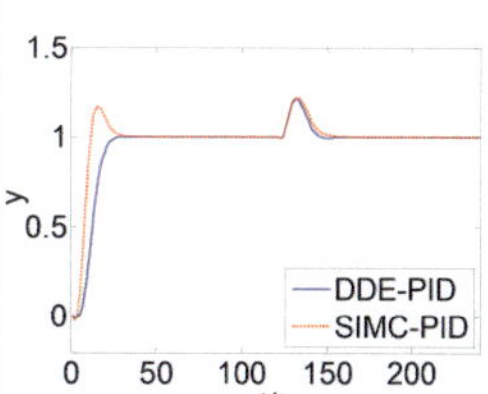	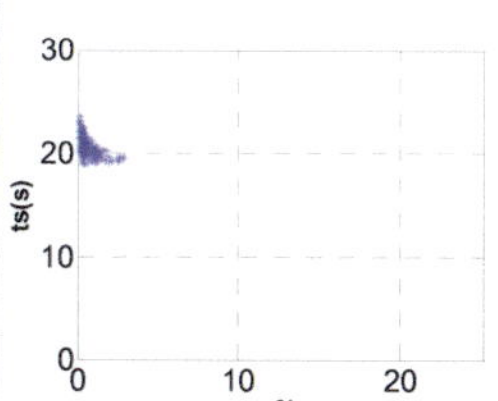	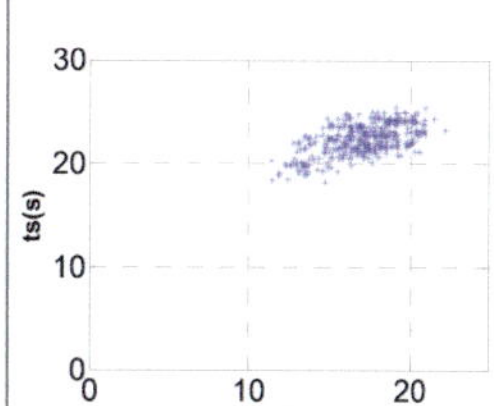
G_{15}	PI	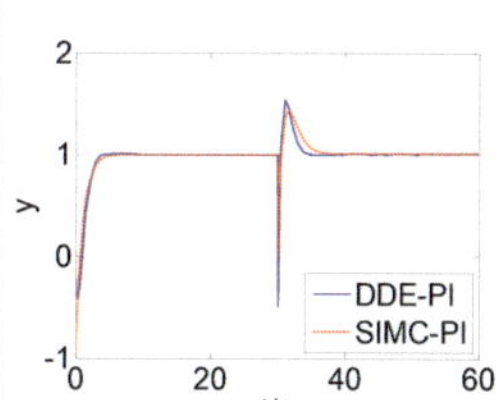	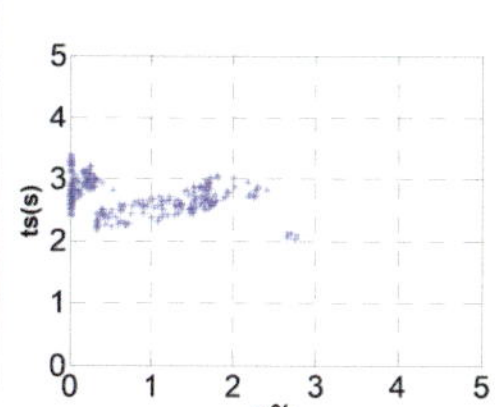	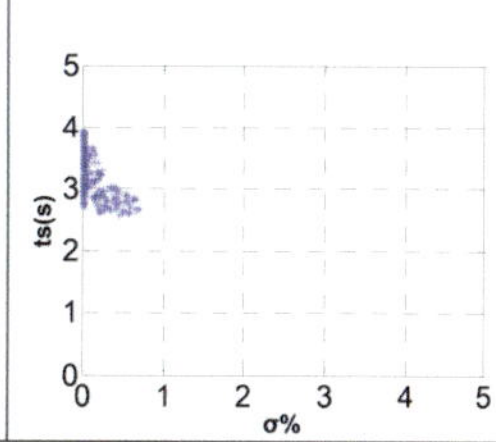

Table 13. Simulation results of Non-minimum model

Model	Method	Overshoot (%)			Adjustment time (s)		
		Scope	Mean	Variance	Scope	Mean	Variance
G_2	DDE-PI	0~1.24	0.099	0.058	8.52~9.72	8.92	0.05
	IMC-PI	0.60~7.31	4.20	1.78	4.85~9.14	5.87	2.01
	DDE-PID	1.91~6.09	3.76	0.967	5.65~9.07	6.21	0.91
	IMC-PID	6.36~13.7	10.1	2.51	6.52~8.31	7.55	0.18
G_7	DDE-PI	0~8.38	2.10	5.24	9.47~16.0	11.1	2.49
	IMC-PI	0~6.53	2.30	2.74	10.1~18.1	11.6	3.47
	DDE-PID	0.06~12.1	2.93	9.71	6.09~12.3	8.37	3.54
	IMC-PID	0~12.1	3.49	10.7	5.62~12.3	7.87	4.03
G_{11}	DDE-PI	0~5.53	1.46	1.93	22.6~32.6	23.7	1.16
	IMC-PI	3.22~11.7	7.47	2.98	15.4~32.6	27.1	15.4
	DDE-PID	0.043~3.0	0.507	0.272	19.0~23.3	20.7	0.91
	IMC-PID	11.6~21.8	16.7	4.99	18.7~25.7	22.4	2.31
G_{15}	DDE-PI	0~2.82	0.66	0.567	2.07~3.45	2.73	0.007
	IMC-PI	0~0.69	0.07	0.015	2.50~3.95	3.23	0.112

Table 14. Performance index of Non-minimum model

For Non-minimum model, the two method has similar step response, but the undershoot is smaller with DDE method. DDE method also has good performance robustness.

Integral is the typical element in control system. If a system contains an integral, it will not be a self-balancing system. It is open-loop unstable and easy to oscillate in close-loop. So it is hard to obtain a good control effect. The simulation results are shown in table 15 and 16.

The simulation results of Model with integral shows that the overshoot of IMC method is much larger than DDE method, and DDE method is much quicker than IMC method. The performance robustness of DDE method is better than IMC method.

The comprehensive comparison is shown in table 17.

Model	Controller	Step response	Performance robustness: DDE	Performance robustness: IMC
G_6	PI	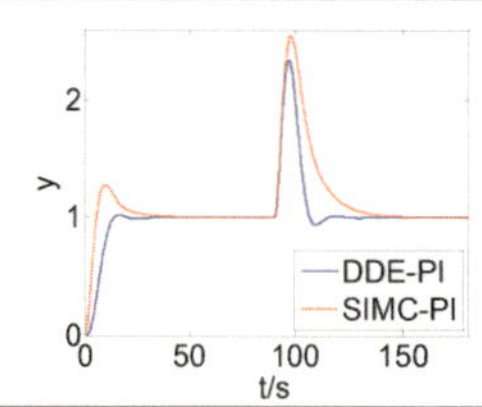	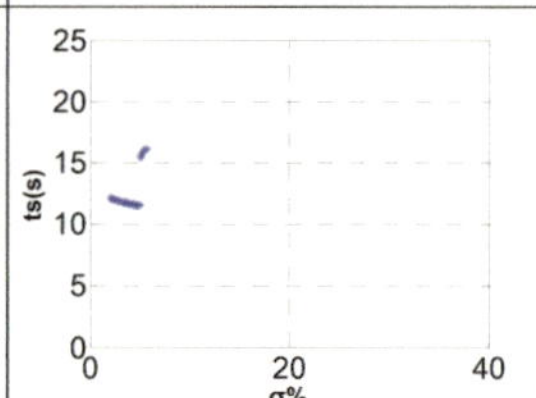	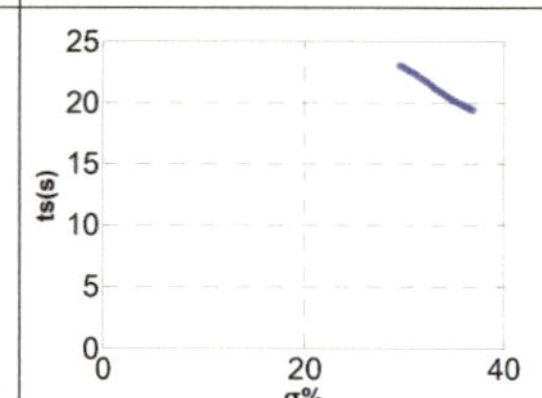
	PID	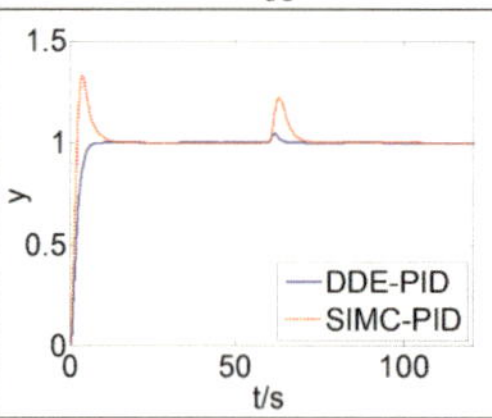	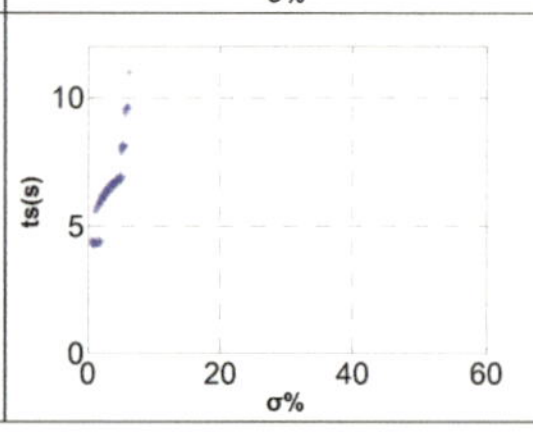	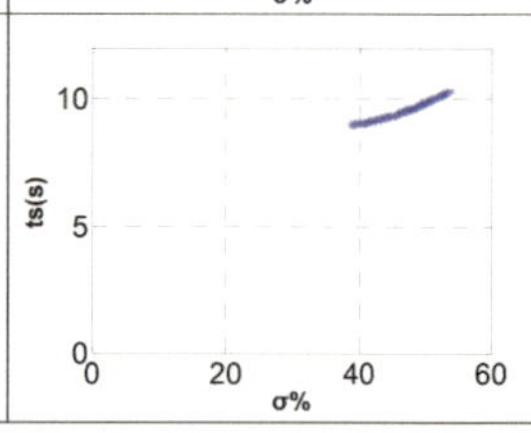

<table>
<tr><th rowspan="2">Model</th><th rowspan="2">Controller</th><th rowspan="2">Step response</th><th colspan="2">Performance robustness</th></tr>
<tr><th>DDE</th><th>IMC</th></tr>
<tr><td rowspan="2">G₈</td><td>PI</td><td>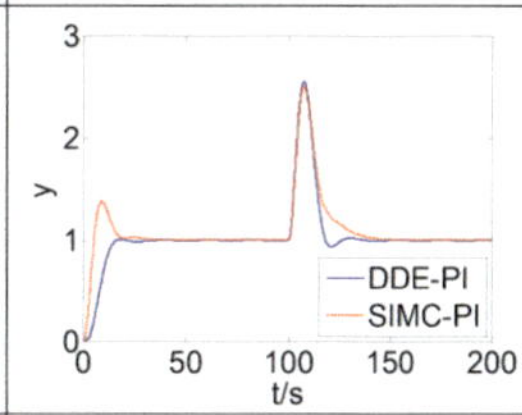
</td><td>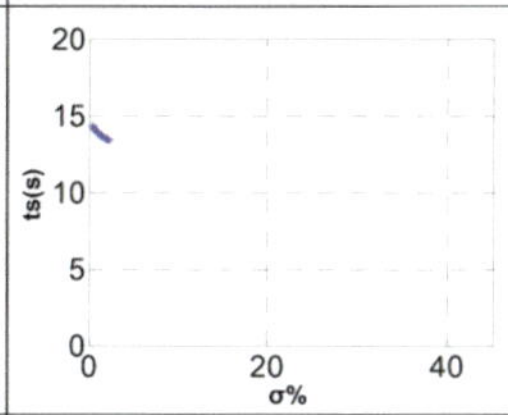
</td><td>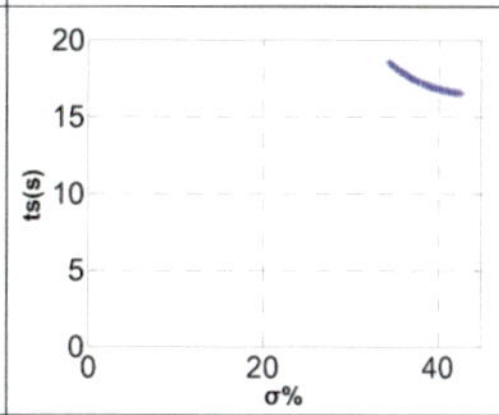
</td></tr>
<tr><td>PID</td><td>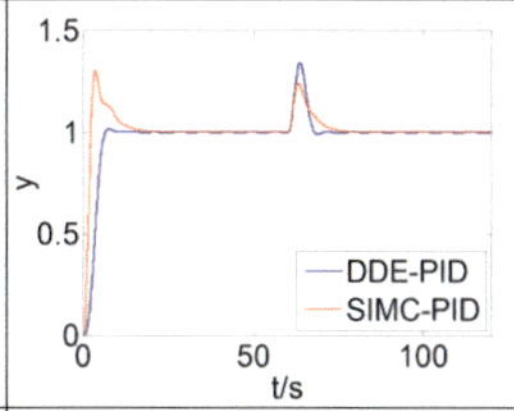
</td><td>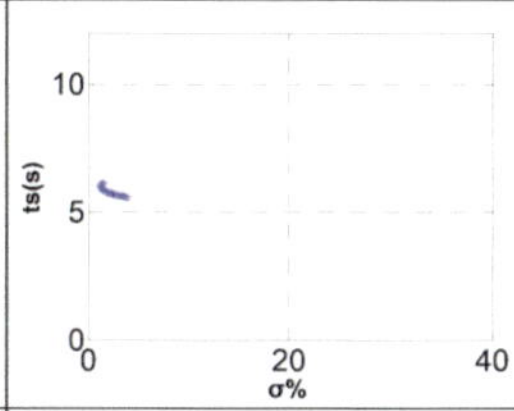
</td><td>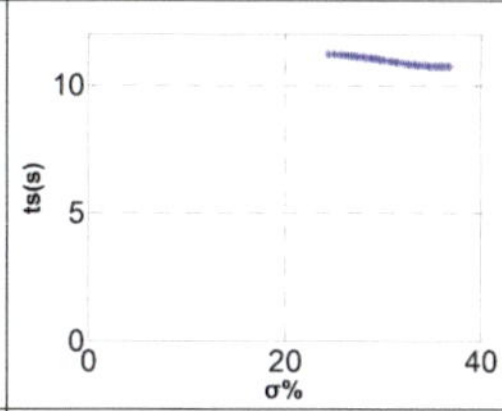
</td></tr>
<tr><td>G₁₄</td><td>PI</td><td>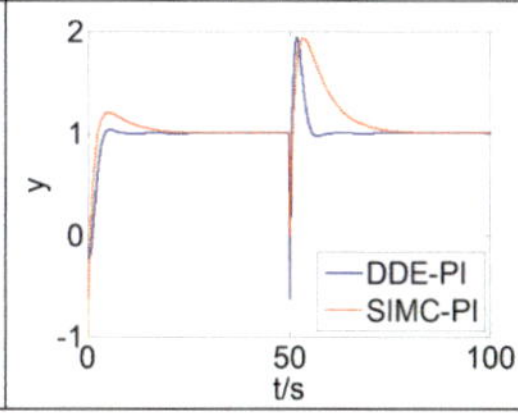
</td><td>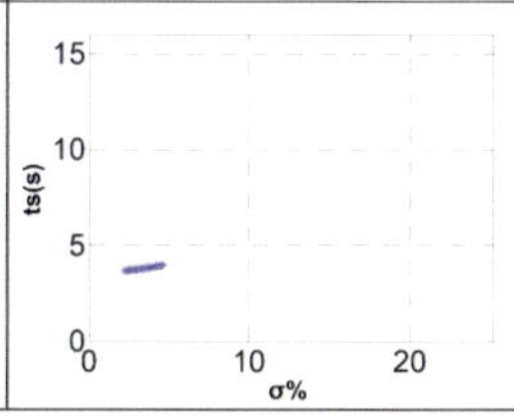
</td><td>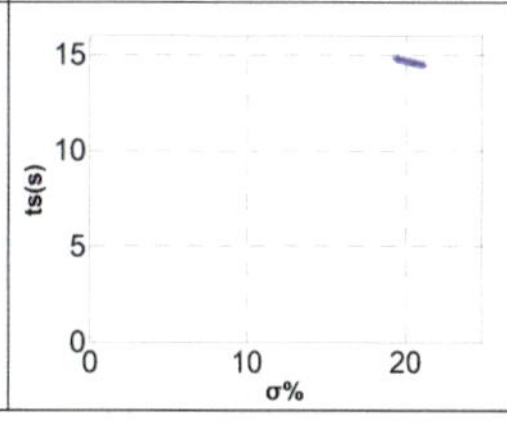
</td></tr>
</table>

Table 15. Simulation results of Model with integral

Model	Method	Overshoot(%)			Adjustment time(s)		
		Scope	Mean	Variance	Scope	Mean	Variance
G_6	DDE-PI	1.99~5.64	3.68	1.15	11.6~16.2	12.5	2.09
	IMC-PI	29.3~37.0	33.2	5.24	19.4~23.1	21.2	1.29
	DDE-PID	0.544~6.33	2.70	2.21	4.29~11.0	6.07	1.63
	IMC-PID	38.4~53.5	46.0	17.8	8.97~10.3	9.53	0.139
G_8	DDE-PI	0.353~2.3	1.19	0.33	13.4~14.4	13.9	0.0756
	IMC-PI	34.1~42.3	38.5	6.41	16.6~18.6	17.3	0.358
	DDE-PID	1.19~3.95	1.97	0.69	5.58~6.15	5.83	0.028
	IMC-PID	24.2~36.8	30.3	13.8	10.8~11.2	11.0	0.029
G_{14}	DDE-PI	2.13~4.66	3.64	0.53	3.71~4.03	3.88	0.008
	IMC-PI	19.2~21.2	20.1	0.33	14.5~14.8	14.7	0.011

Table 16. Performance index of Model with integral

	DDE method	IMC method
Rise time	Slow	Fast
Adjustment time	Relatively fast	Relatively fast
Overshoot	Small	Large
Performance robustness	Good	General
IAE	Large	Small
Demand of model	Relative order	Precise

Table 17. Comparison of DDE method and IMC method

3.3 Performance robustness comparison of DDE and GPM

In this section, we also consider the four typical models shown in table 18.

No.	Types of models	Mathematical form	Examples	Parameters perturbation
1	FOPTD model	$G_{p1}(s)=\frac{K}{1+s\tau}e^{-sL}$	$\frac{[0.9,\ 1.1]}{1+[0.9,\ 1.1]s}e^{-[0.9,\ 1.1]s}$	[Min, Max] Min: the minimum of parameters perturbation. Max: the maximum of parameters perturbation. The parameters are uniformly selected in the scope.
2	SOPTD model	$G_{p2}(s)=\frac{K}{(1+s\tau_1)(1+s\tau_2)}e^{-sL}$	$\frac{[0.9,\ 1.1]}{(1+[0.9,\ 1.1]s)(1+[0.45,\ 0.55]s)}e^{-[0.9,\ 1.1]s}$	
3	High-order model	$G_{p3}(s)=\frac{K}{(1+\tau s)^n}$	$\frac{[0.9,\ 1.1]}{(1+[0.9,\ 1.1]s)^5}$	
4	Non-minimum model	$G_{p4}(s)=\frac{K(a-s)}{(1+\tau s)^n}$	$\frac{[0.9,\ 1.1]([0.9,\ 1.1]-s)}{(1+[0.9,\ 1.1]s)^3}$	

Table 18. Four types of typical model

According to desired adjustment time and prospective gain margin ~ phase margin to design controller in each DDE and GPM methods. Within nominal parameter, design PI controller for FOPTD model, design PID controller for SOPTD model, high-order model and non-minimum model. Proceed performance robustness experiment within ±10% parameter perturbation. In order to keep the comparison impartial, select adjustment time of GPM method as the desired adjustment time. Controller parameters are shown in table 19, results of Monte-Carlo simulation are shown in table 20, comparison of performance indices is shown in table 21.

Simulation results show that DDE method has better performance robustness than GPM method generally. Apparently, the points on overshoot ~ adjustment time plane of DDE method concentrate more together near the bottom left corner than GPM method. Except the G_{P3} result, the points on gain margin ~ phase margin plane of DDE method are more concentrated than GPM method.

Types of models	DDE method									GPM method							
	Settings	PID parameters								Settings		PID parameters					
	t_{sd}	h_0	h_1	l	k	K_p	K_i	K_d	b	A_m	P_m	K_c	T_i	T_d	K_p	K_i	K_d
FOPTD G_{p1}	12.5	0.8		11.6	10	0.93	0.68	0	0.86	3	60°	0.52	1	0	0.52	0.52	0
SOPTD G_{p2}	7.7		2.6	21.6	10	1.28	0.78	0.58	1.2	3	60°	0.52	1	0.5	0.78	0.52	0.26
High-order G_{p3}	20.8		0.96	6.5	10	1.51	0.36	1.69	1.48	3	60°	0.57	1.89	1.89	1.14	0.3	1.08
Non-minimum G_{p4}	13		1.54	13.4	10	1.19	0.44	0.86	1.15	3	60°	0.33	1	1	0.66	0.33	0.33

Table 19. Controller parameters

Types of models	Time-domain performance robustness		Frequency-domain performance robustness	
	DDE method	GPM method	DDE method	GPM method
G_{P1}	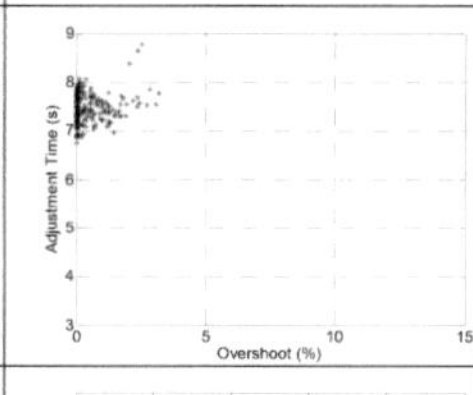	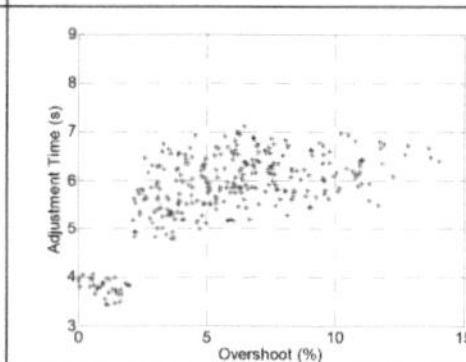	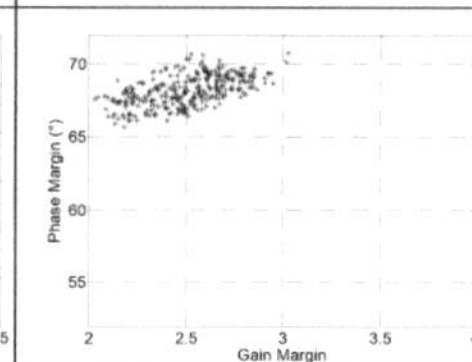	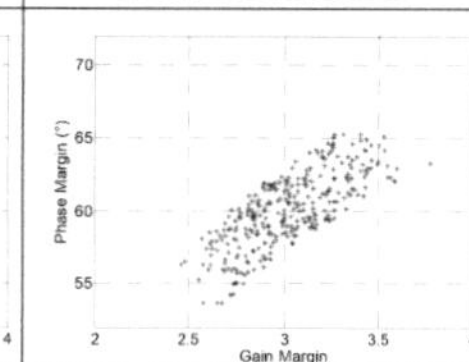
G_{P2}	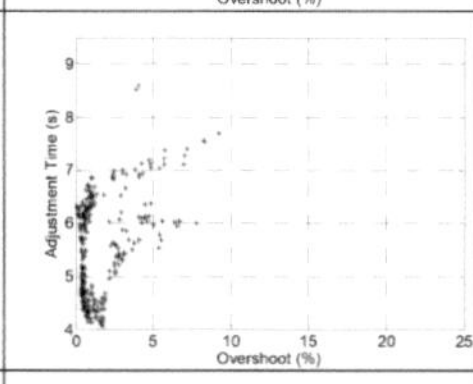	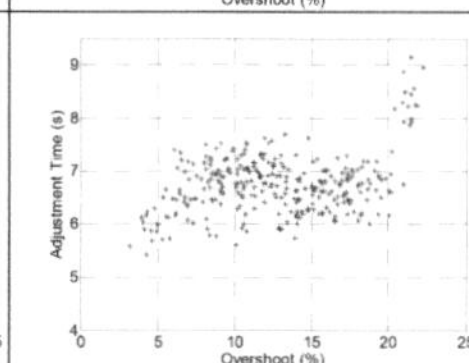	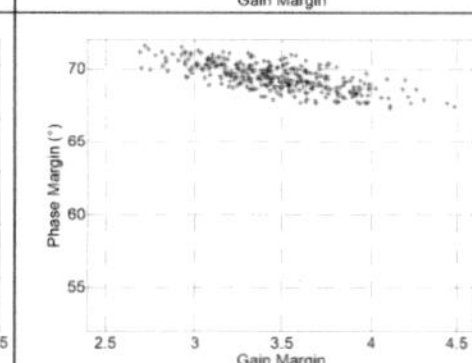	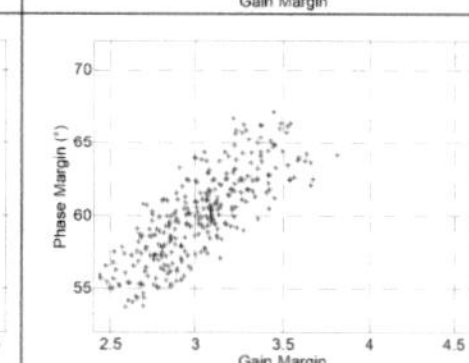
G_{P3}	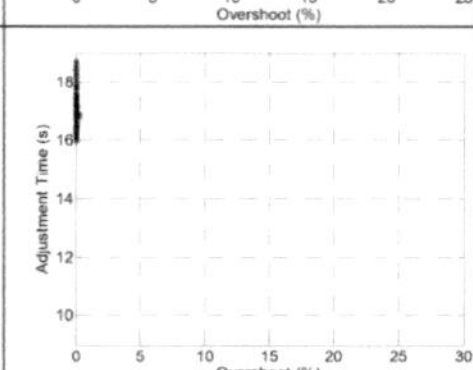	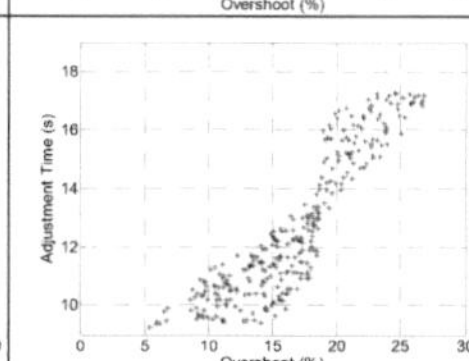	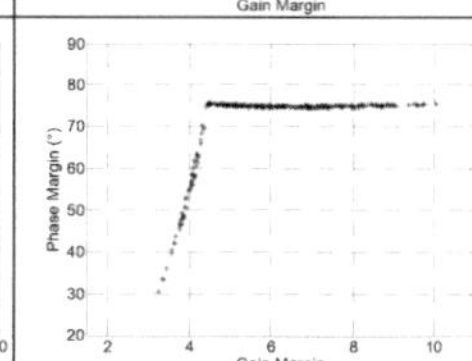	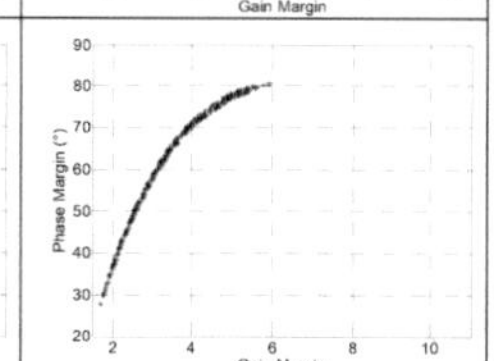
G_{P4}	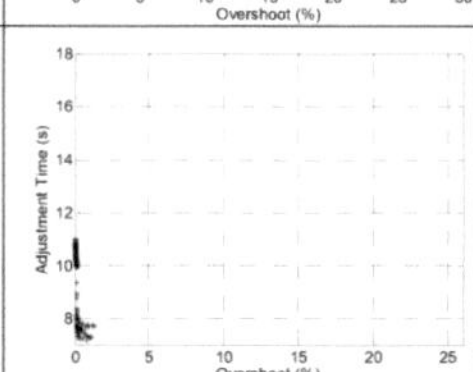	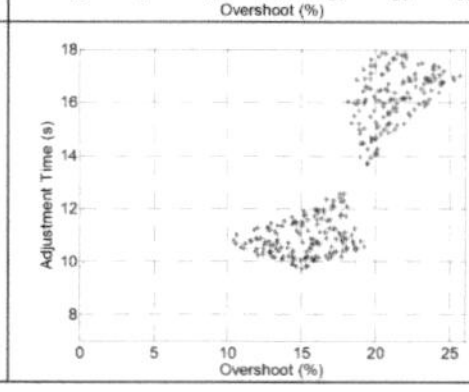	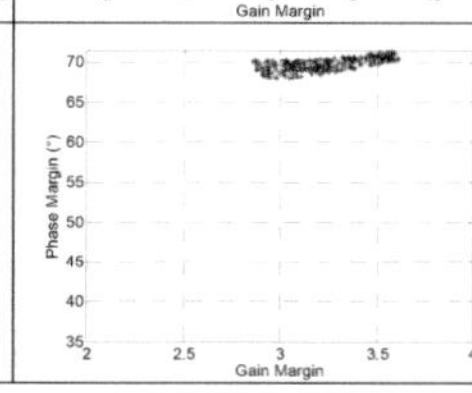	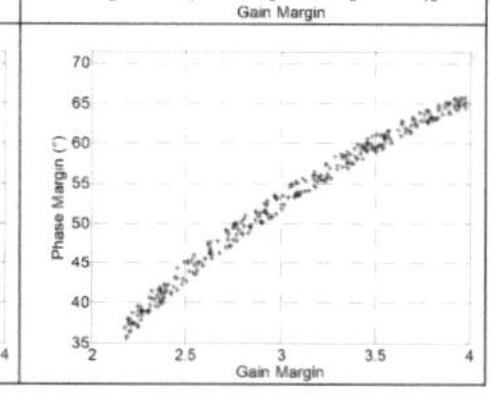

Table 20. Monte-Carlo simulations

Types of models	Tuning method	Overshoot (%)			Adjustment time (s)			Gain margin			Phase margin (°)		
		Scope	Mean	Variance	Scope	Mean	Variance	Scope	Mean	Variance	Scope	Mean	Variance
G_{P1}	DDE method	0.00-3.20	0.35	0.0000	6.76-8.77	7.47	0.08	2.03-3.03	2.50	0.05	65.72-70.76	68.23	1.08
	GPM method	0.00-14.04	5.75	0.0009	3.43-7.11	5.75	0.73	2.46-3.78	3.02	0.06	53.68-65.33	60.11	6.52
G_{P2}	DDE method	0.05-9.19	1.61	0.0003	4.08-8.61	5.53	0.89	2.69-4.48	3.46	0.12	67.33-71.59	69.46	0.85
	GPM method	3.14-22.26	12.79	0.0021	5.43-9.16	6.77	0.33	2.44-3.81	3.02	0.08	53.78-67.15	60.07	9.24
G_{P3}	DDE method	0.00-0.35	0.01	0.0000	15.94-18.79	17.13	0.41	3.23-10.05	6.04	2.65	30.71-75.76	71.34	84.1
	GPM method	5.38-26.78	16.63	0.0023	9.25-17.28	12.66	5.83	1.70-5.94	3.52	1.19	27.95-80.70	61.67	196
G_{P4}	DDE method	0.00-1.28	0.09	0.0000	7.19-11.02	9.78	1.54	2.86-3.60	3.23	0.04	68.07-71.38	69.80	0.58
	GPM method	10.35-25.67	18.10	0.0013	9.73-17.88	13.45	7.60	2.17-3.98	3.06	0.30	35.61-65.80	52.75	76.8

Table 21. Comparison of performance index

The detailed comparison is shown in table 22. Obviously, DDE method has better performance than GPM method. Especially in time-domain, DDE method has nearly zero overshoot and equivalent adjustment time compared with GPM method. In most industry field, the unknown model is inevitable, the simple tuning method, small overshoot and good performance robustness are needed. So the 2-DOF DDE method is available for industry field to meet the high performance requirement.

		DDE Method	GPM Method
Controller Structure		2-DOF	1-DOF
Approximation of Model		No	Yes
Demand of Model		Relative Order	Precise
Complicacy of Tuning Method		Simple	Simple
Design Basis		Time-domain	Frequency-domain
Overshoot		Small	Large
Performance Robustness	Time-domain	Good	Bad
	Frequency-domain	Mostly Good	Mostly Bad

Table 22. Comparison of DDE method and GPM method

4. Conclusions

Combined the Monte-Carlo method, this chapter gives a new method to test the performance robustness of PID control system. This method do not need complex mathematical reasoning, but the simple simulations and visible results are easy to be accepted by engineers. The large numbers of simulations have been done to study the performance robustness of different PID tuning method with the proposed criterion. We can see that the IMC method and GPM method are superior to other classical method. Then the DDE method which does not base on precise model is compared with IMC method and GPM method. The simulation results show that the DDE method perform better than the other two methods in general, especially on the models which the IMC method and GPM method have to design controllers based on approximate model. So, the proposed performance robustness criterion is effective to test PID type controller.

Although PID control is the most popular control method in the industry field, the advanced control theory is developing all the time. We are making effort to apply proposed performance robustness criterion on other type controller.

5. Acknowledgment

This research is supported by the National Natural Science Foundation of China #51176086. The authors would like to thank Ms. Irena Voric for disposing the chapter proposal issue and thank reviewers for their useful comments.

6. References

Åström, K.J. & Hägglund, T. (1984). Automatic Tuning of Simple Regulators with Specifications on Phase and Amplitude Margins. *Automatica*, Vol.20, No.5, (September 1984),pp. 645-651, ISSN 0005-1098

Chien, I.L. & Fruehauf, P.S. (1990). Consider IMC tuning to improve controller performance. *Chemical Engineering Progress*, Vol.86, No.10, (October 1990),pp. 33-41, ISSN 0360-7275

Chien, K.L.; Hrones, J.A. & Reswick, J.B. (1952). On the automatic control of generalized passive systems. *Transaction of the ASME*, Vol.74, No.2, (February 1952),pp. 175-185

Cui, G.; Cai, Z. & Li, M. (2000). Simulation for Heat-transfer and Flow of Continuous Fluid by Direct Simulation Monte Carlo Method. *Journal of Engineering Thermophysics*, Vol.21, No.4, (July 2000),pp. 488-490, ISSN 0253-231X

Ding, M. & Zhang, R. (2000). Monte-Carlo Simulation of Reliability Evaluation for Composite Generation and Transmission System. *Power System Technology*, Vol.24, No.3, (March 2000),pp. 9-12, ISSN 1000-3673

Ho W.K.; Hang C.C. & Cao L.S. (1995). Tuning of PID Controllers Based on Gain and Phase Margin Specifications. *Automatica*, Vol.31, No.3, (March 1995),pp. 497-502, ISSN 0005-1098

Lu, G. & Li, R. (1999). Monte Carlo Computer Simulation and Its Application in Fluid Theory. *Journal of the University of Petroleum China*, Vol.23, No.3, (June 1999),pp. 112-116, ISSN 1673-5005

Quevedo, J. & Escobet, T. (2000). Digital Control: Past, Present and Future of PID Control (PID'00). *Proceedings volume from the IFAC Workshop*, ISBN 0-08-043624-2, Terrassa, Spain, 5-7 April 2000

Sun, F.; Xia, X. & Liu, S. (2001). Calculation of Spacecraft Temperature Field by Monte Carlo Method. *Journal of Harbin Engineering University*, Vol.22, No.5, (October 2001),pp. 10-12, ISSN 1006-7043

Wang, W.; Li, D.; Gao, Q. & Wang, C. (2008). Two-degree-of-freedom PID Controller Tuning Method. *Journal of Tsinghua University*, Vol.48, No.11, (November 2008),pp 1962-1966, ISSN 1000-0054

Xue, D. (2000). *Feedback control system design and analysis*, Tsinghua University, ISBN 7-302-00853-1, Beijing, China

Yuan, M. (1999). Monte Carlo Evaluation Method for Robustness of Heat Exchanger. *Chemical Equipment Technology*, Vol.20, No.2, (April 1999),pp. 33-35, ISSN 1007-7251

Ziegler, J.G. & Nichols, N.B. (1943). Optimum setting for automatic controllers. *Journal of Dynamic Systems Measurement and Control-Transactions of the ASME*, Vol.115, No.2B, (June 1993),pp. 220-222, ISSN 0022-0434

10

Robust LMI-Based PID Controller Architecture for a Micro Cantilever Beam

Marialena Vagia and Anthony Tzes
University of Patras, Electrical and Computer Engineering Department
Greece

1. Introduction

Electrostatic Micro-Electro-Mechanical Systems (MEMS), are mechanical structures, consisting of mechanical moving parts actuated by externally induced electrical forces (Towfighian et al., 2011). The use of electrostatic actuation, is interesting, because of the high energy densities and large forces developed in such microscale devices (Chu et al., 2009; Vagia & Tzes, 2010b). For that reason, electrostatic micro actuators have been used in the fabrication of many devices in recent years, such as capacitive pressure sensors, comb drivers, micropumps, inkjet printer heads, RF switches and vacuum resonators.

Amongst different types of electrostatic micro actuators (Towfighian et al., 2010), electrostatic micro cantilever beams ($E\mu Cbs$) are considered as the most popular resonators. They can be extremely useful for a wide variety of tuning applications such as atomic force microscope (AFM), sensing sequence-specific DNA, detection of single electron spin, mass and chemical sensors, hard disk drives etc.

Accurate modeling of $E\mu Cb$s can be a challenging task, since such micro-systems suffer from nonlinearities that are due to the structural characteristics, the electrostatic force and the mechanical-electrical effects that are present. In addition, there exist more effects that play a dominant role especially in systems of narrow micro cantilever beams undergoing large deflections (Rottenberg et al., 2007). In such structures, the effects of the fringing fields on the electrostatic force are not negligible because of the non zero thickness and finite width of the beam (Gorthi et al., 2004; Younis et al., 2003). Thus, the incorporation of the fringing field capacitance, while modeling $E\mu Cb$s is mandatory. In that case the inclusion of the effects of the fringing field capacitance gives a more complicated but on the other hand a more accurate model of the cantilever beams.

Another important phenomenon appears with the interaction of the nonlinear electrostatic force with the linear elastic restoring one, and is called the "pull-in" phenomenon preventing the electrodes from being stably positioned over a large distance. The "pull-in" phenomenon restricts the allowable displacement of the moving electrodes in $E\mu Cb$'s systems operating in open-loop mode. For that reason, extending the travel range of $E\mu Cb$s is essential, in many practical applications including optical switches, tunable laser diodes, polychromator gratings, optical modulators and millipede data storage systems (Cheng et al., 2004; Towfighian et al., 2010). In order to achieve this extension in attracting mode beyond the conventional one-third of the capacitor beam's gap, researchers have used various methods including charge and current control, and leveraged bending. However, despite the different control approaches proposed until now (Nikpanah et al., 2008; Vagia & Tzes, 2010b), the control scheme to be applied on a micro-structure needs to be simple enough in order to be realizable in CMOS technology so that it can be fabricated on the same chip, next to switch.

In the present study, rather than relying on the design of non–linear control schemes, simplified linear optimal robust controllers (Sung et al., 2000; Vagia et al., 2008) are proposed. The design relies on the linearization of the $E\mu CB$'s nonlinear model at various multiple operating points, prior to the design of the control technique. For the resulting multiple linearized models of the $E\mu Cb$ a combination of optimal robust advanced control techniques in conjunction with a feedforward compensator are essential, in order to achieve high fidelity control of this demanding structure of the $E\mu Cb$ system. The proposed control architecture, relies on a robust time-varying PID controller. The controller's parameters are tuned within an LMI framework. A set of linearized neighboring sub–systems of the nonlinear model are examined in order to calculate the controller's gains. These gains guarantee the local stability of the overall scheme despite any switching between the linearized systems according to the current operating point. In order to enhance the performance of the closed–loop system, a set of PID controllers can be provided. The switching amongst members of the set of the PID controllers depends on the operating point. Each member of this set stabilizes the current linearized system and its neighboring ones. Through this overlapping stabilization of the linearized systems, the $E\mu Cb$'s stability can be enhanced even if the dwell time is not long enough.

The rest of this article is organized as follows, the modelling procedure for a $E\mu Cb$ is presented in Section 2. In Section 3, the proposed controller design procedure is described while in Section 4 simulation studies are carried, in order to prove the effectiveness of the proposed control technique. Finally in Section 5 the Conclusions are drawn.

2. Modeling of the electrostatic micro cantilever beam with fringing effects

The electrostatically actuated $E\mu CB$ is an elastic beam suspended above a ground plate, made of a conductive material. The cantilever beam moves under the actuation of an externally induced electrostatic force. The conceptual geometry of an electrostatic actuator composed of a cantilever beam separated by a dielectric spacer of the fixed ground plane is shown in Figure 1.

In the above Figure, ℓ, w, h are the length, the width and the thickness of the beam, η is the vertical displacement of the free end end from the relaxed position, and $\eta^{\max}$ is the initial thickness of the airgap between the moving electrode and the ground and F_{el} is the electrically-induced force between the two electrodes (Sun et al., 2007; Vagia & Tzes, 2010a).

The governing equation of motion of the $E\mu CB$ presented in Figure 1, is obtained, if considering that the mechanical force of the beam is modeled in a similar manner to that of a parallel plate capacitor with a spring and damping element (Batra et al., 2006; Pamidighantam et al., 2002).

The dynamical equation of motion due to the mechanical, electrostatic and damping force is equal to:

$$m\ddot{\eta} + b\dot{\eta} + k\eta = F_{el} \tag{1}$$

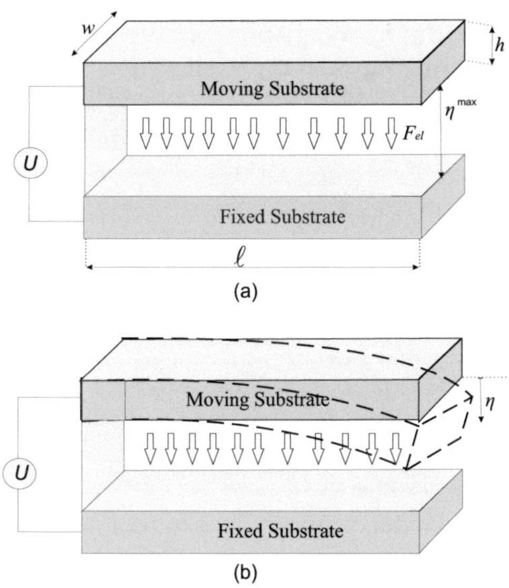

Fig. 1. Electrostatic Micro Cantilever Beam architecture

where m is the beam's mass, k is the spring's stiffness, b is the damping caused by the motion of the beam in the air.

2.1 Electrical force model

In a system of an $E\mu Cb$ composed of a cantilever beam separated by a dielectric spacer from the ground, the developed electrostatic force pulls the beam towards to the fixed ground plane as presented in Figure 1(b). The electrostatic attraction force F_{el} can be found by differentiating the stored energy between the two electrodes with respect to the position of the movable beam and can be expressed as (Batra et al., 2006; Chowdhury et al., 2005):

$$F_{el} = -\frac{d}{d\eta}\left(\frac{1}{2}CU^2\right) \tag{2}$$

where C is the $E\mu Cb$'s capacitance and U is the applied voltage between the beam's two surfaces.

The cantilever beam shown in Figure 1 can be viewed as a semi-infinitely VLSI on-chip interconnect separated from a ground plane (substrate) by a dielectric medium (air). If the bandwidth-airgap ratio is smaller than 1.5, the fringing field component becomes the dominant one.

The capacitance C in Equation (2), can be written as (Rottenberg et al., 2007):

$$C = e_0 e_r \ell \left(\frac{w}{\eta^{\max}}\right) + 0.77 e_0 \ell + 1.06 e_0 e_r \ell \left(\frac{w}{(\eta^{\max} - \eta)}\right)^{0.25} + \\ 1.06 e_0 e_r \ell \left(\frac{h}{(\eta^{\max} - \eta)}\right)^{0.5} + 1.06 e_0 e_r w \left(\frac{\ell}{\eta^{\max} - \eta}\right)^{0.25} \tag{3}$$

where e_0 is the permittivity of the free space and e_r is the dielectric constant of the air.

The first term on the right-hand side of Equation (3), describes the parallel-plate capacitance, the third term expresses the fringing field capacitance due to the interconnect width w, the fourth term captures the fringing field capacitance due to the interconnect thickness h and the fifth expresses the fringing field capacitance due to the interconnect length ℓ as shown in Figure 2.

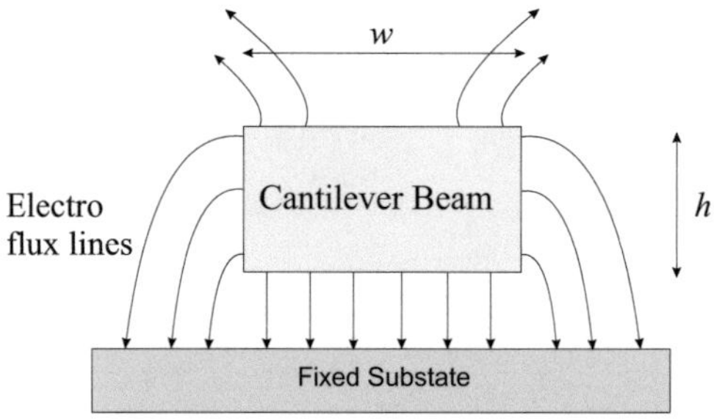

Fig. 2. Electric flux lines between the cantilever beam and the ground plane

After performing the differentiation of Equation (2) the electrical force is equal to:

$$F_{el} = \frac{e_0 w \ell U^2}{2(\eta^{\max} - \eta)^2} + \frac{0.1325 e_0 w^{0.25} \ell U^2}{(\eta^{\max} - \eta)^{1.25}} + \frac{0.265 e_0 h^{0.5} \ell U^2}{(\eta^{\max} - \eta)^{1.5}} + \frac{0.1325 e_0 w \ell^{0.25} U^2}{(\eta^{\max} - \eta)^{1.25}}. \tag{4}$$

The nonlinear equation of motion incorporating the expressions of the electrical and mechanical forces applied on the beam, is presented in Equation (5) as follows:

$$m\ddot{\eta} + b\dot{\eta} + k\eta = \frac{e_0 w \ell U^2}{2(\eta^{\max} - \eta)^2} + \frac{0.1325 e_0 w^{0.25} \ell U^2}{(\eta^{\max} - \eta)^{1.25}} + \frac{0.265 e_0 h^{0.5} \ell U^2}{(\eta^{\max} - \eta)^{1.5}} + \frac{0.1325 e_0 w \ell^{0.25} U^2}{(\eta^{\max} - \eta)^{1.25}}. \tag{5}$$

2.2 Linearized equations of motion

Equation (5) is a nonlinear equation due to the presence of the parameters η and U. All possible "equilibria"-points η_i^o, $i = 1, \ldots, M$ depend on the applied nominal voltage U_o. Equation (5) for $\ddot{\eta}_i^o = \dot{\eta}_i^o = 0$ and η_i^o yields:

$$k\eta_i^o = \underbrace{\frac{e_0 \ell w}{2(\eta^{\max} - \eta_i^o)^2}}_{k_{11}} U_o^2 + \underbrace{\frac{0.1325 e_0 \ell w^{0.25}}{(\eta^{\max} - \eta_i^o)^{1.25}}}_{k_{22}} U_o^2 + \underbrace{\frac{0.265 e_0 \ell h^{0.5}}{(\eta^{\max} - \eta_i^o)^{1.5}}}_{k_{33}} U_o^2 + \underbrace{\frac{0.1325 e_0 \ell^{0.25} w}{(\eta^{\max} - \eta_i^o)^{1.25}}}_{k_{44}} U_o^2 \Leftrightarrow$$

$$U_o = \pm \left[\frac{k\eta_i^o}{k_{11} + k_{22} + k_{33} + k_{44}} \right]^{1/2} \tag{6}$$

This nominal Uo-voltage must be applied if the beam's upper electrode is to be maintained at a distance $\eta_i^o \leq \frac{\eta^{\max}}{3}$ from its un-stretched position and equals to the feedforward compensator. This fact must be taken into account, as in the presented system, the "pull-in" phenomenon exists resulting to a single bifurcation point at $\eta^b = \frac{\eta^{\max}}{3}$. The resulting linearized systems that exist below this point are stable, while the linearized sub-systems above this limit are unstable. In the sequel Uo voltage, that keeps the system at η^b and will be referred to as the "bifurcation parameter".

The linearized equations of motion around the equilibria points $\left(U_o,\ \eta_i^o, \text{and } \ddot{\eta}_i^o = \dot{\eta}_i^o = 0\right)$ can be found using standard perturbation theory for the variables U and η_i where $U = U_o + \delta u$ $\eta_i = \eta_i^o + \delta\eta_i$. The linearized equation can be described as:

$$m\delta\ddot{\eta}_i + b\dot{\eta}_i + k\delta\eta_i + k\eta_i^o = \frac{e_0\ell w U_o^2}{2(\eta^{\max} - \eta_i^o)^2} + \frac{e_0 w\ell U_o^2}{(\eta^{\max} - \eta_i^o)^3}\delta\eta_i + \frac{e_0 w\ell U_o}{(\eta^{\max} - \eta_i^o)^2}\delta u + \frac{0.1325 e_0\ell w^{0.25}U_o^2}{(\eta^{\max} - \eta_i^o)^{1.25}} +$$

$$\frac{0.165 e_0 w^{0.25}\ell U_o^2}{(\eta^{\max} - \eta_i^o)^{2.25}}\delta\eta_i + \frac{0.265 e_0 w^{0.25}\ell U_o}{(\eta^{\max} - \eta_i^o)^{1.25}}\delta u + \frac{0.265 e_0\ell h^{0.5}U_o^2}{(\eta^{\max} - \eta_i^o)^{1.5}} + \frac{0.397 e_0 h^{0.5}\ell U_o^2}{(\eta^{\max} - \eta_i^o)^{2.5}}\delta\eta_i + \frac{0.53 e_0 h^{0.5}\ell U_o}{(\eta^{\max} - \eta_i^o)^{1.5}}\delta u +$$

$$\frac{0.1325 e_0\ell^{0.25}wU_o^2}{(\eta^{\max} - \eta_i^o)^{1.25}} + \frac{0.1625 e_0 w\ell^{0.25}U_o^2}{(\eta^{\max} - \eta_i^o)^{2.25}}\delta\eta_i + \frac{0.265 e_0 w\ell^{0.25}U_o}{(\eta^{\max} - \eta_i^o)^{1.25}}\delta u,\ i = 1,\ldots,M.$$

Substitution of:

$$k_i^a = k - \frac{e_0 w\ell U_o^2}{(\eta^{\max} - \eta_i^o)^3} - \frac{0.165 e_0 w^{0.25}\ell U_o^2}{(\eta^{\max} - \eta_i^o)^{2.25}} - \frac{0.397 e_0 h^{0.5}\ell U_o^2}{(\eta^{\max} - \eta_i^o)^{2.5}} - \frac{0.1625 e_0 w\ell^{0.25}U_o^2}{(\eta^{\max} - \eta_i^o)^{2.25}}$$

$$\beta_i = \frac{e_0 w\ell U_o}{(\eta^{\max} - \eta_i^o)^2} + \frac{0.265 e_0 w^{0.25}\ell U_o}{(\eta^{\max} - \eta_i^o)^{1.25}} + \frac{0.53 e_0 h^{0.5}\ell U_o}{(\eta^{\max} - \eta_i^o)^{1.5}} + \frac{0.265 e_0 w\ell^{0.25}U_o}{(\eta^{\max} - \eta_i^o)^{1.25}}, i = 1,\ldots,M.$$

yields to the final set of linearized equations describing the nonlinear system, for all different operating points:

$$m\delta\ddot{\eta}_i + b\delta\dot{\eta}_i + k_i^a\delta\eta_i = \beta_i\delta u, M = 1,\ldots,M. \tag{7}$$

The equations of motion describing the linearized subsystems, in state space form are equal to:

$$\begin{bmatrix}\delta\dot{\eta}_i\\ \delta\ddot{\eta}_i\end{bmatrix} = \begin{bmatrix}0 & 1\\ \frac{-k_i^a}{m} & \frac{-b}{m}\end{bmatrix}\begin{bmatrix}\delta\eta_i\\ \delta\dot{\eta}_i\end{bmatrix} + \begin{bmatrix}0\\ \frac{\beta_i}{m}\end{bmatrix}\delta u = \tilde{A}_i\begin{bmatrix}\delta\eta_i\\ \delta\dot{\eta}_i\end{bmatrix} + B_i\delta u,\ i = 1,\ldots,M$$

$$\delta\eta_i = [1\ \ 0]\begin{bmatrix}\delta\eta_i\\ \delta\dot{\eta}_i\end{bmatrix} = C\begin{bmatrix}\delta\eta_i\\ \delta\dot{\eta}_i\end{bmatrix}, i = 1,\ldots,M. \tag{8}$$

3. Switching robust control design

The design aspects of the used robust switching (Ge et al., 2002; Lam et al., 2002) LMI–based PID–controller comprised of $N+1$ "switched" PID sub-controllers, coupled to a feedforward controller (FC), as shown in Figure 3, will be presented in this Section.

The feedforward term provides the voltage U_0 from Equation (6) while the robust switching PID controller for the set of the M–linearized systems in Equation (8) is tuned via the utilization of LMIs (Boyd et al., 1994) and a design procedure based on the theory of Linear Quadratic Regulators (LQR).

This robust switching PID–controller is specially designed to address the case where multiple–system models have been utilized (Chen, 1989; Cheng & Yu, 2000; Hongfei & Jun, 2001; Narendra et. al., 1995; Pirie & Dullerud, 2002; Vagia et al., 2008) in order to describe the uncertainties that are inherent from the linearization process of the nonlinear system model.

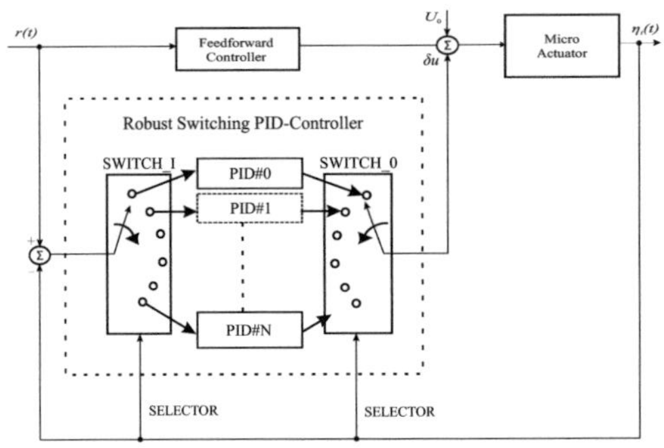

Fig. 3. Feedforward and Switching PID Control Architecture

The nature of the PID–structure in the controller design can be achieved if the linearized system's state vector $\delta\bar{\eta}_i = [\delta\eta_i, \delta\dot{\eta}_i]^T$ is augmented with the integral of the error signal $\int e_i dt = \int (r(t) - \eta_i(t))\, dt$. In this case, the augmented system's description is

$$\begin{bmatrix} \delta\dot{\bar{\eta}}_i \\ -e_i \end{bmatrix} = \hat{A}_i \begin{bmatrix} \delta\bar{\eta}_i \\ -\int e_i dt \end{bmatrix} + \begin{bmatrix} B_i \\ 0 \end{bmatrix} \delta u + \begin{bmatrix} -1 \\ 0 \end{bmatrix} r, \tag{9}$$

where $\hat{A}_i = \begin{bmatrix} \tilde{A}_i & 0 \\ 1 & 0 \end{bmatrix}$.

The LQR–problem for each system $(i = 0, \ldots, M)$ described in Equation (9) can be cast in the computation of δu in order to minimize the following cost:

$$J(\delta u) = \int_0^\infty (\delta\tilde{\eta}_i{}^T\, Q\, \delta\tilde{\eta}_i + \delta u^T\, R\, \delta u) dt \tag{10}$$

where $\delta\tilde{\eta}_i = \left[\delta\bar{\eta}_i, -\int e_i dt\right]^T$ is the state vector of the augmented system, and Q, R are semidefinite and definite matrices respectively. If a single PID-controller was desired $(N = 0)$, then the solution to the LQR problem relies on computing a common Lyapunov matrix that satisfies the Algebraic Ricatti Equations (AREs):

$$\hat{A}_i{}^T P + P\hat{A}_i - PB_i R^{-1} B_i{}^T P + Q = 0, i = 0, \ldots, M. \tag{11}$$

Rather than using the $\hat{A}_i$–matrices in the LQR–problem, the introduction of the auxiliary matrices $A_i = \hat{A}_i + \Lambda\mathbf{I}$, where $\Lambda > 0$ and $\mathbf{I}$ the identity matrix generates an optimal control $\delta u = -S\delta\tilde{\eta}$ such that the closed–loop's poles have real part less than $-\Lambda$, or $\Re(\mathrm{eig}(\hat{A}_i - B_i S)) < -\Lambda\ \forall i \in \{0, \ldots, M\}$.

The switching nature of the PID–controller is based on the following principle. Under the assumption of $M + 1$ linearized systems and $N + 1$ available PID controllers $(N \leq M)$, the objective of $\jmath$th PID–controller is to stabilize the $\jmath$–th system $\jmath \in \{0, \ldots, N\}$ and its 2Δ-neighboring ones $\jmath - \Delta, \ldots, \jmath - 1, \jmath, \jmath + 1, \ldots, \jmath + \Delta$, where Δ is an ad-hoc designed parameter related to the range of the affected neighboring subsystems.

If the stability-issue is the highest consideration, thus allowing for increased conservatism, only one $(N + 1 = 1)$ controller is designed for all $M + 1$ subsystems, or $\jmath - \frac{M}{2}, \ldots, \jmath, \ldots, \jmath + \frac{M}{2}$ (under the assumption that $\Delta = \frac{M}{2}$). This fixed time–invariant controller $(\delta u = -S\delta\tilde{\eta})$

stabilizes any linearized system (A, B) within the convex hull defined by the $(A_i, B_i), i = 0, \ldots, M$ vertices, or $(A, B) \in \text{Co}\{(A_0, B_0), \ldots, (A_M, B_M)\}$.

There is no guarantee, that this fixed linear time–invariant controller when applied to the nonlinear system will stabilize it, nor that it can stabilize the set of all linearized systems when switchings of the control occur. The promise is that when there is a slow switching process, then this single PID controller will stabilize any switched linear system $(A(t), B(t)) \in \text{Co}\{(A_i, B_i), i = 0, \ldots, M\}$.

Furthermore if M increases then the approximation of the nonlinear system by a large number of linearized systems is more accurate. This allows the interpretation of the solution to the system's nonlinear dynamics

$$\delta\dot{\tilde{\eta}} = f(\delta\tilde{\eta}) + g(\eta, \delta u) \tag{12}$$

as a close match to the solution of the system's time-varying linearized dynamics

$$\delta\dot{\tilde{\eta}} = A(t)\delta\tilde{\eta} + B(t)\delta u. \tag{13}$$

The increased conservatism stems from the need to stabilize a large number of systems with a single controller, thus limiting the performance of the closed loop system.

In order to enhance the system's performance, multiple controllers can be used; each controller needs not only to stabilize the current linearized system but also its neighboring ones thus providing increased robustness against switchings at the expense of sacrificing the system's performance.

In a generic framework, the $\jmath$th robust switching–PID controller's objective is to optimize the cost in (10) while the $\jmath$th–linearized system is within

$$\text{Co}\{(A_i, B_i), i \in \{\jmath - \Delta, \ldots, \jmath + \Delta\}\}. \tag{14}$$

Henceforth , the needed modification to (11) is the adjustment of the spam of the systems from $\{0, \ldots, M\}$ to $\{\jmath - \Delta, \ldots, \jmath + \Delta\}$. It should be noted that the optimal cost at Equation (10) is equal to $\delta\tilde{\eta}^T(0)\hat{P}^{-1}\delta\tilde{\eta}(0)$ for a P-matrix satisfying (11). An efficient alternative solution for the optimal control $\delta u = -S\delta\tilde{\eta}$ can be computed by transforming the aforementioned optimization problem, subject to the concurrent satisfaction of the AREs in Equation (11), into an equivalent LMI–based algorithm, where a set of auxiliary matrices $\hat{P}$, Y and an additional variable γ $(\gamma > 0)$ have been introduced.

The γ–variable is used as an upper bound of the cost, or

$$\delta\tilde{\eta}^T(0)\hat{P}^{-1}\delta\tilde{\eta}(0) \leq \gamma. \tag{15}$$

Therefore the optimal control problem amounts to the minimization of γ subject to the satisfaction of the AREs in (11). The optimal control $\delta u = -S\delta\tilde{\eta}$ is encapsulated in the following formulation which is amenable for solution via classical LMI–based algorithms; relying on Schur's complement (Boyd et al., 1994), and the introduction of a set of auxiliary matrices $\hat{P}$, Y and an additional variable γ $(\gamma > 0)$ the controller computation problem is transformed to:

$$\min \gamma$$

$$\text{subject to} \begin{cases} \begin{bmatrix} \gamma & \delta\tilde{\eta}^T(0) \\ \delta\tilde{\eta}(0) & \hat{P} \end{bmatrix} \leq 0 \\ \begin{bmatrix} A_i\hat{P} + \hat{P}A_i^T + B_iY + Y^TB_i^T & \hat{P} & Y^T \\ \hat{P} & -Q^{-1} & 0 \\ Y & 0 & -R^{-1} \end{bmatrix} \leq 0\,, \\ \text{for } i = \jmath - \Delta, \ldots, \jmath + \Delta \\ \hat{P} > 0\,. \end{cases}$$

The feedback control can be computed based on the recorded values of $\hat{P}^*$ and Y^* for the last feasible solution:

$$\delta u = Y^*(\hat{P}^*)^{-1}\delta\tilde{\eta} = -S\delta\tilde{\eta} = -\left[\, s_p | s_d | s_i \,\right] \begin{bmatrix} \delta\eta_i \\ \delta\dot{\eta}_i \\ -\int edt \end{bmatrix}$$
$$= \left[s_p e + s_d \dot{e} + s_i \int edt \right] + \left[s_p \left(\eta_i^o - r\right) - s_d \dot{r} \right]. \tag{16}$$

The first portion of the controller form in (16) is equivalent to that of a PID–controller. It should be noted that the operating points are

$$\eta_i{}^o = \eta_0{}^{\min} + \frac{\eta_M{}^{\max} - \eta_0{}^{\min}}{M} \cdot i = \eta_0{}^{\min} + W \cdot i, \;\; i = 0, \ldots, M, \tag{17}$$

where W is the distance related to the separation of the operating points. The $\jmath$th locally stabilizing PID controller stabilizes the linearized systems that are valid over the interval

$$\left[\eta_\jmath{}^{\min}, \eta_\jmath{}^{\max}\right) = \left[\eta_{j-\Delta}^o - \frac{W}{2}, \eta_{j+\Delta}^o + \frac{W}{2}\right) \bigcup \cdots \bigcup$$
$$\left[\eta_j^o - \frac{W}{2}, \eta_j^o + \frac{W}{2}\right) \bigcup \cdots \bigcup \left[\eta_{j+\Delta}^o - \frac{W}{2}, \eta_{j+\Delta}^o + \frac{W}{2}\right). \tag{18}$$

Essentially the resulting PID structure is equivalent to that of an overlapping decomposition controller. The region of validity for each controller with respect to the available travel distance of the $E\mu Cb$ appears in Figure 4. Small number of W and Δ lead to smaller regions of validity with insignificant overlapping (i.e., when $\Delta = 0$ there is no overlapping and each controller is responsible for the region $\left[\eta_j^o - \frac{W}{2}, \eta_j^o + \frac{W}{2}\right)$)

For the travel-distances where there is overlapping the PID-controller maintains its gains, and when the beam moves out of the boundaries of that region the PID controller readjusts its gains. To exemplify this issue, consider the motion of the $E\mu Cb$ as shown in Figure 5.

The controller's switching mechanism starts with the set of gains of the $(\jmath - 1)$th controller for $\eta(t) \in \left[\eta_{\jmath-2}^{\max}, \eta_{\jmath-1}^{\max}\right)$. At time $t = t_1$, when $\eta(t) = \eta_{\jmath-1}^{\max}$ the controller switches to its new $(\jmath)$th controller and maintains this set of gains till time t_2. For $t \geq t_2$, or when $\eta(t) \geq \eta_\jmath^{\max}$ the $(\jmath + 1)$th controller is activated, until time instant t_3 at which $\eta(t) = \eta_{\jmath+1}^{\min}$. For $t_3 < t \leq t_4$, or $\eta_{\jmath+1}^{\min} < \eta(t) \leq \eta_\jmath^{\min}$ the $(\jmath)$th controller is activated. It should be noted that each controller

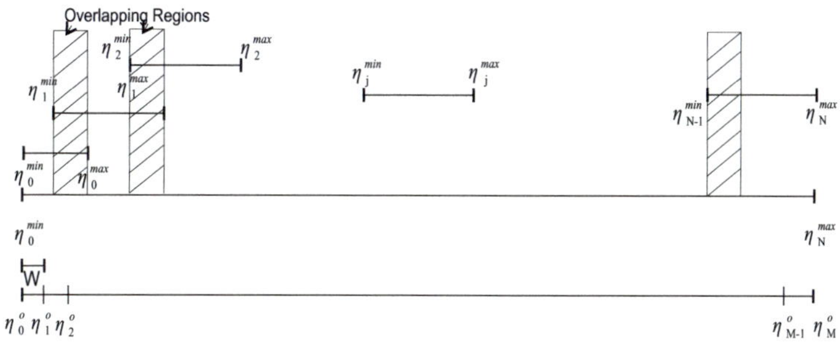

Fig. 4. Controllers' Regions of Validity

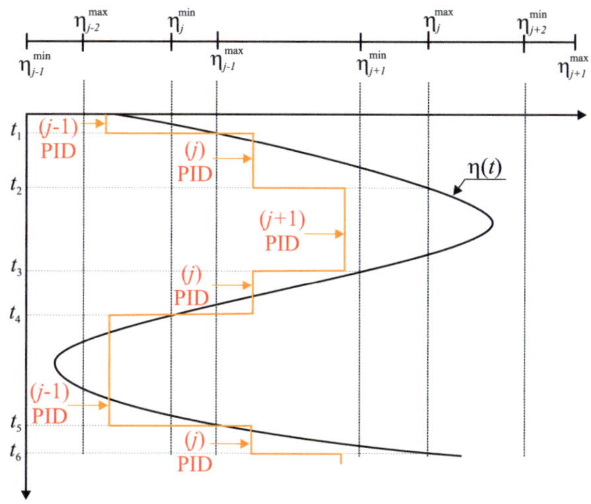

Fig. 5. PID- controller Gain Switching Example

is activated in a manner that resembles a "hysteresis"–effect. In the noted example, as $\eta(t)$ increases, the jth controller operates when $\eta(t) \in \left[\eta_{j-1}^{\max}, \eta_{j}^{\max}\right)$, while as $\eta(t)$ decreases the same controller operates when $\eta(t) \in \left[\eta_{j}^{\min}, \eta_{j+1}^{\min}\right)$.

In the suggested framework the control design needs to select the number of:

1. the number $M + 1$ of linearized systems (partitions)
2. the number $N + 1$ of the switched controllers
3. the "width" Δ of the "overlapping system stabilizations" of each controller and
4. the cost Q, R parameters and the Λ factor used to "speed up" the system's response.

In general Q and R are given, and ideally M is desired to be as large as possible. As far as the three parameters N, Δ and Λ there is a trade–off in selecting their values. Large N-values lead to superfluous controller switchings which may destabilize the system; small N typically leads to a slow–responding system thus hindering its performance. Large values of Δ increase the system's stability margin while decreasing the system's bandwidth (due to the need to simultaneously stabilize a large number of systems). The parameter Λ directly affects the speed of the system's response. From a performance point of view, large Λ-values are desired; however this may lead to an infeasibility issue in the controller design.

It should be noted that a judicious selection of these parameters is desired, since there are contradicting outcomes behind their selection. As an example, large values of N leads to a faster performance at the expense of causing significant switchings caused by the transition of the controller's operating regime. Similarly, large values of Δ increase the systems's stability margin at the expense of decreasing its bandwidth which is also affected by the parameter Λ.

Practical considerations ask for an a priori selection of N and Δ while computing the largest Λ that generates a feasible controller.

4. Simulation studies

Simulation studies were carried on a $E\mu Cb$'s non–linear model. The parameters of the system unless otherwise stated are equal to those presented in the following Table.

parameter (Unit)	Description	Value
$w(m)$	Beam Width	7.5×10^{-6}
$h(m)$	Beam Height	1.2×10^{-6}
$\ell\ (m)$	Beam Length	100×10^{-6}
$\eta^{max}(m)$	Maximum Distance	4×10^{-6}
$\mu\ (kg\ m/sec^2)$	Viscosity Coefficient	18.5×10^{-6}
$\rho\ (kg/m^3)$	Density	1.155
$e_0\ (Coul^2/Nm^2)$	Dielectric constant of the air	8.85×10^{-12}
$P_a\ (N/m^2)$	Ambient Pressure	10^5
k (N/m)	Stiffness of the spring	0.249

The allowable displacements of the $E\mu Cb$ in the vertical axis: $\eta \in [0.1, 1.33]$ μm=$[\eta_o^{\min}, \eta_M^{\max}]$. This is deemed necessary in order to guarantee the stability of the linearized open–loop system and retain it, below the well known-bifurcation points. These are the points where the behavior of the system changes from stable to unstable and vice versa and can be easily found by setting the derivative of $\frac{\partial U_e}{\partial \eta}$ of the expression in Equation (6) equal to zero. It should be noted that as presented at Figure 6, the bifurcation point is equal to the extrema of the graph presented, at $\eta^b = 1.33\mu m = \frac{\eta^{\max}}{3}$.

As far as the controller's design parameters are concerned, different test cases were examined in order to prove the effectiveness of the suggested control scheme. Different test cases, regarding the values of M, N, Δ, Λ are examined in order to prove the relevance between them and the system's performance.

Each set of the parameters of the controller switches at the instants, when: a) there is a movement of the upper plate from its initial to its final position, and b) at the crossings of the boundaries $\eta_i^{\min}, \eta_i^{\max}$ where each linearized model is valid.

Figure 7 presents the nonlinear system's responses for different Λ-values when a single robust PID controller is designed. The goal of the controller was to move the beam's upper plate from an initial position to a new desired one (set–point regulation). In this case, 5-linearized subsystems were used in each case for the controller's design, and thus $M = 5$ and $N = 0$. As expected, the system responds faster in the cases where the Λ-value is higher, since it is guaranteed that its closed–loop poles will be deeper in the LHP.

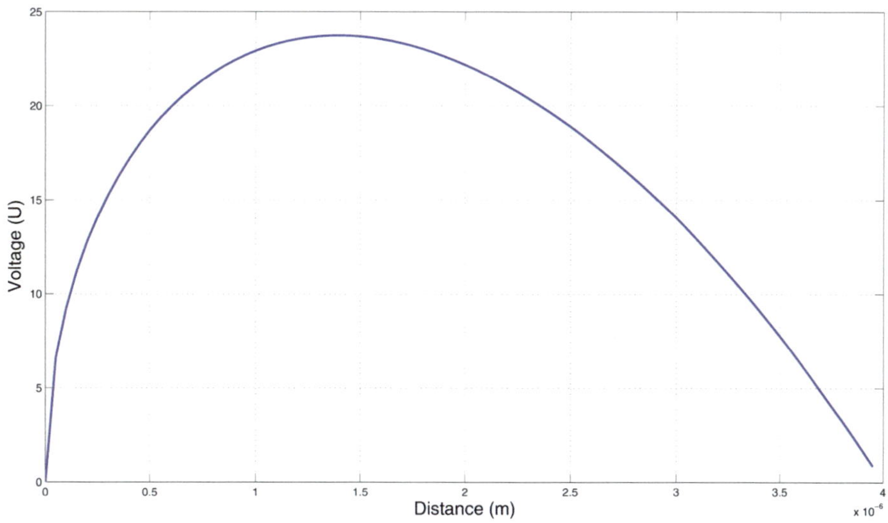

Fig. 6. Bifurcation points of the System

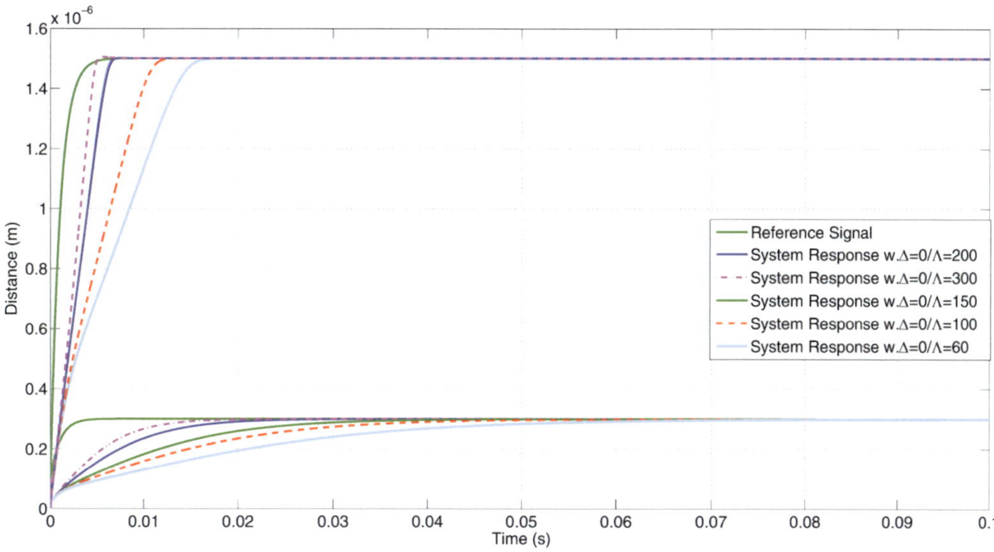

Fig. 7. Systems' Responses for different Λ-values

Another parameter to be examined is the number of the operating points (M value), and its effect on the system's performance. Figure 8 (9) presents the responses (control efforts) of the system when $M = 1, 5, 10$ and $N = 0$. Comparing the systems' responses in an apparent performance improvement is observed when using more operating points. However, due to the continuous switchings between the operating regimes, the control effort in the latter case ($M = 10$) is quite "noisy" and might cause significant aging on the beam's moving electrode.

In the sequel Figure 10 presents the responses of the system for different N-values ($N + 1 = 1, 4, 10$). The other parameters of the controller equal to: $\Delta = 0$ and $M = 5$ for all the three cases. The number of the switchings between the different designed PID-controllers

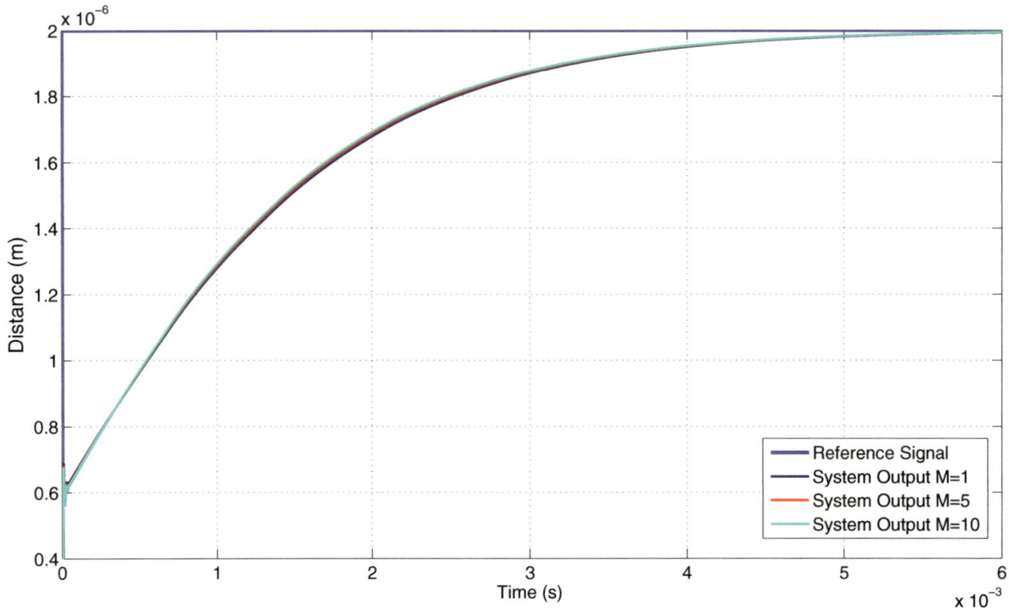

Fig. 8. Systems' Responses for different M values

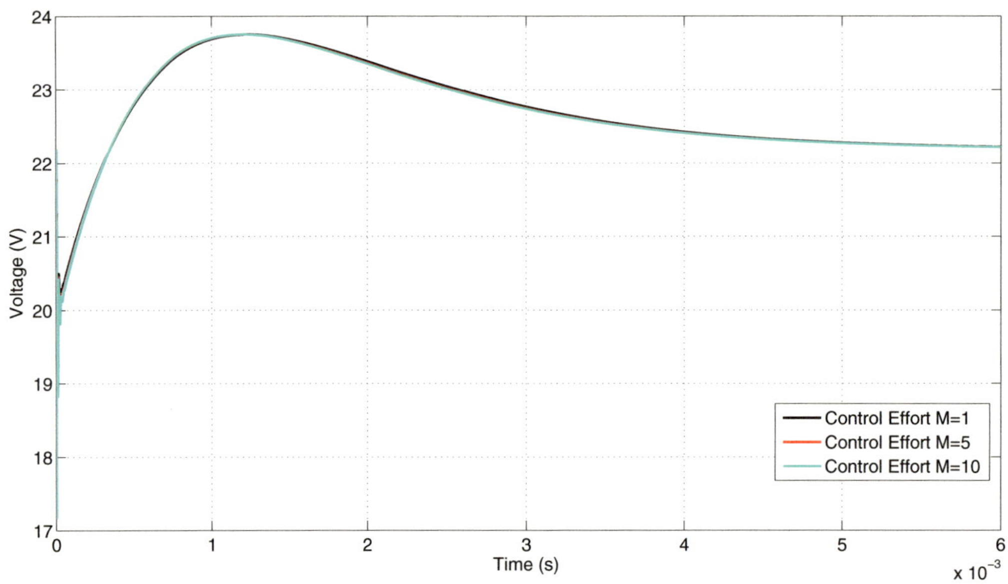

Fig. 9. Control Efforts for different M values

has a great impact on the system's output. The grater the number of N the faster the system becomes. On the other hand, an increase of N-values makes the system's response more oscillatory. Therefore the control law designed need to take into consideration, the trade off that exists between the velocity and the performance of the system when more controllers are used. Figure 11 presents the control efforts of the system that are in full harmony with the previous mentioned results.

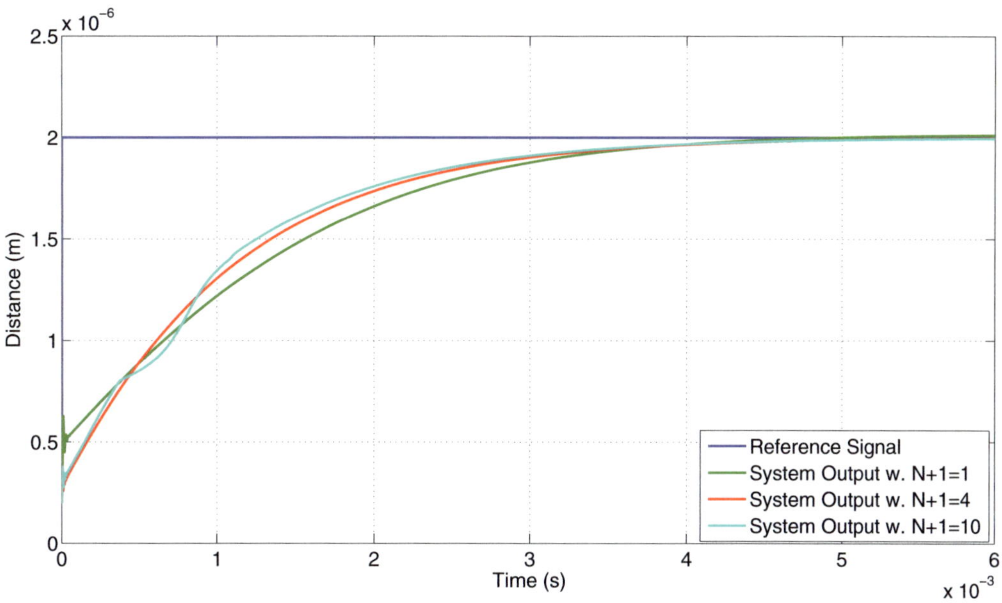

Fig. 10. Systems' Responses for different N-values

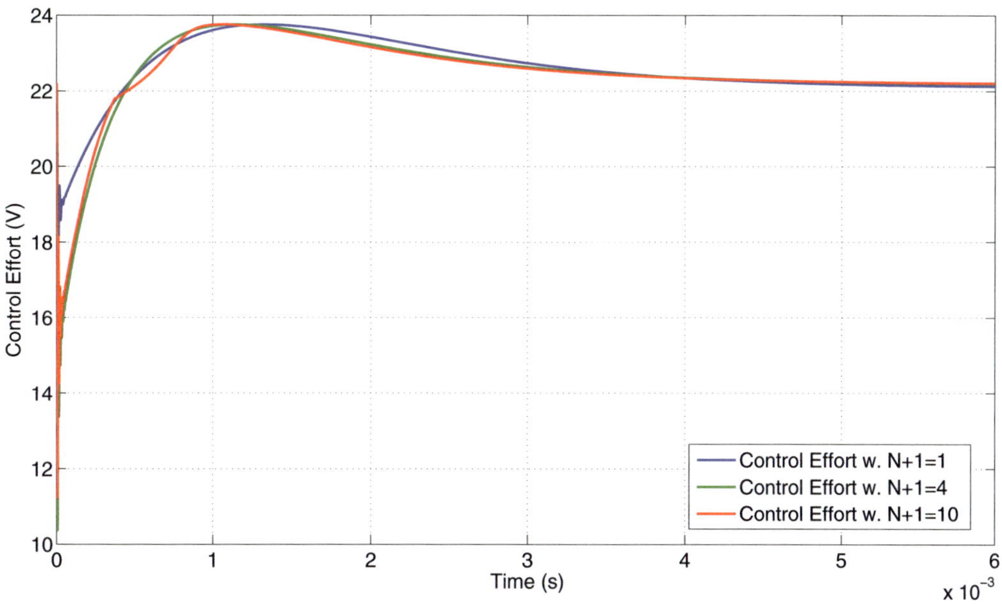

Fig. 11. Control Efforts for different N values

In Figure 12, the responses of the $E\mu Cb$'s non-linear model are presented for different values of Δ. Thirteen ($M+1$=14) operating points were selected at $\eta_i^o = \eta_0^{\min} + W \cdot i$, where $W = 0.1\mu m$ and $i \in \{0, \ldots, 13\}$. Three test-cases were examined as far as the number of the switched controllers: a) $N+1 = 13$, b) $N+1 = 9$ and c) $N+1 = 1$. For the first case there are no overlapping regions, thus ($\Delta = 0$). For the second case, there are three overlapping regions

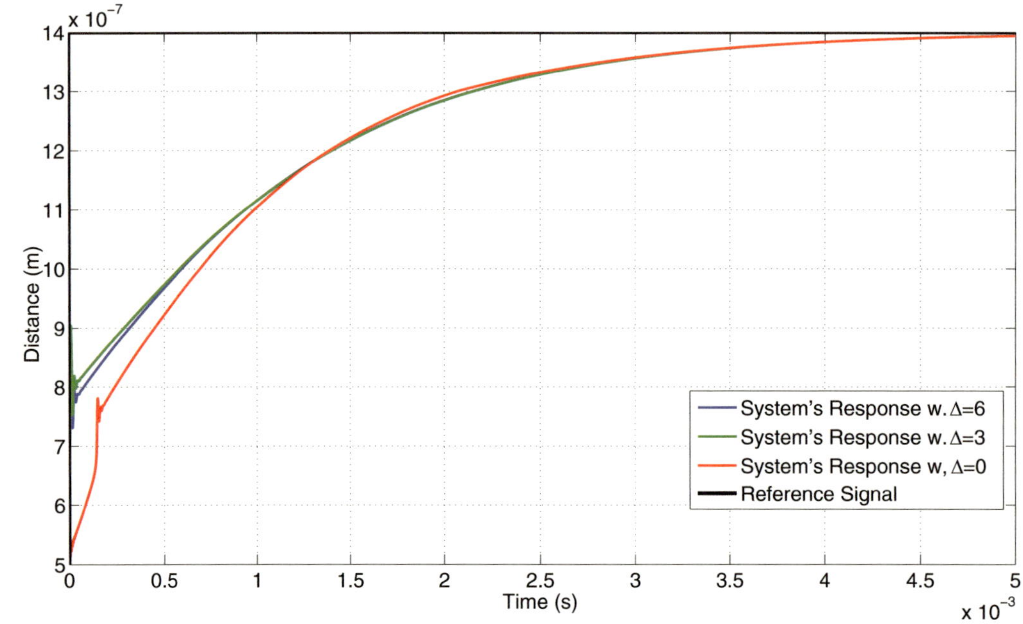

Fig. 12. Systems' Responses for different Δ values

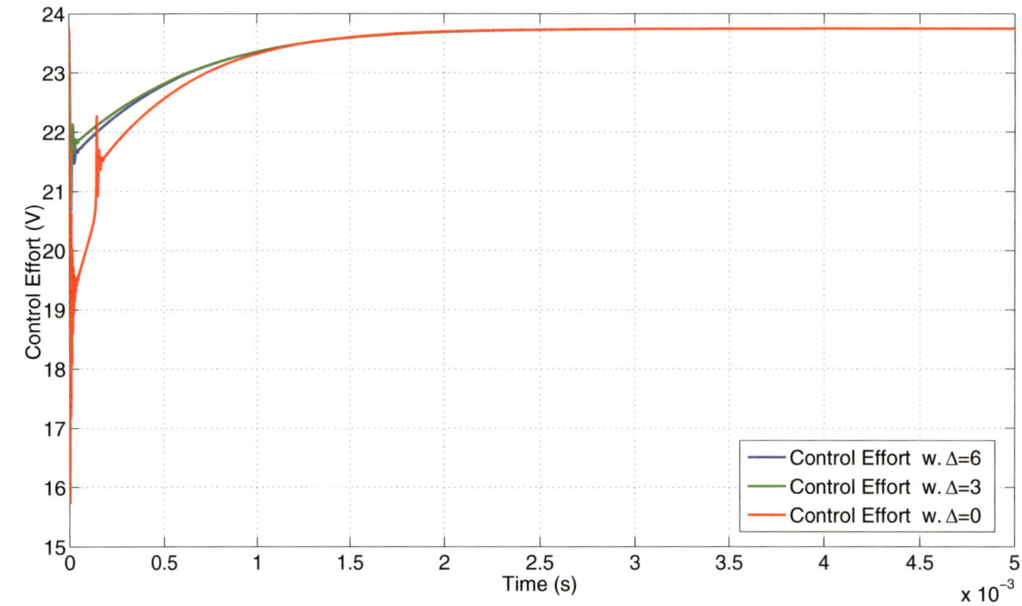

Fig. 13. Control Efforts for different Δ values

($\Delta = 3$) around each operating point. For the last case, where only one controller is used ($\Delta = 6$) this controller's region of validity is:

$$\left(\eta_6^0 - \Delta W - \frac{W}{2}, \eta_6^0 + \Delta W + \frac{W}{2}\right). \tag{19}$$

.

Figures 12 and 13 present the corresponding systems' responses and control efforts. In the cases where the Δ value is higher, the system's response becomes slower but the oscillations are diminished. This is also apparent from a direct comparison between the control effort shown in Figure 13.

5. Conclusion

In this article a robust switching control scheme is firstly designed, and then applied on the system of an $E\mu Cb$. The control architecture consisting of several robust switching PID controllers tuned with the utilization of the LMI technique, in conjunction with a feedforward term, is applied on the nonlinear beam's system. In an attempt to address the performance, the switching PID-controllers are designed in order to push the poles deep inside the LHP. The resulting scheme relies on a minimization procedure subject to the satisfaction of several LMI-constraints. Several test cases are provided in order to find any possible relevance between the different values used during the controller design procedure. Simulation studies prove the efficiency of the suggested scheme and highlight the provoked indirect effects caused by the frequency switchings of the time-varying control architecture.

6. References

Batra, R.C.; Porfiri, M. & Spinello, D. (2006). Capacitance estimate for electrostatically actuated narrow microbeams. *Micro & Nano Letters IET* Vol 1., No. 2., pp. 71-73.

Boyd, S.; Ghaoui, L.El; Feron, E. & Balakrishnan, V. (1994). Linear Matrix Inequalities in System and Control Theory. *SIAM*, Philadelphia, PA.

Chen C., (1989). A simple method for on-line identification and controller tuning. *AIChE*, Vol. 35., No. 12., pp. 2037–2039.

Cheng, Y. & Yu, C. (2000). Nonlinear process control using multiple models: relay feedback approach. *Industrial Eng.Chem*, Vol. 39., No. 2., pp. 420–431.

Cheng, J.; Zhe, J. & Wu, X. (2004). Analytical and finite element model pull-in study of rigid and deformable electrostatic microactuators. *Journal of Microelectronics and Microengineering*, Vol. 14., No. 1., pp. 57-68.

Chowdhury, S.; Ahmadi, M. & Mille, W. C. (2005). A closed-form model for the pull in voltage of electrostatically actuated cantilever beams. *Journal of Microelectronics and Microengineering*, Vol. 15., No. 4., pp. 756-763.

Chu, C, H.; Shih, W. P.; Chung, S. Y.; Tsai, H. C.; Shing, T. K. & Chang, P.Z. (2005). A low actuation voltage electrostatic actuator for RF MEMS switch applications. *Journal of Micromechanics and Microengineering*, Vol. 17., No.8., pp. 1649-1656.

Ge, M.; Chiu, M. & Wang, Q. (2002). Robust PID controller Design via LMI approach. *Journal of Process Control*, Vol. 12., No.1., pp. 3–13.

Gorthi, S.; Mohanty, A. & Chatterjee, A. (2004). Cantilever beam electrostatic MEMS actuators beyond pull-in. *Journal of Microelectronics and Microengineering*, Vol.14., No. 9., pp. 1800-1810.

Hongfei, S. & Jun, Z. (2001). Control Lyapunov functions for switched control systems. *Proceedings of the 2001 American Control Conference*, pp. 1890–1891, Arlington, Virginia, June 2001.

Lam, H. K.; Leung, F. H. F. & Tam, P. K. S. (2002). A switching controller for uncertain nonlinear systems. *IEEE Control System Magazine*, Vol. 22., pp. 7–14.

Narendra, K. S.; Balakrishnan, J. & Ciliz, K. (1995). Adaptation and Learning using multiple models, switching and tuning. *IEEE Control Systems Magazine*, Vol. 15., No. 3., pp. 37–51.

Nikpanah, M.; Wang, Y.; Lewis, F. & Liu, A. (2008). Real time controller design to solve the pull-in instability of MEMS actuator. *Proceedings of 10th Int. Conf. on Control, Automation, Robotics and Vision*, pp. 1724-1729, Hanoi, Vietnam.

Pamidighantam, S.; Puers, R.; Baert, K. & Tilmans, H. A. C. (2002). Pull in voltage analysis of electrostatically actuated beam structures with fixed-fixed and fixed-free end conditions. *Journal of Microelectronics and Microengineering*. Vol. 12., pp. 458-464.

Pirie, C. & Dullerud, G. (2002). Robust Controller synthesis for uncertain time-varying systems. *SIAM Journal on Control and Optimization*, Vol. 40. No. 4., pp. 1312-1331.

Rottenberg, X.; Brebels, S.; Ekkels, P.; Czarnecki, P.; Nolmans, P.; Mertens, R. P.; Nauwelaers, B.; Puers, R.; De Wolf, I.; De Raedt, W. & Tilmans, H. A. C. (2007). An electrostatic fringing-field actuator (EFFA): application towards a low-complexity thin-film RF-MEMS technology. *Journal of Micromechanics and Microengineering*. Vol. 17., No. 7., pp. 204-210.

Sun, D. M.; Dong, W.; Liu, C. X.; Chen, W. Y. & Kraft, M. (2007). Analysis of the dynamic behaviour of a torsional micro-mirror. *Journal of Microsystem Technology*. Vol. 13., No. 1., pp. 61-70.

Sung, L.; Yongsang, Q. K. & Dae-Gab, G. (2000). Continuous gain scheduling control for a micro-positioning system: Simple, robust and no overshoot responser. *Control Engineering Practice* Vol. 8., No. 2., pp. 133–138.

Towfighian, S.; Seleim, A.; Abdel-Rahman, E. M. & Heppler, G. R. (2010) Experimental validation for an extended stability electrostatic actuator. *Proceedings of ASME 2010 Int. Design Engineering Technical Conference.*, Montreal, QC, Canada .

Towfighian, S.; Seleim, A.; Abdel-Rahman, E. M. & Heppler, G. R. (2011) A large-stroke electrostatic micro-actuator. *Journal of Micromechanics and Microengineering*, Vol. 21., No. 7., pp. 1-12.

Vagia, M.; Nikolakopoulos, G., & Tzes, A. (2008). Design of a robust PID-control switching scheme for an electrostatic micro-actuator. *Control Engineering Practice*. Vol. 16., No. 11., pp. 1321-1328.

Vagia, M. & Tzes, A. (2010a). LMI-region Pole Placement Control Scheme for an Electrostatically Actuated micro Cantilever Beam. *the 18th Mediterranean Conference on Control and Automation (MED 10).*, Congress Palace, Marrakech, Morocco, pp. 231-236.

Vagia, M. & Tzes, A. (2010b). A Literature Review on Modeling and Control Design for Electrostatic Microactuators with Fringing and Squeezed Film Damping Effects. *Proceedings of American Control cofnerence 2010.*, Baltimore, Maryland, US, pp. 3390-3402.

Younis, M.; Abdel Rahman, E. M. & Nayfeh, A. H. (2003). A reduced order model for Electrically Actuated Microbeam-Based MEMS. *Journal of Microelectromechanical Systems.*, Vol. 12., No. 5., pp. 672-680.

Part 4

Disturbance Rejection for PID Controller Design

11

The New Design Strategy on PID Controllers

Wei Wang
Information School, Renmin University of China
P. R. China

1. Introduction

PID control is a classical control technique. Because of its simplicity and robustness, it is still extensively used in the control of many dynamical processes. The dominative status of PID control in engineering applications is unchanged even with the advances of modern control theories. However, owing to the uncertainty or complexity of the controlled systems, and the randomness of the external disturbances, PID control still faces a great challenge. How to design an effective PID controller as well as with simple architecture? To meet the requirements for the control of practical systems, a breakthrough should be made in the design strategy of PID controllers.

In this chapter, a brief summing-up will be made on the basic ideas which have been considered in the tuning of PID controllers. It is also intended to provide a summary on the new design strategy of PID controllers, which has been proposed in recent decades. It will be emphasized that the further improvements to the design method will be based on the advance of modern control theory rather than just based on a certain technique.

The chapter is organized as follows: In section 2, some of the conventional methods which have been extensively used on the tuning of PID controllers will be discussed. In section 3, it will be reviewed on the breakthrough made for PID controller designing, especially in China in recent years. It includes three aspects: Firstly, it is the method of signal processing by using tracking-differentiators (TDs), which is fundamental for improving PID control; Secondly, it is the nonlinear PID controller, which uses nonlinear characteristics to improve the performance of PID control; Thirdly, it is the method of active disturbance rejection control (ADRC) which can reject the uncertainties and disturbances by using an extended state observer (ESO), and can implement the output regulation effectively. In section 4, the new separation principle on PID controller tuning is discussed, in which the design of disturbance rejection and the design of effective output regulation can be carried out separately, and disturbances can be rejected without using any observer. Therefore it makes the design of PID controllers more effective and simple. And another new method based on the stripping principle is also discussed, which is an extension of the new separation principle to networked control systems. In section 5, some applications are provided to indicate the efficiency of the methods discussed in section 4. Followed by some conclusions.

2. A brief summary on the conventional methods of PID controller tuning

As been pointed out above, PID control is still being widely used in process control because of its robustness and with relatively simple structure, thousands of books and papers have been published (see (Åström & Hagglund, 1995), (Cong & Liang, 2009), (Datta et al., 2000), (Han, 2009), (Heertjes et al., 2009), (Hernandez-Gomez et al., 2010), (Leva et al., 2010), (Nusret & Atherton, 2006), (Santibanez et al., 2010), (Yaniv & Nagurka, 2004), etc.). However, the main problem in PID control is how to tune the parameters to adapt the controller to different situations ((Han, 2009), (Leva et al., 2010), (Liu & Hsu, 2010), (Sekara et al., 2009), (Toscano & Lyonnet, 2009)). In this section, the conventional methods on the tuning of PID controllers will be summarized.

2.1 The conventional PID control and the requirements

Classical PID control is an implementation of error-based feedback control. The basic principle is particularly rather primitive and simplified. It can be described by the following formula.

Suppose that $y_r(t)$ is the set output or reference input, and $y(t)$ is the actual output of the system. Then the error is $e(t) = y(t) - y_r(t)$, and the classical PID control input is the one as follows:

$$u(t) = -a_0 \int_{t_0}^{t} e(\tau)d\tau - a_1 e(t) - a_2 \dot{e}(t) \tag{1}$$

where a_0, a_1, and a_2 are the design parameters or the gains of integral, proportional, and derivative, respectively.

Before going into the discussion on conventional tuning methods, let's look at what are the purposes of the controller tuning. If possible, it would like to have both of the following for the control system:

- Fast responses, and
- Good stability (the overshoot should be limited to a certain extent).

Unfortunately, for most practical processes being controlled with a PID controller, these two criteria can not be achieved simultaneously ((Han, 1994), (Han, 2009), (Haugen, 2010)). In other words, the result will be

- The faster responses, the worse stability, or
- The better stability, the slower responses.

For a practical control system, it often shows that the output response will sway due to a step change of the setpoint.

And in most cases, it is important that having good stability is better than being fast. So, the acceptable stability (good stability, but not so good, as it gives too slow of a response) should be specified. That is to say, a way of trade-off between fastness and overshoot should be found ((Han, 1994), (Han, 2009)).

From the viewpoint of practical implementations, the most important of all is that the structure of the controller should be as simple as possible.

2.2 The tuning methods based on the knowledge of systems

There are a large number of tuning methods, but for covering most practical cases, there are three kinds of methods for calculating proper values of PID parameters, i.e. controller tuning. These methods are as follows ((Haugen, 2010)):

• The good gain method

It is a simple experimental method which can be used without any knowledge about the process to be controlled. It aims at obtaining acceptable stability as explained above. The method is a simple one which has proven to give good results on laboratory processes and on simulators.

However, if a process model can be obtained, the good gain method can be used on a simulator instead of in the physical process.

• Skogestad's method

It is a model-based tuning method. It is assumed that the mathematical model (a transfer function) of the process can be obtained. It does not matter how to derive the transfer function - it can stem from a model derived from physical principles, or from the calculation of model parameters (e.g. gain, time-constant and time-delay, etc.) from an experimental response, typically a step response experiment with the process.

With this tuning method, the controller parameters should be expressed as functions of the process model parameters.

• Ziegler-Nichols' methods

It is the ultimate gain method (or closed-loop method) and the process reaction curve method (the open-loop method).

The ultimate gain method has actually many similarities with the good gain method, but the former method has one serious drawback. Namely, it requires the control loop to be brought to the limit of stability during the tuning, while the good gain method requires a stable loop during the tuning. The Ziegler-Nichols' open-loop method is similar to a special case of Skogestad's method, and Skogestad's method is more applicable.

2.3 The methods based on improving the control performance

From another point of view, or according to the attention paid to improving control performance ((Åström & Hagglund, 2004), (Heertjes et al., 2009) ,(Sekara et al., 2009)), the tuning methods can be summarized as follows.

Firstly, the intelligent methods, such as neural networks, fuzzy theory, particle swarm optimization, etc., have been employed to tune the PID controllers, and some rules have been obtained by technical analysis and a series of experiments. The rules based on certain logical relationships are popularly used, especially with the applications of computer technology.

Secondly, the separation method of PID controller tuning was proposed ((Åström & Hagglund, 1995)). The parameters can be determined based on features of the step response, for example, K, T_i, T_d and T_f are determined to deal with disturbances and robustness, and the parameters b and c can then be chosen to give the desired set-point response. However, the trouble with the method is that the tuning for disturbance rejection and for robustness are mixed, so there are some difficulties for choosing the parameters.

Thirdly, the appearance of integration method, the main trends are that the effectiveness of most tuning methods is enhanced by the combination of PID control with other control methods such as, variable structure control, artificial neural networks, intelligent control, etc., see (Cong & Liang, 2009), (Haj-Ali & Ying, 2004), (Li & Xu, 2010), (Liu & Hsu, 2010).

In spite of the improvements on the tuning methods mentioned above, most of them are still empirical. The trouble with the existing PID control methods is that the more requirements, the more complicated the structure and more parameters to be tuned ((Han, 2009), (Oliveira et al., 2009), (Sekara et al., 2009)).

3. The new improvements to conventional PID controllers

In this section, some new efforts which have been made in recent decades will be reviewed. They are the proposition and development of a group of methods relevant with the improvement of PID control. Because they have essential effects on the tuning of PID controllers, they will be discussed with more details.

3.1 The nonlinear tracking-differentiators

As it is known, the implementation of PID controllers needs to obtain the derivative of the error between the actual output and reference input. Unfortunately, the errors often contain measurement noise, or are usually in discrete form. That is to say, they are discontinuous, let alone differentiable. Usually, people use the difference to replace the derivative. It may amplify the affects of noise, and result in worse control results.

At the same time, the reference input or the setpoint being encountered may also be nondifferentiable, i.e., perhaps the tracked patterns are unusual ones, such as square waves, sawtooth waves, etc.

To avoid setpoint jump, it is necessary to construct a transient profile which the output of the plant can reasonably follow. While this need is mostly ignored in any typical control textbook, engineers have devised different motion profiles in servo systems.

To overcome these drawbacks, or to meet these needs for implementing PID control, and to obtain the derivatives of signals which are nondifferentiable, or contain noise, the nonlinear tracking-differentiator (TD) was introduced. And the general forms and theoretical results of TDs can be found in (Han & Wang, 1994). One of the specific forms of TDs is a nonlinear system as follows:

$$\begin{cases} \dot{y}_1 = y_2 \\ \dot{y}_2 = -R_1 sat\left(y_1 - y_r(t) + \frac{|y_2|y_2}{2R_1}, \delta\right) \end{cases} \tag{2}$$

where sat is the saturation function. From (Han & Wang, 1994), it is proved that, as $R_1 \to \infty$, the following equality holds,

$$\lim_{R_1 \to \infty} \int_0^T |y_1(t) - y(t)|dt = 0 \tag{3}$$

That is to say, $y_1 \to y_r$ as $R_1 \to \infty$. So, y_2 can be regarded as the (generalized) derivative of y_r. It can be used as the derivative for PID control.

At the same time, higher-order derivatives can be obtained by using cascading of TDs, and the transient process can also be arranged by using TDs to overcome overshoots in the responses of PID control, especially for the set inputs which are nondifferentiable.

And because TDs use integrators to process the signal, they also have the ability to filer the noise, especially for dealing with measurement noise.

3.2 The nonlinear PID controllers

Usually, conventional PID control, as a control law, employs a linear combination of proportional (present), integral (accumulative), and derivative (predictive) forms of the tracking error. And, for a long time, other possibilities of combinations, which may be much more effective, are ignored. As a result, it often needs the method on trade-off between overshoot and fastness of the control response.

On the other hand, based on the inspiration of intelligent control, the response curve of neural networks is often a S-type curve, i.e. it is nonlinear. And the optimal control also uses feedback with nonlinear form. Can the control performance be improved by using certain kinds of nonlinear error feedback?

In order to avoid the contradiction between overshoot and fastness in output response of conventional PID controllers, the nonlinear PID (NPID) controller was introduced (see (Han, 1994) and the references therein). Then the control input can be chosen as follows:

$$u(t) = -a_0 \left| \int_{t_0}^{t} e(\tau)d\tau \right|^{\alpha} sgn\left(\int_{t_0}^{t} e(\tau)d\tau\right) - a_1|e(t)|^{\alpha} sgn(e(t)) - a_2|\dot{e}(t)|^{\alpha} sgn(\dot{e}(t)) \quad (4)$$

where $0 < \alpha \leq 1$ is also a design parameter. It has been proved that, the proper choice of α can improve the damping ability for overshoot ((Han, 1994), (Han, 1995)). As to the nonlinear forms in (4), the general expression for these functions is selected heuristically based on experimental results, such as the following one,

$$g(e_1, \alpha, \delta) = \begin{cases} |e_1|^{\alpha} sgn(e_1), & as \ |e_1| > \delta, \\ \frac{e_1}{\delta^{1-\alpha}}, & as \ |e_1| \leq \delta \end{cases} \quad (5)$$

An important property of the function $g(\cdot,\cdot,\cdot)$ is that, for $0 < \alpha < 1$, it yields a relatively high gain when the error is small, and a small gain when the error is large. The constant δ is a small number used to limit the gain in the neighborhood of the origin and defines the range of the error corresponding to high gain ((Han, 1998), (Talole et al., 2009)).

At the same time, the nonlinear feedback usually provides surprisingly better results to the output response in practice. For example, with linear feedback, the tracking error, at best, approaches zero in infinite time; Otherwise, with nonlinear feedback of the following form

$$u = |e|^{\alpha} sgn(e)$$

where $\alpha < 1$, the error can reach zero much more quickly. Such α can also help reduce steady state error significantly, to the extent that an integral control, together with its downfalls, can be avoided. In the extreme case, $\alpha = 0$, i.e., bang-bang control, the nonlinear feedback can bring the system with zero steady state error, even if without the integral term in PID control. It is because of such efficacy and unique characteristics of nonlinear feedback that people proposed a systematic and experimental investigation. There are also some nonlinear feedback functions, such as fal and $fhan$, which play an important role in the newly proposed control framework ((Han, 1998), (Han, 2009)).

Furthermore, by using the tracking-differentiator (TD) in the nonlinear PID controller, the following needs can be satisfied, for example, to deal with measurement noise and to make a powerful tracking for targets which are nondifferentiable or even discontinuous, such as sawtooth wave, square wave, or even certain random signals, etc.

The trouble in the nonlinear PID controller is how to find the proper parameters when dealing with different situations.

3.3 Active disturbance rejection control

With the advances mentioned above, another improvement to PID control technique is the proposition of active disturbance rejection control (ADRC) ((Han, 1998), (Han, 2009)), which inherited the essence from conventional PID controllers and observers. The basic principle of ADRC is that it uses the extended state observer (ESO) ((Han, 1995), (Talole et al., 2009)) to estimate the total disturbances and uncertainties, and then it forces the system to change in a canonical way. Then it only needs to construct a control input for the canonical system.

3.3.1 The extended state observer (ESO)

One of the main troubles in system synthesis is the uncertainties and disturbances. How to find an effective way to deal with these factors? One of the reasonable methods is to construct an estimator.

In order to do that, it is necessary to introduce a new concept: total disturbance, i.e. in a system which contains uncertainties and disturbance, all the nonlinear parts including uncertainties and disturbance, and even the external noise can be regarded as a whole of the total disturbance. Although such a concept is, in general, applicable to most nonlinear multi-input-multi-output (MIMO) time varying systems, for the sake of simplicity and clarity, only a second-order single-input-single-output (SISO) system is used.

The idea of ESO can be explained as follows ((Han, 1995)). For the following system, for instance,

$$\begin{cases} \dot{x}_1 = x_2 \\ \dot{x}_2 = f(x_1, x_2, \omega(t), t) + bu \qquad t \geq t_0. \\ y = x_1 \end{cases} \tag{6}$$

where $f(x_1, x_2, \omega(t), t)$ refers to the total disturbance. Let $F(t) = f(x_1, x_2, \omega(t), t)$, and introduce a new variable x_3 as an additional state variable such that $x_3 = F(t)$. Then, the system (6) can be rewritten in the following extended way,

$$\begin{cases} \dot{x}_1 = x_2 \\ \dot{x}_2 = x_3 + bu \\ \dot{x}_3 = G(t) \\ y = x_1 \end{cases} \quad t \geq t_0. \tag{7}$$

where $G(t)(= \dot{F}(t))$ is unknown to us, and it is the representative of total disturbance.

To estimate the total disturbance, the extended state observer (ESO) ((Han, 1995), (Han, 2009)) for (7) can be built as follows

$$\begin{cases} \dot{x}_1 = x_2 - \beta_0 e \\ \dot{x}_2 = x_3 - \beta_1 e + bu \qquad t \geq t_0. \\ \dot{x}_3 = -\beta_2 e \end{cases} \tag{8}$$

Then the system (6) will be forced to change in the following way

$$\begin{cases} \dot{x}_1 = x_2 \\ \dot{x}_2 = u_0 \end{cases} \tag{9}$$

which can be obtained by using $u = \frac{u_0 - x_3}{b}$ in (6), and u_0 is the proper combination of the derivative and proportional parts of error variables, which is only for the canonical system (9).

It can be shown that, by properly choosing the parameters in ESO, it can estimate the total disturbance changing to a large extent.

3.3.2 The Implementation of active disturbance rejection control

Based on the methods of TD, NPID, and ESO, which are used for the generation of transient profiles, the nonlinear combination of errors, and the estimation and rejection of total disturbances, respectively, the proposition of ADRC would be a natural thing, and ADRC takes the form as shown in Fig. 1. The corresponding control algorithm and the observer gains can be found in (Han, 1998) or (Han, 2009). As to the tuning of parameters in ADRC, r is the amplification coefficient that corresponds to the limit of acceleration, c is a damping coefficient to be adjusted in the neighborhood of unity, h_1 is the precision coefficient that determines the aggressiveness of the control loop, and it is usually a multiple of the sampling period h by a factor of at least four, and b_0 is a rough approximation of the coefficient b in the plant within a ±50% range.

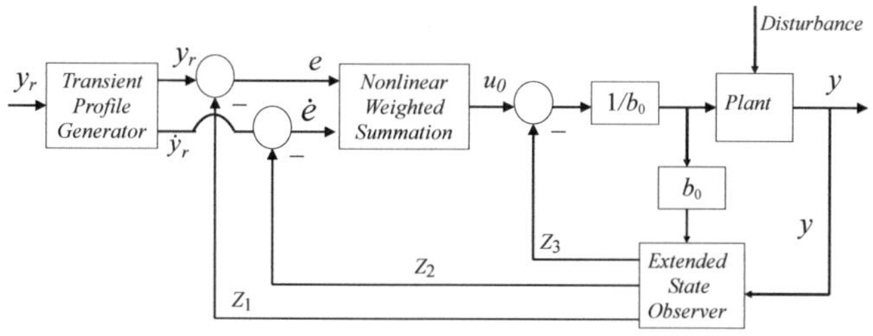

Fig. 1. The control block based on the ADRC

Because of its strong ability to estimate or reject disturbances, ADRC has been widely expanded and it has become a practical and popular control technique, especially in China.

Even though ADRC is presented mainly for a second-order system, it is by no means limited to that. In fact, many complex control systems can be reduced to first or second order systems, and ADRC makes such simplification much easier by lumping many untrackable terms into "total disturbance". Still, the proper problem formulation and simplification is perhaps the most crucial step in practice, and some suggestions are offered as follows ((Han, 1998) or (Han, 2009)).

• Identify the control problem

For a physical process which may contain many variables, one should identify which is the input that can be manipulated and which is the output to be controlled. Maybe this is not very clear, and the choice may not be unique in digital control systems when there are many variables being monitored and many different types of commands can be executed. That is to say, at the beginning, the control problem itself sometimes is not well defined. And it is needed to identify what the control problem is, including its input and output.

• Determine the system structure

According to the relative degree of the system, the structure of the system should be determined. In linear systems, it is easy to obtain the order in terms of its transfer function. For others, such as nonlinear and time varying systems, it might not be straightforward. From the diagram of the plant, the order of the system can be determined simply by counting the number of integrators in it. In the same diagram, however, the relative degree can be found as the minimum number of integrators from input to output through various direct paths.

• Lump the factors which affect the performance

The key in a successful application of ADRC is how well one can reformulate the problem by lumping various known and unknown quantities that affect the system performance into total disturbance. This is a crucial step in transforming a complex control problem into a simple one.

• Proper use of the pseudo control variables

Another effective method for problem simplification is the intelligent use of the pseudo control variable, as shown previously. ADRC shows an obviously quite different way of going about control design. This is really a paradigm shift.

Therefore, the problems of ADRC are that, with the use of ESO, the structure is more complex and it also has more parameters to be tuned. And the implementation needs more skills, especially for some complex industrial processes.

4. The new separation strategy on tuning of PID controllers

In this section, some of the improvements and extensions, which have been done to the active disturbance rejection control (ADRC) in recent years, will be summarized. Firstly, the improvement to ADRC is the proposition of a new separation principle (NSP) such that disturbance rejection and high precision output regulation can be implemented separately. Based on NSP, without using any observer or estimator, disturbances can be rejected, and the output regulation can be carried out effectively even for higher-order systems or multi-input-multi-output interconnected systems. Secondly, in order to meet the needs for the control of uncertain networked systems or complex systems, the NSP is extended and the stripping principle (SP) is proposed. The method based on SP can remove all the interconnected parts, uncertainties and disturbances from the related subsystems, and make the synthesis of complex systems easier and effective. Thirdly, the methods on the stability analysis of the proposed methods will be summed up.

4.1 The new separation principle (NSP) to PID controllers

Based on the necessity analysis above, and from the viewpoint of control system synthesis, if an effective way to reject disturbances and uncertainties can be found, the design for the remaining system will be easier, and the synthesis will have more choices.

4.1.1 The proposition of the new separation principle

It is well known that the variable structure control (VSC) or sliding mode (SM) method has played an important role for disturbance or uncertainty rejection ((Utkin, 1992), (Emelyanov et al., 2000), (Li & Xu, 2010)). But the associated control switching will lead to chattering. PID-type control has a strong ability to reject the uncertainty or disturbance, the tuning of its parameters will need more skills ((Åström & Hagglund, 1995), (Heertjes et al., 2009), (Luo et al., 2009), (Sekara et al., 2009)). At the same time, although active disturbance rejection control (ADRC) can avoid chattering, the need of the extended state observer (ESO) will result in new trouble.

In order to overcome the shortcomings stated above, inspired by the idea of ADRC and the separation principle (SP) ((Blanchini et al., 2009)), it is intended to find a systematic method so that disturbance rejection and high accuracy output regulation can be implemented separately.

• The basic idea of NSP

It is hoped that the PID control input with the form of (1) can be divided into two parts: one is for the rejection of uncertainties and disturbances; the other is for the control of the remaining system which is the one without uncertainties and disturbances. That is to say, it is hope to separate the PID control input into two parts, and each of them has the function of its own. That will reduce the difficulty for the tuning of the parameters.

As a matter of fact, based on the binary control system theory ((Emelyanov et al., 2000)), or by introducing a dynamic mechanism to adjust the gain of the integral feedback, an effective strategy which can powerfully reject disturbances and/or uncertainties will be found. At the same time, by using proper feedback of proportional and derivative (PD) for the remaining parts, the efficient output regulation can be carried out. That is the new separation principle which has been proposed ((Wang, 2010a), (Wang, 2010b), (Wang, 2010e)). The block of the control system based on the new separation principle is shown in Fig.2.

• The implementation of NSP

The dynamic mechanism $\mu(t)$ can be chosen based on the ideas and results of (Emelyanov et al., 2000). One of the particular forms of $\mu(t)$ is as follows, which is described by the following differential equation with a discontinuous right-hand side ((Emelyanov et al., 2000))

$$\dot{\mu}(t) = \begin{cases} -\gamma sgn(\sigma(e)), & \text{as } |\mu(t)| \leq 1, \\ -\omega\mu(t), & \text{as } |\mu(t)| > 1, \quad \mu(t_0) = sgn(\sigma(e(t_0))), \end{cases} \tag{10}$$

where $\sigma(e)$ is a compound function of the regulation error e, γ is a positive parameter, and $\omega(> 0)$ is a given positive constant. The purpose of $\mu(t)$ is to adjust the gain of the

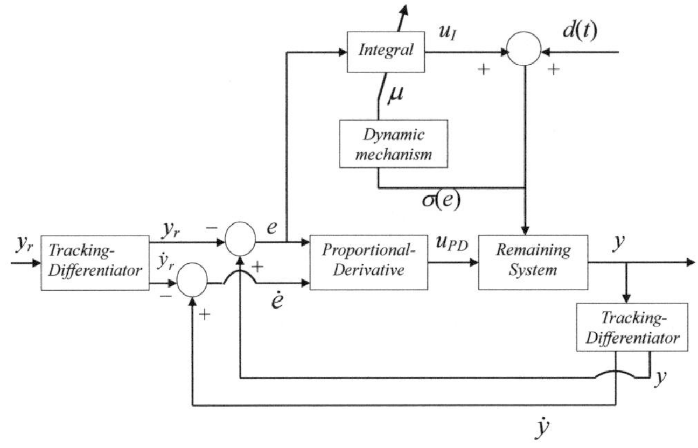

Fig. 2. The control block based on the new separation principle

integral part such that the uncertainties or disturbances can be rejected, in which $\sigma(e)$ can be determined by the parts which should be removed from the system, or equivalently, it should be the part which will be left after the rejection.

It has been proved that, using $\mu(t)$ as (10), the idea of NSP can be realized. The most important of all is that the two procedures (disturbance rejection and output regulation) can be implemented separately. This is the new strategy for PID controller tuning that has been discussed.

Therefore, different from ADRC which uses the following form of control input

$$u = -a_0|e_0|^\alpha sgn(e_0) - a_1|e_1|^\alpha sgn(e_1) - a_2|e_2|^\alpha sgn(e_2) \tag{11}$$

the total control input based on the NSP is the following one

$$u = a_0\mu|e_0|^\alpha - a_1|e_1|^\alpha sgn(e_1) - a_2|e_2|^\alpha sgn(e_2) \tag{12}$$

where e_0 is the integration of the error e_1, a_0 can be determined by the ranges of disturbance and the disturbance's derivative, and the parameters a_1, a_2 can be obtained by assigning, which should maintain the stability of the state (0, 0) of the following remaining error system

$$\begin{cases} \dot{e}_1 = e_2 \\ \dot{e}_2 = -a_1|e_1|^\alpha sgn(e_1) - a_2|e_2|^\alpha sgn(e_2) \end{cases} \tag{13}$$

and in (12), μ can be determined by the following differential equation with a discontinuous righthand side

$$\dot{\mu} = \begin{cases} -\gamma sgn(\dot{e}_2 + a_1|e_1|^\alpha sgn(e_1) + a_2|e_2|^\alpha sgn(e_2)), & \text{as } |\mu| \leq 1, \\ -\omega\mu, & \text{as } |\mu| > 1, \quad |\mu(t_0)| \leq 1 \end{cases} \tag{14}$$

From the results stated above, it can be seen that the function of integral part is to reject the uncertainties and disturbances, the functions of the proportional and derivative parts are to maintain the stability of the remaining system. Therefore, the tuning of the parameters will be more purposeful. And the most important of all is that the choice of a_0 and a_1, a_2 can be implemented separately.

The benefits of the improvements are as follows: (i) It can transform the synthesis of PID controllers into two separate parts. So it will make the synthesis of systems or the tuning of parameters more convenient; (ii) The functions of the design parameters have the new explanations. It will make the tuning more purposeful; (iii) Without any online estimation of uncertainties or disturbances, or with less design parameters or simple structure, the proposed method can effectively reject uncertain factors; (iv) The method has nothing to do with the structure or the relative degree of the system. That is to say, it can also be extended to the output regulation of higher-order systems, or of multi-input-multi-output (MIMO) interconnected systems; (v) It is because of the improvements that a rigorous theoretic foundation will be laid down for the tuning of PID controllers rather than just solve the tuning problem by rules of thumb or experiments. And this will make it possible that the controllable ranges can be extended to more general uncertain systems. It is because of the robustness to the uncertainties or disturbances that it will be effective for engineering applications.

At the same time, based on the using of tracking-differentiators (TDs) ((Han & Wang, 1994)), a powerful control can be made for tracking targets which are nondifferentiable or even discontinuous, such as sawtooth waves, square waves, or certain random signals etc, even if the output contains measurement noise. And the method also has the ability to deal with random noise, especially of measurement noise. The most important of all is that the two procedures (disturbance rejection and output regulation) stated above can be implemented separately. This is the NSP for PID controller tuning that has been considered.

4.1.2 The comparative analysis

Here some brief comparisons will be given about the differences between the tuning method based on NSP and other typical methods.

First, the method based on the NSP is different from the separation method of PID controller tuning discussed in (Åström & Hagglund, 1995), (Åström & Hagglund, 2004). In (Åström & Hagglund, 2004), the parameters can be determined based on features of the step response, for example, K, T_i, T_d and T_f are determined to deal with disturbances and robustness, and the parameters b and c can then be chosen to give the desired set-point response. And the relevant tuning method is based on rules obtained by a series of experiments. However, for the method from the NSP, the parameters can be chosen based on the empirical estimation of disturbances and the analysis to the structure of the remaining part. There also have some theoretical results to guide the choice. So the tuning will be more purposeful and active. And it is easy to extend the controllable ranges.

Second, the difference between NSP and ADRC ((Han, 2009)) is that, in ADRC the rejection of disturbances is realized by constructing ESO, which may be difficult for some systems with complex structures or that are interrupted seriously by disturbances, and this will need more parameters. But here the rejection of disturbances can be realized by using dynamic feedback, which is just based on estimating the boundaries of the disturbances. The rejection has nothing to do with the relative degree of the controlled systems. So there is a need for small number of design parameters by using NSP.

It seems that the NSP has more parameters to be tuned than the conventional ones, or the structure is more complex than former PID controllers. As a matter of fact, the main parameters in the NSP are a_0 and γ, which can be determined by the theorems given in (Wang, 2010b) or (Wang, 2010e). And the key is that the method of disturbance rejection is independent of the order of the system. The other parameters such as a_1 and a_2 can be chosen by assigning. And the parameters in TDs can be found in (Han, 2009), (Han & Wang, 1994) or (Han & Yuan, 1999). So the tuning is much easier here than that of former ones. It should be emphasized that the choice of those parameters can be accomplished separately.

Of course, the problem of the method based on NSP is that, in u_I (see Fig. 2), the integration about the absolute value of errors is used. It may cause the amplification to the error in disturbances' rejection. This can be avoided by using $|\int_{t_0}^{t} e(\tau)d\tau|$ instead of $\int_{t_0}^{t} |e(\tau)|d\tau$. But such change will have very little influence on the results of output regulation.

As to the computational complexity on obtaining the design parameters, from the results in (Wang, 2010b) or (Wang, 2010e), it is known that the parameters should be estimated only by knowing the boundaries of $d(t)$ and $\dot{d}(t)$. In fact, they can also be obtained by a empirical estimation of the disturbance. It should be noticed that, once certain values of the parameters are satisfactory, the proper changes of them are permissible.

4.2 The Extensions of NSP—the proposition of the stripping principle

In recent years, the control for networked systems or complex systems has attracted increasing attention of scientists or engineers from many fields, especially in some large scale engineering, such as electric grids, communication, transportation, biology systems, and so on, see (Hespanha et al., 2007), (Tipsuwan & Chow, 2003), (Wang, 2010a) or (Wang, 2010b), and the references therein. From a system-theoretical point of view, a networked system can be considered as a large-scale system with special interconnections among its dynamical nodes. For such a class of systems, both the synthesis and the implementation of a centralized controller are often not feasible in practice. Techniques aimed at investigating decentralized or distributed controller architecture have been studied. And many interesting results have been established, for example, decentralized fixed modes, decentralized controller design, diagonal Lyapunov function method, M-matrix method, and the model-based approach, etc. (see (Wang, 2010b) and the references therein). Decentralized control has many advantages for its lower dimensionality, easier implementation, lower cost, etc.

However, due to the structural restrictions in decentralized control, it is very difficult to develop a unified and effective design strategy. As a result, many theoretical and practical problems remain unsolved in this field. And many papers are devoted to stability analysis and decentralized controller design for networked systems with linear forms or the nonlinear terms are restricted to the Lipschitz condition (see (Emelyanov et al., 1992), (Vrancic et al., 2010) and the references therein). At the same time, uncertainties or disturbances are often encountered in many control problems. There is no exception for networked control systems ((Wang, 2010a), (Wang, 2010b)). The existence of uncertainty and/or disturbance will result in more difficulty for the system synthesis. The method of ADRC will face great challenges in such cases.

4.2.1 The basic idea of the stripping principle

Based on the philosophic thinking of 'On Contradiction' ((Mao, 1967)) that one should pay more attention to the principal contradiction or the main factors when dealing with some complicated problems, the stripping principle (SP) has been proposed ((Wang, 2011a)). That is to say, to overcome the difficulties caused by interconnected terms, uncertainties and disturbance, it is hoped to find a systematic method such that the networked control system can get rid of the influence of interconnected parts, uncertainties and disturbances completely. And then the control of networked systems will become a decentralized control for the completely independent subsystems.

Suppose that a dynamical network has N nodes, and different nodes may have different forms of structure or dimensionalities. Each node i $(i = 1, \ldots, N)$ in the network is a continuous-time nonlinear uncertain system described by

$$\begin{cases} \dot{x}_{1i} = x_{2i} \\ \quad \vdots \\ \dot{x}_{r_i i} = f_i(x_i) + \triangle f_i(x_i) + g_i(x^\tau) + d_i(t) + h_i(t)u_i(t) \\ y_i = x_{1i} \end{cases} \tag{15}$$

where $x_i = (x_{1i}, \ldots, x_{r_i i})^\tau$, A^τ means the transpose of A, $f_i(x_i)$ is the main component of node i which is well modelled, $\triangle f_i(x_i)$ is the uncertainty or unmodelled part of the model of node i, and $g_i(x^\tau) = g_i(x_1^\tau, \ldots, x_N^\tau)$ is the interconnected parts which indicate that the subsystem i is affected by other nodes, it may be also unknown to us, $d_i(t)$ is the disturbance, $u_i(t)$ is the control input of the subsystem i, y_i is the output of node i. Suppose that $h_i(t)(\geq h_{0i} > 0)$ is the coefficient of the control input, in which h_{0i} is a constant. Usually, it is supposed that $\triangle f_i, d_i, g_i$ are unknown to us.

In such circumstances, similar to the method based on NSP, for each of the nodes, by introducing integral feedback with a variable gain, a new control method, which can remove the interconnected parts, uncertainties and disturbances, will be obtained. At the same time, the proper nonlinear feedback will be used to realize efficient control for each of the subsystems even if the subsystems are with non-smooth components ((Wang, 2011a)).

4.2.2 The implementation of the stripping principle

As to the implementation of SP, integral feedback with a variable gain will be introduced, or for each subsystem i $(i \in \{1, \ldots, N\})$, let

$$u_i(t) = a_{0i}\mu_i(t) \int_{t_0}^{t} \sum_{i=1}^{N} |x_{1i}(\tau)| d\tau + v_i(t) \tag{16}$$

where $\mu_i(t)$ is a dynamic mechanism, $v_i(t)$ is the new control input for the remaining parts of the subsystem i. It has been proved that, through the proper choice of $\mu_i(t)$ similar to (10), all the non-principal factors, such as the interconnected parts, uncertainties and disturbances, can be removed from the subsystem i by using the control input with the form of (16).

If the interconnected parts, uncertainties and disturbances are taken away from the subsystem i, the new control input $v_i(t)$ can be chosen as the one which depends on the control performance of the subsystem i. The simple one of $v_i(t)$ can be obtained by using the principle of poles' placement when the subsystems are linear systems, i.e., $v_i(t)$ can be taken as the following form:

$$v_i(t) = -\sum_{j=1}^{r_i} a_{ji}x_{ji}(t) \tag{17}$$

where $a_{ji}(> 0)$ can be chosen according to the requirement of poles or the performance of the subsystem i.

For the subsystems which are with nonlinear forms, and in order to improve the efficiency of the output regulation, based on the idea of nonlinear PID controllers ((Han, 1994)), the control variable $v_i(t)$ can be chosen as follows ((Wang, 2011a)):

$$v_i(t) = -\sum_{j=1}^{r_i} a_{ji}|x_{ji}(t)|^{\alpha} sgn(x_{ji}(t)) \tag{18}$$

where $0 < \alpha \leq 1$ and $a_{ji}, (j = 1, \cdots, r_i)$ are the design parameters as that of (17).

As far as the existence of γ_i and a_{0i} are concerned, it is needed only to assume that, for certain constants $\bar{k}_i$ and $\bar{\bar{k}}_i$, the following inequalities

$$|\max_{0<\tau\leq t}\{d_i, \triangle f_i, g_i\}| \leq \int_0^t \sum_1^N |x_{1i}(\tau)|d\tau + \bar{k}_i \tag{19}$$

and

$$|\max_{0<\tau\leq t}\{\dot{d}_i, \frac{d}{d\tau}[\triangle f]_i, \dot{g}_i\}| \leq \int_0^t \sum_1^N |x_{1i}(\tau)|d\tau + \bar{\bar{k}}_i \tag{20}$$

hold in the meaning of supremum.

In fact, from the basic conclusions of ordinary differential equations (Emelyanov et al., 2000) and (Filippov, 1988), it is known that, for a positive constant k, if a function $v(t)$ satisfies the following inequality

$$|v(t)| \leq k + \int_0^t \kappa(s)|v(s)|ds \tag{21}$$

then the usual Gronwall inequality can be obtained as follows:

$$|v(t)| \leq k\, exp\left(\int_0^t \kappa(s)ds\right) \tag{22}$$

Therefore, if $|x_1(t)| \leq k_0 + \int_0^t |x_1(\tau)|d\tau$, then $x_1(t)$ will be dominated by a exponential function. On the contrary, it can be proved that if the variation of $x_1(t)$ does not exceed certain exponential functions, the inequality of (21) holds. Then, as long as (19) and (20) hold, the γ_i and a_{0i} can be found so that the disturbance can be rejected. And from (19) and (20), it is known that there are a large number of functions satisfying the restrictions. Of course, step-like functions belong to the set which can be rejected in this way. It also indicates that the Lipshitz condition, which will be required for the analysis of uncertain systems by using some of the other methods ((Emelyanov et al., 1992), (Vrancic et al., 2010)), will not be necessary.

In (20), the derivatives of the uncertainties will be needed. It can be replaced by the existence of the generalized derivatives when those uncertainties are nondifferentiable. And, in (12) and (16), the integration of absolute value of states or errors will be used, then the gains of those expressions will be infinite as $t \to \infty$, so it is capable of rejecting the disturbance such as random noise, even though the derivative of random noise does not exist.

From the above results it is known that, in contrast to other methods, the improvements and extensions on disturbance rejection have nothing to do with the relative degree or the interconnected terms of systems. Therefore the method has less requirements for the systems' structure. That will greatly reduce the complexity for obtaining the design parameters.

4.2.3 The benefits of the stripping principle

The benefits of the method based on the stripping principle are as follows: (i) It can transform the control of networked systems into decentralized control. So it can make the synthesis more convenient; (ii) Without any online estimation of uncertainties, disturbances, or interconnected parts, or with less design parameters or simple structures, the proposed method can effectively reject the troublesome factors; (iii) The functions of the design parameters also have the new explanations. It will make the synthesis of the control system more purposeful; (iv) It is the method that a rigorous theoretic foundation will be laid down for the control method rather than just solve the problem by rules of thumb or experiments. And this will make it possible that the controllable ranges can be extended to more general uncertain networked systems. It is the benefits of the control system that a controller with a simple structure and less parameters to be tuned will be obtained. And it is because of the robustness to the uncertainties or disturbances, then it is easy for engineering implementation.

Theoretical analysis is provided to guarantee the possibility of the method and the stability of the relevant control systems.

4.3 The stability analysis of the relevant control systems

Similar to the synthesis method which uses two separated parts, the stability analysis of the control systems can also be divided into two parts. The first one is to determine the condition to keep the states on a 'variable sliding mode'. The second one is to find the stability condition for the remaining parts (Wang, 2010e).

Firstly, to the stability condition for the disturbance rejection, the following Lyapunov function can be used:

$$L_i(\sigma_i(x_i)) = \frac{1}{2}\sigma_i^2(x_i), \; i = 1, \ldots, N \tag{23}$$

where $\sigma_i(x_i)$ is the part of the non-principal factors of node i which will be stripped, or equivalently, the remaining part after the stripping of the non-principal factors. Based on the condition discussed in (Emelyanov et al., 2000) and (Filippov, 1988), it is known that, under the restrictions of (19) and (20), $\sigma_i(x_i)$ will converge to zero asymptotically.

Secondly, the stability condition for the remaining parts is easy to be obtained. In fact, the parameters such as a_1, a_2 in (13) or $a_1, \cdots, a_n$ in (17) can be chosen by using placement of poles. Especially, for linear error systems, the parameters can be chosen based on the Hurwitz rules ((Emelyanov et al., 2000)), or the optimal control method.

If the remaining system is a second-order linear system, the nonlinear feedback as (13) can be chosen. In such a case, the Lyapunov function can be chosen as follows:

$$L_2(x_1, x_2) = \frac{a_1}{\alpha+1}|x_1|^{\alpha+1} + \frac{1}{2}x_2^2 \tag{24}$$

From the Krasovskii theorem, it is known that the remaining system is stable ((Han & Wang, 1994)). For other kinds of remaining systems, the Lyapunov function can be chosen similarly.

So with the Lyapunov functions of (23) and (24), the stability of the whole system can be proved. It means that the stability of the whole control system can be easily guaranteed.

5. Some applications

In this section, some of the applications are provided to demonstrate the effectiveness of the new methods discussed in section 3 and 4. The applications include the control for tracking unusual motion patterns, the safety and comfort control of vehicles, the active control for noise suppressing, and the synchronization control of networked systems, etc.

5.1 The control for tracking unusual control patterns

The control on output tracking for some unusual control patterns was considered in (Wang, 2010d) by using ADRC. Here, just some simulation results are selected. The simulation can be carried out by using Matlab. In all simulations, let $\omega = 0.5$ in (14). Other parameters can be chosen based on certain given situations.

Consider the following second-order nonlinear system

$$\begin{cases} \dot{x}_1 = x_2 \\ \dot{x}_2 = x_1^2 + 1.5x_2^2 + 4.9sin(t) + u \\ y = x_1 \end{cases} \tag{25}$$

where $4.9sin(t)$ can be regarded as a term of disturbance. It is hoped that the output $y = x_1$ can follow the reference signal y_r, in which the reference inputs y_r are chosen in the form of square and sawtooth waves, respectively. Owing to the nondifferentiable of the reference signals, a TD can be used to smooth y_r. Here, for the purpose of precision, the discrete form of TDs ((Han & Yuan, 1999)) is used. The parameters in TD are chosen as $r = 30, h = 0.01, T = 0.01$. The parameters in (12) are given by $a_0 = 20, a_1 = 90, a_2 = 90$. And the parameters of ω, γ are 0.5 and 10 respectively. The results on output tracking of the two cases are shown in Fig. 3 and Fig. 4 respectively.

The results indicate that, with the using of TDs, the method can deal with the signals which are nondifferentiable, generate the proper transient profile, and implement the output regulation effectively.

5.2 The control of higher-order systems

The method based on NSP will demonstrate great potentialities to the control of higher-order systems, for example, the control for the safety and comfort of vehicles ((Wang, 2010a)).

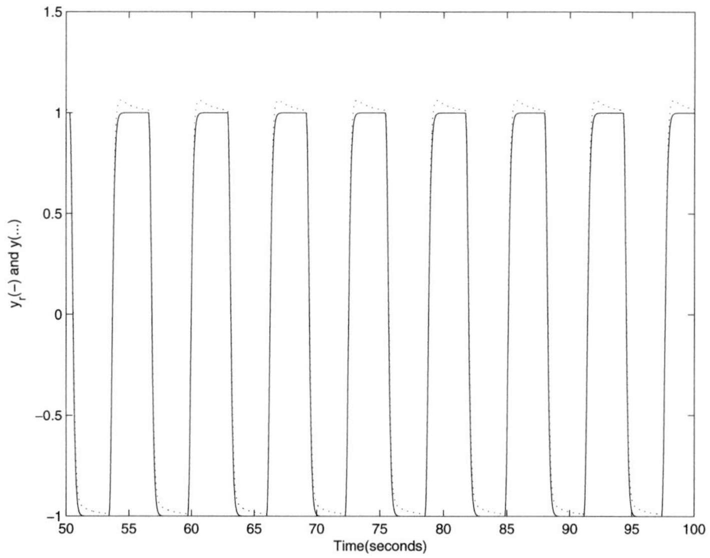

Fig. 3. The output result of tracking the square wave

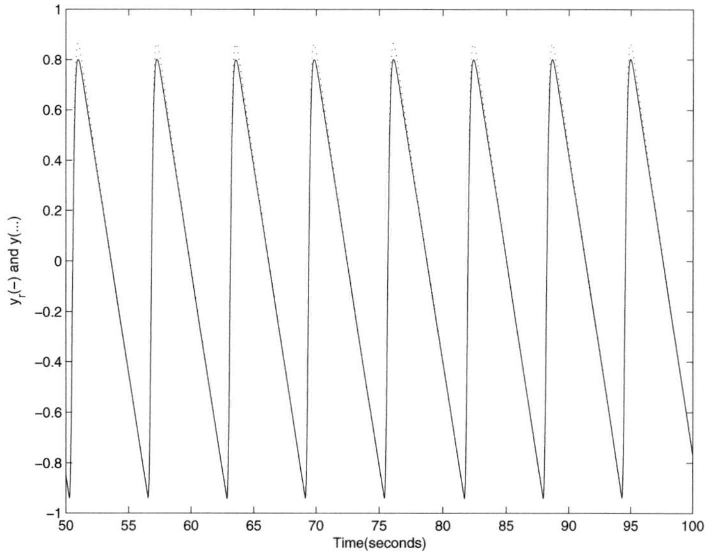

Fig. 4. The output result of tracking the sawtooth wave

Usually, the model of vehicles is a three-order differential equation, so it is easy to be controlled by using the method based on NSP. As an example, the three-order system is considered as follows:

$$\begin{cases} \dot{x}_1 = x_2 \\ \dot{x}_2 = x_3 \\ \dot{x}_3 = 0.3x_1 - 0.2x_2x_3 - x_3 + 0.4x_2^2 + 3sin(t) + 2 + d_1(t) + u \\ y = x_1 + \delta n(t) \end{cases} \tag{26}$$

where $d_1(t)$ refers to the disturbance, which is the possible influence of the road, it can be simulated by the stochastic process obtained from signal generator (SG) of Matlab with the type of random and the parameter of amplitude is 5. The measurement noise can be obtained by random number (RN) with the coefficient of δ. The output $y = x_1$, and its derivatives $\dot{y} = \dot{x}_1$, $\ddot{y} = \ddot{x}_1$ and $\dddot{y} = \dddot{x}_1$ refer to the location, velocity, acceleration and jerk of the vehicle, respectively. The satisfactory control results for all the state variables can be obtained ((Wang, 2010a)).

Another example is the control of a fourth-order system. Consider a flexible joint robotic system which is a single-link manipulator with a revolute joint actuated by a DC motor, and the elasticity of the joint is modelled as a linear torsional spring with stiffness K (see Fig.1 in (Talole et al., 2009) for details). By introducing certain coordinates to the system, its mathematical model is the one as follows, which can be found in (Talole et al., 2009)

$$\begin{cases} \dot{x}_1 = x_2 \\ \dot{x}_2 = x_3 \\ \dot{x}_3 = x_4 \\ \dot{x}_4 = a(x) + d_2(t) + bu \\ y = x_1 + \delta n(t) \end{cases} \tag{27}$$

where $x = (x_1\ x_2\ x_3\ x_4)^\tau$, $a(x) = \frac{MgL}{I} sin(x_1)\left(x_2^2 - \frac{K}{J}\right) - \left(\frac{MgL}{I} cos(x_1) + \frac{K}{J} + \frac{K}{I}\right) x_3$ and $b = \frac{K}{IJ}$. The parameters of the system can also be found in (Talole et al., 2009). Different from the method discussed in (Talole et al., 2009) which uses linearization and ESO, the cascade of TDs are used to obtain the estimation of the states, and the disturbance $d_2(t)$, which is the type of random variable obtained from SG of Matlab with the parameter of amplitude 2, is also introduced. The satisfactory results, especially with good performance can still be obtained ((Wang, 2011c)).

These results indicate that, even if the system contains disturbance, the method based on NSP is not only with a simple control structure but also has strong ability to the control of higher-order systems.

5.3 The noise suppression by using active control based on the NSP

To indicate the ability of the control method for disturbance rejection, the results on noise suppression based on the NSP is demonstrated, the example is considered as follows.

For the following form of noise

$$n_1(t) = rand * sawtooth * square + rand + sin(4t) + 30 \quad (28)$$

where *rand* is the the random signal, and *sawtooth*, and *square* are sawtooth and square waves respectively, and the amplitude and frequency of *sawtooth, and square* are 2, 10 and 1, 5 respectively. It is hoped to suppress $n_1(t)$ to zero. The control method based on the NSP is used. The noise suppression result can be shown in Fig.5. The more details and results can be found in (Wang, 2010c).

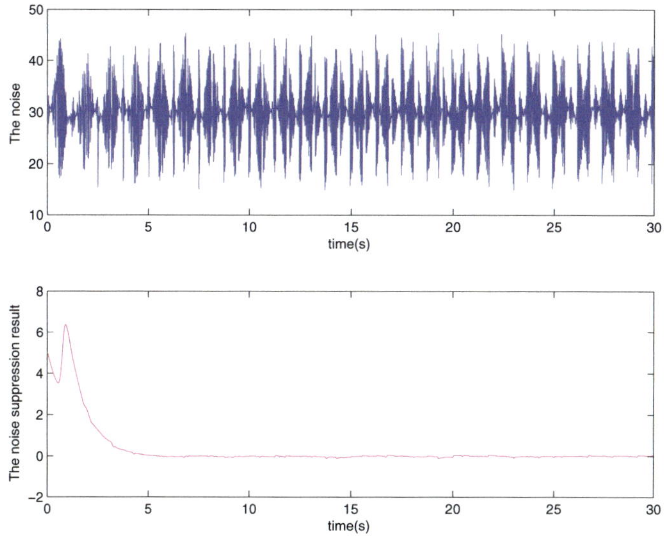

Fig. 5. The noise suppression result based on the new separation principle

The result indicates that the method of NSP has strong ability for rejecting noise, even if the noise is a colored one.

5.4 The control of interconnected dynamical systems

The control results on general unstable interconnected systems can be found in (Talole et al., 2009) and (Foo & Rahman, 2010). Here, an example on the application of the control method based on NSP for some practical MIMO systems is provided.

For an interior permanent-magnet synchronous motor (IPMSM) which has been considered extensively in recent years (see (Foo & Rahman, 2010) and the references therein), the model can be described by the following equation

$$\frac{dx}{dt} = f(x) + g(x)u \quad (29)$$

where $x = (i_d \; i_q \; \omega_r)^{\tau}$, $u = (v_d \; v_q)^{\tau}$, $g(x) = \begin{pmatrix} \frac{1}{L_d} & 0 \\ 0 & \frac{1}{L_q} \\ 0 & 0 \end{pmatrix}$, and

$$f(x) = \begin{pmatrix} f_1(x) \\ f_2(x) \\ f_3(x) \end{pmatrix} = \begin{pmatrix} -\frac{R}{L_d} i_d + P\frac{L_q}{L_d} i_q \omega_r \\ -\frac{R}{L_q} i_q - P\frac{L_d}{L_q} i_d \omega_r - P\frac{\lambda_f}{L_q}\omega_r \\ \frac{3}{2}P\frac{(L_d - L_q)}{J} i_d i_q + \frac{3}{2}P\frac{\lambda_f}{J} i_q - \frac{B}{J} - \frac{T_L}{J} \end{pmatrix} \tag{30}$$

and the meanings and their values of the parameters in (29) and (30) can be found in (Foo & Rahman, 2010).

It is evident that it is a MIMO interconnected system. It is hoped that the outputs i_d and ω_r can track the reference values $i_d^* = \frac{L_d - L_q}{\lambda_f} i_q^2$ and ω_r^* respectively. By introducing the error variables $e_1 = i_d - i_d^*, e_2 = \omega_r - \omega_r^*$, the following error system can be obtained

$$\begin{pmatrix} \dot{e}_1 \\ \ddot{e}_2 \end{pmatrix} = \begin{pmatrix} \tilde{f}_1 \\ \tilde{f}_2 \end{pmatrix} + \tilde{G}u \tag{31}$$

where

$$\tilde{f}_1 = f_1(x) - 2i_q \frac{L_d - L_q}{\lambda_f} f_2(x),$$
$$\tilde{f}_2 = \frac{3}{2}P\left(\frac{L_d - L_q}{J}\right) f_1(x) i_q + \frac{3}{2}P\left(\frac{L_d - L_q}{J}\right) f_2(x) i_d + \frac{3}{2}P\frac{\lambda_f}{J} f_2(x) - \frac{B}{J} f_3(x),$$

and

$$\tilde{G} = \begin{pmatrix} \frac{1}{L_d} & -\frac{2(L_d - L_q)}{L_q \lambda_f} i_q \\ \frac{3}{2}\frac{P}{J}\frac{(L_d - L_q)}{L_d} i_q & \frac{3}{2}\frac{P}{J}\frac{[\lambda_f + (L_d - L_q) i_d]}{L_q} \end{pmatrix}$$

in which $f_1(x), f_2(x)$ and $f_3(x)$ are the ones given in (30).

Let $u = \tilde{G}^{-1}(v_1, v_2)^\tau$, this can realize the decoupling to the error system (31). Then v_1, v_2 can be chosen respectively. From the structure of this error system, for the control of i_d, PI control is used, and for the control of ω_r, PID-type control is used. Compared with the method used in (Foo & Rahman, 2010), without the help of other control such as artificial intelligent (AI), satisfactory results can still be obtained. The structure of the control system here is also a simple one (Wang, 2011c).

It should be emphasized that, if the form of the model or the design parameters are changed in certain ranges, the results on the output regulation may have almost no change at all.

5.5 The synchronization control of complex dynamical networks

In (Wang, 2011b), the output synchronization control is considered for following networked system with three second-order coupled nonlinear subsystems

$$\begin{cases} \dot{x}_{11} = x_{21} \\ \dot{x}_{21} = \sum_{j=1}^{3} f_{j1}(x_1, x_2, x_3) + d_1 + u_1 \\ \dot{x}_{12} = x_{22} \\ \dot{x}_{22} = \sum_{j=1}^{3} f_{j2}(x_1, x_2, x_3) + d_2 + u_2 \\ \dot{x}_{13} = x_{23} \\ \dot{x}_{23} = \sum_{j=1}^{3} f_{j3}(x_1, x_2, x_3) + d_3 + u_3 \\ y = (x_{11}, x_{12}, x_{13}) \end{cases} \tag{32}$$

in which $x_i = (x_{1i}, x_{2i})^{\tau}, i = 1,2,3$ are the states of the subsystems, $f_{11} = sin(x_{11}) + x_{21}^2$, $f_{22} = x_{12}^2 + x_{22}^2$, $f_{33} = -x_{13}^2 + x_{23}^2$, $f_{21} = f_{23} = -sin(x_{12}) - x_{22}$, $f_{31} = f_{32} = sin(x_{13}) + x_{23}$, $f_{12} = f_{13} = sin(x_{11}) + x_{21}$ are the interconnected terms. $d_1(t) = 5sin(t) + 3, d_2(t) = 3sin(t) + 5, d_3(t) = 3cos(t) + 6$ can be regarded as terms of disturbances. Obviously, the subsystems are unstable.

It is hoped that the outputs x_{11}, x_{12}, x_{13} can synchronize with the filtered random signal. Owing to the need of obtaining the higher-order derivatives, the discrete form of TDs ((Han & Yuan, 1999)) is used, and the parameters in the TD are chosen as $r = 30, h = 0.01, T = 0.01$. The parameters in (12) are given by $a_{0i} = 30, a_{1i} = 80, a_{2i} = 80$ and $\alpha = 0.6$. And the parameters of ω, γ_i in (14) are 0.5 and 10 respectively. The results on the synchronization control of the three subsystems are shown in Fig. 6. The more results can be found in (Wang, 2011b).

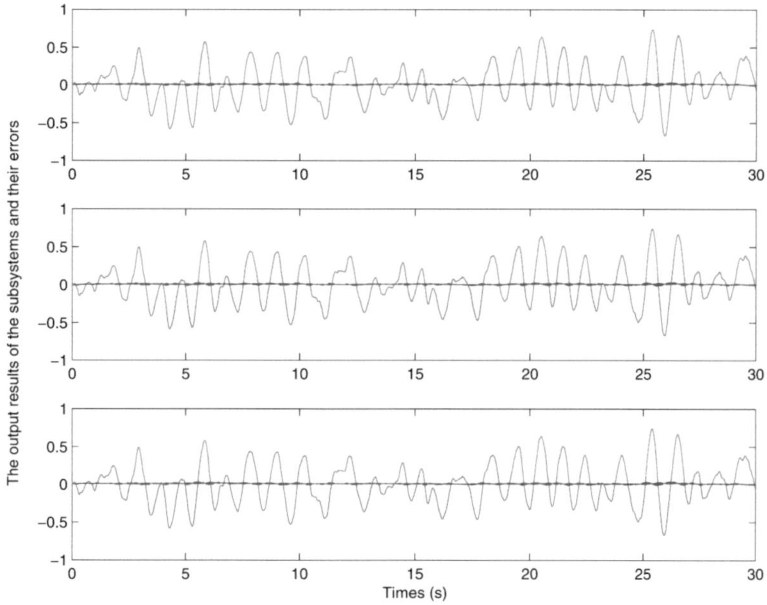

Fig. 6. The output synchronization of networked system

From all these simulation results, it is known that the parameters of γ and ω in the dynamic mechanism $\mu(t)$ can be chosen almost the same. That is to say, they are about 10, and 0.5, respectively.

The results of these applications show that the control methods based on the new strategy can not only have a fast response but also with a strange stability, even if the controlled system contains uncertainty or disturbance, or even has a complex structure. In addition, the methods can also deal with noise effectively.

And, for the usage of these methods, more attention should be paid mainly for choosing the coefficients of the control inputs, and the choice of the coefficients may be carried out separately or purposefully. So it will make the tuning of PID controllers easier.

6. Conclusions

In this chapter, a summary is provided on the improvements and extensions related to the tuning of PID controllers, which have been done from another point of view in recent decades. The development of ADRC is an important step to the improvement of PID control. The new separation principle (NSP) is an another essential improvement of PID control. That is to say, in order to overcome the shortcomings mentioned above, inspired by the idea of the binary control system theory ((Emelyanov et al., 2000)) and the separation method of PID control((Åström & Hagglund, 2004)), a systematic method was proposed so that disturbance rejection and high accuracy output regulation can be implemented separately ((Wang, 2010a)). The extension of NSP is the proposition of the stripping principle (SP) which can provide an effective systematic method for the control of networked systems or complex systems. The method based on SP can remove all the non-principal factors, and it transforms the synthesis of complex systems into decentralized control, so it makes the implementation easier. The efficiency of the related methods is shown by some applications. These methods also have a great potential value for the control of practical complex systems.

As to the problem on how to make the idea more suitable for practical applications, it will be considered in the future.

7. References

Åström, K.J. & Hagglund, T. (1995). *PID Controllers: Theory, Design, and Tuning*, Instrument Society of America, Research Triangle Park, NC.

Åström, K.J. & Hagglund, T. (2004). Revisiting the Ziegler-Nichols step response method for PID control, *J. Process Control*, Vol.14: 635-650.

Blanchini, F., Miani, S. & Mesquine, F. (2009). A separation principle for linear switching systems and parametrization of all stabilizing controllers, *IEEE Trans. Autom. Control*, Vol. 54: 279-292.

Cong, S. & Liang, Y. (2009). PID-Like neural network nonlinear adaptive control for uncertain multivariable motion control systems, *IEEE Trans. Ind. Electron.*, Vol. 56: 3872-3879.

Datta, A., Ho, M.T. & Bhattachoryya, S.P. (2000). *Structure and Synthesis of PID Controllers*, Springer, Berlin.

Emelyanov, S.V. & Korovin, S. K. (2000). *Control of Complex and Uncertain Systems: New Types of Feedback*, Springer, Berlin.

Emelyanov, S.V., Korovin, S.K., Nersisian, A.L. & Nisenzon, Yu. E. (1992). Output feedback stabilization of uncertain plants: a variable structure systems approach, *Int. J. of Control*, Vol.52: 61-81.

Filippov, A.F. (1988). *Differential Equations with Discontinuous Righthand Sides*, Kluwer Academic Publishers, Dordrecht.

Foo, G. & Rahman, M. F. (2010). Sensorless sliding-mode MTPA control of an IPM synchronous motor drive using a sliding-mode observer and HF signal injection, *IEEE Trans. on Ind. Electron.*, Vol.57(4): 1270-1278.

Haj-Ali, Amin & Ying, H. (2004). Structural analysis of fuzzy controllers with nonlinear input fuzzy sets in relation to nonlinear PID control with variable gains, *Automatica*, vol. 40: 1551-1559.

Han, J.Q. (1994). Nonlinear PID controller, *J. Autom.*, Vol. 20(4): 487-490. (in Chinese).

Han, J.Q. (1995). Extended state observer for a class of uncertain plants, *Control & Decis.*, Vol. 10(1): 85-88. (in Chinese).

Han, J.Q. (1998). Auto disturbances rejection controller and its applications, *Control & Decis.*, Vol. 13(1): 19-23. (in Chinese).

Han, J.Q. (2009). From PID to active disturbance rejection control, *IEEE Trans. Ind. Electron.*, vol. 56(3): 900-906.

Han, J.Q. & Wang, W. (1994). Nonlinear Tracking-Differentiators, *J. Syst. Sci. Math. Sci.*, Vol.14: 177–183. (In Chinese).

Han, J.Q. & Yuan, L. (1999). The discrete form of Tracking-Differentiators, *J. Syst. Sci. Math. Sci.*, Vol.19: 268–273. (In Chinese).

Haugen, F. (2010). Basic Dynamics and Control, TechTeach, Skien, Norway, URL: *techteach.no*

Heertjes, M. F., Schuurbiers, Xander G. P., & Nijmeijer, H. (2009). Performance-improved design of N-PID controlled motion systems with applications to wafer stages, *IEEE Trans. on Ind. Electron.*, Vol. 56(5): 1347-1355.

Hernandez-Gomez, M., Ortega, R., Lamnabhi-Lagarrigue, F., & Escobar, G. (2010). Adaptive PI stabilization of switched power converters, *IEEE Trans. Control Syst. Technol.*, Vol. 18(3): 688-698.

Hespanha, J. P., Naghshtabrizi, P. & Xu, Y. (2007). A survey of recent results in networked control systems, *Proc. IEEE*, Vol. 95(1): 138-162.

Hwu, K.I. & Yau, Y.T. (2010). Performance enhancement of boost converter based on PID controller plus linear-to-nonlinear translator, *IEEE Trans. Power Electron.*, Vol. 25(5): 1351-1361.

Leva, A., Negro, S. & Papadopoulos, A.V. (2010). PI/PID autotuning with contextual model parametrisation, *J. Process Control*, Vol. 20 (4): 452-463.

Li, Y. & Xu, Q. (2010). Adaptive sliding mode control with perturbation estimation and PID sliding surface for motion tracking of a piezo-driven micromanipulator, *IEEE Trans. Control Syst. Technol.*, Vol. 18(4): 798-810.

Liu, C.-H. & Hsu, Y.-Y. (2010). Design of a self-tuning PI controller for a STATCOM using particle swarm optimization, *IEEE Trans. on Ind. Electron.*, Vol. 57(2): 702-715.

Luo, A., Tang, C., Shuai, Z., Tang, J., Xu, X. Y. & Chen, D. (2009). Fuzzy-PI-based direct-output-voltage control strategy for the STATCOM used in utility distribution systems, *IEEE Trans. on Ind. Electron.*, Vol. 56(7): 2401-2411.

Mao, Tse-tung. (1967). On Contradiction, in *Selected Works of Mao Tse-tung, I*, Beijing:Foreign Languages Press, pp. 311-347.

Nusret, T. & Atherton, D.P. (2006). Design of stabilizing PI and PID controllers, *Int. J. Syst. Sci.*, Vol. 37: 543-554.

Oliveira, Vilma A., Cossi, Lúcia V., Teixeira, Marcelo C.M. & Silva, Alexandre M.F. (2009). Synthesis of PID controllers for a class of time delay systems, *Automatica*, Vol. 45: 1778-1782.

Santibanez, V., Camarillo, K., Moreno-Valenzuela, J., & Campa, R. (2010). A practical PID regulator with bounded torques for robot manipulators, *Int. J. Control Autom. and Syst.*, Vol. 8(3): 544-555.

Šekara, Tomislav B. & Mataušek, Miroslav R. (2009). Optimization of PID controller based on maximization of the proportional gain under constraints on robustness and sensitivity to measurement noise, *IEEE Trans. Autom. Control*, Vol.54(1): 184-189.

Talole, S. E., Kolhe, J. P. & Phadke, S. B. (2009). Extended-state-observer-based control of flexible-joint system with experimental validation, *IEEE Trans. Ind. Electron.*, Vol. 56(4): 1411-1419.

Tipsuwan, Y. & Chow, M.Y. (2003). Control methodologies in networked control systems, *Control Eng. Practice*, Vol. 11: 1099-1111.

Toscano, R. & Lyonnet, P. (2009). Robust PID controller tuning based on the heuristic Kalman algorithm, *Automatica*, Vol. 45: 2099-2106.

Vrancic, D., Strmcnik, S., Kocijan, J. et al. (2010). Improving disturbance rejection of PID controllers by means of the magnitude optimum method, *ISA Transactions.* Vol. 49 (1): 47-56.

Wang, W. (2010a). The safety and comfort control of vehicles by the separation principle of PID controller tuning, *Proc. IEEE International Conference on Industrial Applications*, IEEE Inc., Vina del Mar, Chile, pp.145-150.

Wang, W. (2010b). The output regulation for a kind of uncertain networked systems, *Proc. of the 8th World Congress on Intelligent Control and Automation*, IEEE Inc., Jinan, China, pp. 4407-4412.

Wang, W. (2010c). A method on noise suppression based on active control by using systems with high-frequency vibrations, *Proc. 2010 3rd International Congress on Image and Signal Processing (CISP 2010)*, IEEE Computer Society, Yantai, China, pp. 3681-3685.

Wang, W. (2010d). The output regulation for tracking unusual motion patterns by active disturbance rejection controls, *in* M. Iskander, V. Kapila, and M.A. Karim (eds.), *Technological Develpoments in Education and Automation*, Springer, pp.243-248.

Wang, W. (2010e). The stability control for a kind of uncertain networked systems by using a new separation principle, *Proc. of the IEEE International Conference on Control and Automation*, IEEE Inc., Xiamen, China, pp.790-795.

Wang, W. (2011a). The stripping principle to the system synthesis for a kind of complex networked systems, *Advanced Meterials Research*, Vol. 220-221: 999-1002.

Wang, W. (2011b). The synchronization control for a kind of complex dynamical networks, *Proc. of 2011 International Conference on Network Computing and Information Security (NCIS'11)*, IEEE Computer Society, Guilin, China, pp.157-161.

Wang, W. (2011c). A double-mechanism strategy of control systems for improving the accuracy of output regulations, *Proc. of the 6th IEEE Conference on Industrial Electronics and Apllications (ICIEA2011)*, IEEE Inc., Beijing, China, pp.2062-2067.

Utkin, V.I. (1992). *Sliding Modes in Control and Optimization*, Springer-Verlag, Berlin, New York.

Yaniv, O. & Nagurka, M. (2004). Design of PID controllers satisfying gain margin and sensitivity constraints on a set of plants, *Automatica*, Vol. 40: 111-116.

12

IMC Filter Design for PID Controller Tuning of Time Delayed Processes

M. Shamsuzzoha[1] and Moonyong Lee[2]
[1]*Department of Chemical Engineering, King Fahd University of Petroleum and Minerals, Daharan,*
[2]*School of Chemical Engineering, Yeungnam University, Kyongsa,*
[1]*Kingdom of Saudi Arabia*
[2]*Korea*

1. Introduction

The proportional integral derivative (PID) control algorithm is widely used in process industries because of its simplicity, robustness and successful practical application. Although advanced control techniques can show significantly improved performance, a PID control system can suffice for many industrial control loops.

A survey by Desborough and Miller (2002) of over 11,000 controllers in the process industries found that over 97% of regulatory controllers use the PID algorithm. Kano and Ogawa (2010) reported that PID control, conventional advanced control and model predictive control are used in an approximately 100:10:1 ratio. Although, a PID controller has only three adjustable parameters, finding appropriate settings is not simple, resulting in many controllers being poorly tuned and time consuming plant tests often being necessary to obtain process parameters for improved controller settings.

There are several approaches for controller tuning, with that based on an open-loop model (g) being most popular. This model is typically given in terms of the plant's gain (K), time constant (τ) and time delay (θ). For a given a plant model, g, controller settings are often obtained by direct synthesis (Seborg et al., 2004). The IMC-PID tuning method of Rivera et al. (1986) is also popular. Original direct synthesis approaches, like that of Rivera et al. (1986), give very good performance for set-point changes, but show slow responses to input (load) disturbances for lag-dominant (including integrating) processes with $\theta/\tau < 0.1$. To improve load disturbance rejection, Shamsuzzoha and Lee (2007a and b, 2008a and b) proposed modified IMC-PID tuning methods for different types of process. For PI tuning, Skogestad (2003) proposed a modified SIMC method where the integral time is reduced for processes with large process time constants, τ. PI/PID tuning based on the IMC and direct synthesis approaches has only one tuning parameter: the closed-loop time constant, τ_c.

Tuning approaches based on an open-loop plant require an open-loop model (g) of the process to be obtained first. This generally involves an initial open-loop experiment, for example a step test, to acquire the required process data. This can be time consuming and may result in undesirable output changes. Approximations can then be used to obtain the process model, g, from the open-loop data.

A two-step procedure, based on a closed-loop set-point experiment with a P-controller, was originally proposed by Yuwana and Seborg (1982). They developed a first-order with delay model by matching the closed-loop set-point response with a standard oscillating second-order step response that results when the time delay is approximated by a first-order Pade approximation. From the set-point response, they identified the first overshoot, the first undershoot and the second overshoot and then used the Ziegler-Nichols (1942) tuning rules for the final PID controller settings, which shows aggressive responses.

Controller tuning based on closed-loop experiments was initially proposed by Ziegler-Nichols (1942). It can simply and directly obtain controller setting from closed-loop data, without explicitly obtaining an open-loop model, g. This approach requires very little information about the process; namely, the ultimate controller gain (K_u) and the oscillations' period (P_u), which can be obtained from a single experiment. The recommended settings for a PI-controller are $K_c=0.45K_u$ and $\tau_I=0.83P_u$. However, there are several disadvantages; the system needs to be brought to its limit of instability, which may require many trials. This problem can be circumvented by inducing sustained oscillations with an on-off controller using the relay method of Åström and Hägglund (1984), though this requires the system to have installed the ability to switch to on/off-control. Ziegler-Nichols (1942) tuning does not work well with all processes. Its recommended settings are aggressive for lag-dominant (integrating) processes (Tyreus and Luyben, 1992) and slow for delay-dominant process (Skogestad, 2003). To improve robustness for lag-dominant (integrating) processes, Tyreus and Luyben (1992) proposed the use of less aggressive settings ($K_c=0.313K_u$ and $\tau_I=2.2P_u$), though this further slowed responses for delay-dominant processes (Skogestad, 2003). This is a fundamental problem of the Ziegler-Nichols (1942) method because it considers only two pieces of information about the process (K_u, P_u), which correspond to the critical point on the Nyquist curve. Therefore, distinguishing between, for example, a lag-dominant and a delay-dominant process is not possible. Additional closed-loop experiments may fix this problem, for example an experiment with an integrating controller (Schei, 1992). A third disadvantage of the Ziegler-Nichols (1942) method is that it can only be used with processes for which the phase lag exceeds -180 degrees at high frequencies. For example, it is inapplicable to a simple second-order process.

Shamsuzzoha and Skogestad (2010) proposed a simple tuning method for a PI/PID controller of an unidentified process using closed-loop experiments. It requires one closed-loop step set-point response experiment using a proportional only controller, and mainly uses information about the first peak (overshoot), which is easily identified. The set-point experiment is similar to that of Ziegler-Nichols (1942) but the controller gain is typically about one half, so the system is not at the stability limit with sustained oscillations. Simulations of a range of first-order with delay processes allow simple correlations to be derived to give PI controller settings similar to those of the SIMC tuning rules (Skogestad, 2003). The recommended controller gain change is a function of the height of the first peak (overshoot); the controller integral time is mainly a function of the time to reach the peak. The method includes a detuning factor that allows the user to adjust the final closed-loop response time and robustness. The proposed tuning method, originally derived for first-order with delay processes, has been tested with a range of other typical processes for process control applications and the results are comparable with SIMC tunings using the open-loop model.

The IMC-PID tuning rules have the advantage of using only a single tuning parameter to achieve a clear trade-off between closed-loop performance and robustness against model inaccuracies. The IMC-PID controller provides good set-point tracking but shows slow responses to disturbances, especially for processes with small time-delay/time-constant ratios (Chen and Seborg, 2002; Horn et al., 1996; Shamsuzzoha and Lee, 2007 and 2008; Shamsuzzoha and Skogestad, 2010; Morari and Zafiriou, 1989; Lee et al., 1998; Chien and Fruehauf, 1990; Skogestad, 2003). However, as disturbance rejection is often more important than set-point tracking, designing a controller with improved disturbance rejection rather than set-point tracking is an important problem that much current research aims to address.

The IMC-PID tuning methods of Rivera et al. (1986), Morari and Zafiriou (1989), Horn et al. (1996), Lee et al. (1998) and Lee et al. (2000), and the direct synthesis method of Smith et al. (1975) (DS) and Chen and Seborg (2002) (DS-d) are examples of two typical tuning methods based on achieving a desired closed-loop response. These methods obtain the PID controller parameters by computing the controller which gives the desired closed-loop response. Although this controller is often more complicated than a PID controller, its form can be reduced to that of either a PID controller or a PID controller cascaded with a first- or second-order lag by some clever approximations of the dead time in the process model.

Regarding disturbance rejection for lag time-dominant processes, Ziegler and Nichols' (1942) established design method (ZN) shows better performance than IMC-PID design methods based on the IMC filter $f=1/(\tau_c s+1)^r$. Horn et al. (1996) proposed a new type of IMC filter that includes a lead term to cancel process-dominant poles. Based on this filter, they developed an IMC-PID tuning rule that leads to the structure of a PID controller with a second-order lead-lag filter. The resulting controller showed advantages over those based on the conventional IMC filter. Chen and Seborg (2002) proposed a direct synthesis design method to improve disturbance rejection in several popular process models. To avoid excessive overshoot in the set-point response, they employed a set-point weighting factor. To improve set-point performance with a set-point filter, Lee et al. (1998) proposed an IMC-PID controller based on both the filter suggested by Horn et al. (1996) and a two-degrees-of-freedom (2DOF) control structure. Lee et al. (2000) extended the tuning method to unstable processes such as first- and second-order delayed unstable process (FODUP and SODUP) models and for set-point performance, they used a 2DOF control structure.

Veronesi and Visioli (2010) reported another two-step approach that assesses and possibly retunes a given PI controller. From a closed-loop set-point or disturbance response of the existing PI controller, a first-order with delay model and time constant are identified and used to assess the closed-loop performance. If performance is worse than expected, the controller is retuned using, for example, the SIMC method. This method has only been developed for integrating processes. Seki and Shigemasa (2010) proposed a controller retuning method based on comparing closed-loop responses obtained with two different controller settings.

The IMC-PID approach determines the performance of the PID controller mainly through the IMC filter structure. Most previous reports of IMC-PID design have the IMC filter structure designed as simple as possible while satisfying the necessary performance requirements of the IMC controller. For example, the order of the lead term in the IMC filter is designed small enough to cancel out the dominant process poles and the lag term is set simply to make the IMC controller realizable. However, the performance of the resulting PID controller is determined both by the performance of the IMC controller performance and by how closely the PID controller approximates the ideal controller equivalent to the IMC controller, which mainly depends on the structure of the IMC filter. Therefore, in IMC-PID design, the optimum IMC filter structure has to be selected considering the performance of the resulting PID controller rather than that of the IMC controller.

2. The IMC-PID approach for PID controller design

The block diagram of IMC control (Figure 1-a) has G_p represent the process, $\tilde{G}_p$ the process model, and q the IMC controller. In the IMC control structure, the controlled variable is given by:

$$C = \frac{G_p q}{1 + q\left(G_p - \tilde{G}_p\right)} f_R R + \left[\frac{1 - \tilde{G}_p q}{1 + q\left(G_p - \tilde{G}_p\right)}\right] G_d d \tag{1}$$

In the nominal case of, $G_p = \tilde{G}_p$ the set-point and disturbance responses are simplified to:

$$\frac{C}{R} = \tilde{G}_p q f_R \tag{2}$$

$$\frac{C}{d} = \left[1 - \tilde{G}_p q\right] G_d \tag{3}$$

In the classical feedback control structure (Figure (1-b)), the set-point and disturbance responses are respectively:

$$\frac{C}{R} = \frac{G_c G_p f_R}{1 + G_c G_p} \tag{4}$$

$$\frac{C}{d} = \frac{G_d}{1 + G_c G_p} \tag{5}$$

where G_c denotes the equivalent feedback controller.

IMC parameterization (Morari and Zafiriou, 1989) allows the process model $\tilde{G}_p$ to be decomposed into two parts:

$$\tilde{G}_p = P_M P_A \tag{6}$$

where P_M and P_A are the portions of the model that are inverted and not inverted, respectively, by the controller (P_A is usually a non-minimum phase and contains dead times and/or right half plane zeros); $P_A(0) = 1$.

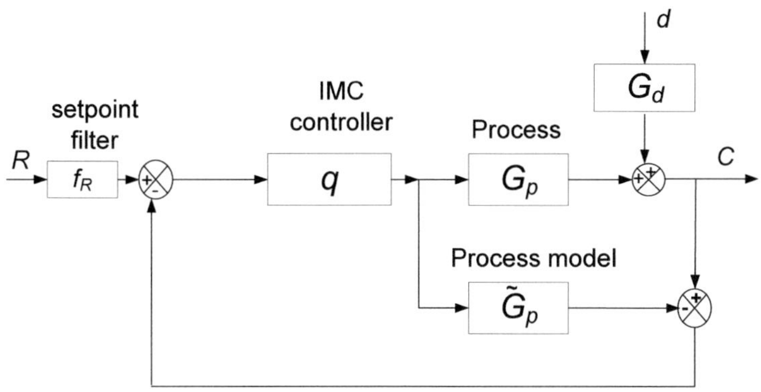

(a) The IMC structure

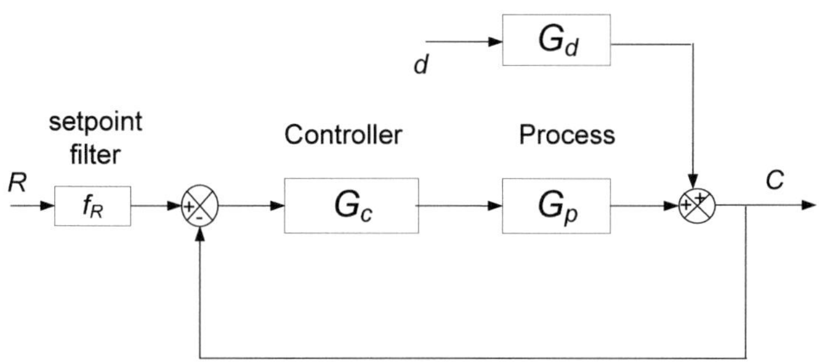

(b) Classical feedback control structure

Fig. 1. Block diagrams of IMC and classical feedback control structures

The IMC controller is designed by:

$$q = P_M^{-1} f \tag{7}$$

The ideal feedback controller that is equivalent to the IMC controller can be expressed in terms of the internal model, $\tilde{G}_p$, and the IMC controller, q :

$$G_c = \frac{q}{1 - \tilde{G}_p q} \tag{8}$$

Since the resulting controller does not have the form of a standard PID controller, it remains to design the PID controller to approximate the equivalent feedback controller most closely. Lee et al. (1998) proposed an efficient method for converting the ideal feedback controller, G_c , to a standard PID controller. Since G_c has an integral term, it can be expressed as:

$$G_c = \frac{g(s)}{s} \tag{9}$$

Expanding G_c in Maclaurin series in s gives:

$$G_c = \frac{1}{s}\left(g(0) + g'(0)s + \frac{g''(0)}{2}s^2 + \ldots \right) \tag{10}$$

The first three terms of which can be interpreted as the standard PID controller which is given by:

$$G_c = K_c\left(1 + \frac{1}{\tau_I s} + \tau_D s \right) \tag{11}$$

where

$$K_c = g'(0) \tag{12a}$$

$$\tau_I = g'(0)/g(0) \tag{12b}$$

$$\tau_D = g''(0)/2g'(0) \tag{12c}$$

3. IMC-PID tuning rules for typical process models

This section proposes tuning rules for several typical time-delayed processes.

3.1 First-Order Plus Dead time Process (FOPDT)

The most commonly used approximate model for chemical processes is the FOPDT model:

$$G_p = G_d = \frac{Ke^{-\theta s}}{\tau s + 1} \tag{13}$$

where K is the gain, τ the time constant and θ the time delay. The optimum IMC filter structure is $f = (\beta s + 1)^2 / (\tau_c s + 1)^3$ and the resulting IMC controller becomes $q = (\tau s + 1)(\beta s + 1)^2 / K(\tau_c s + 1)^3$. Therefore the ideal feedback controller, which is equivalent to the IMC controller, is:

$$G_c = \frac{(\tau s + 1)(\beta s + 1)^2}{K\left[(\tau_c s + 1)^3 - e^{-\theta s}(\beta s + 1)^2 \right]} \tag{14}$$

The analytical PID formula can be obtained from Eq. (12) as:

$$K_c = \frac{\tau_I}{K(3\tau_c - 2\beta + \theta)} \tag{15a}$$

$$\tau_I = (\tau + 2\beta) - \frac{\left(3\tau_c^2 - \theta^2/2 + 2\beta\theta - \beta^2\right)}{(3\tau_c - 2\beta + \theta)} \tag{15b}$$

$$\tau_D = \frac{(2\tau\beta + \beta^2) - \dfrac{\left(\tau_c^3 + \theta^3/6 - \beta\theta^2 + \beta^2\theta\right)}{(3\tau_c - 2\beta + \theta)}}{\tau_I} - \frac{\left(3\tau_c^2 - \theta^2/2 + 2\beta\theta - \beta^2\right)}{(3\tau_c - 2\beta + \theta)} \tag{15c}$$

The value of the extra degree of freedom, β, is selected so that it cancels the open-loop pole at $s = -1/\tau$ that slows responses to load disturbance. Thus, β is chosen so that the term $[1 - Gq]$ has a zero at the pole of G_d, so that $[1 - Gq]\big|_{s=-1/\tau} = 0$ and $\left[1 - (\beta s + 1)^2 e^{-\theta s} / (\tau_c s + 1)^3\right]\Big|_{s=-1/\tau} = 0$. After simplification, β becomes:

$$\beta = \tau\left[1 - \left(\left(1 - \frac{\tau_c}{\tau}\right)^3 e^{-\theta/\tau}\right)^{1/2}\right] \tag{16}$$

3.2 Delayed Integrating Process (DIP)

$$G_p = G_d = \frac{Ke^{-\theta s}}{s} \tag{17}$$

A delayed integrating process (DIP) can be modeled by considering the integrator as a stable pole near zero because the above IMC procedure is not applicable to DIPs, since the β term disappears at $s = 0$. A controller based on a model with a stable pole near zero can give more robust closed-loop responses than one based on a model with an integrator or unstable pole near zero, as suggested by Lee et al. (2000). Therefore, a DIP can be approximated to a FOPDT as follows:

$$G_p = G_d = \frac{Ke^{-\theta s}}{s} = \frac{Ke^{-\theta s}}{s + 1/\psi} = \frac{\psi Ke^{-\theta s}}{\psi s + 1} \tag{18}$$

where Ψ is a sufficiently large arbitrary constant. Accordingly, the optimum filter structure for a DIP is same as that for a FOPDT, i.e. $f = (\beta s + 1)^2 / (\tau_c s + 1)^3$.

Therefore, the resulting IMC controller becomes $q = (\psi s + 1)(\beta s + 1)^2 / K\psi(\tau_c s + 1)^3$ and the ideal feedback controller is $G_c = \dfrac{(\psi s + 1)(\beta s + 1)^2}{K\psi\left[(\tau_c s + 1)^3 - e^{-\theta s}(\beta s + 1)^2\right]}$. The resulting PID tuning rules are listed in Table 1.

Process	K_c	τ_I	τ_D	β
$\frac{Ke^{-\theta s}}{\tau s+1}$	$\frac{\tau_I}{K(3\tau_c-2\beta+\theta)}$	$(\tau+2\beta)-\frac{\left(3\tau_c^2-\theta^2/2+2\beta\theta-\beta^2\right)}{(3\tau_c-2\beta+\theta)}$	$\frac{(2\tau\beta+\beta^2)-\frac{\left(\tau_c^3+\theta^3/6-\beta\theta^2+\beta^2\theta\right)}{(3\tau_c-2\beta+\theta)}}{\tau_I}-\frac{\left(3\tau_c^2-\theta^2/2+2\beta\theta-\beta^2\right)}{(3\tau_c-2\beta+\theta)}$	$\beta=\tau\left[1-\left(\left(1-\frac{\tau_c}{\tau}\right)^3 e^{-\theta/\tau}\right)^{1/2}\right]$
$\frac{Ke^{-\theta s}}{s}$	$\frac{\tau_I}{\psi K(3\tau_c-2\beta+\theta)}$	$(\psi+2\beta)-\frac{\left(3\tau_c^2-\theta^2/2+2\beta\theta-\beta^2\right)}{(3\tau_c-2\beta+\theta)}$	$\frac{(2\psi\beta+\beta^2)-\frac{\left(\tau_c^3+\theta^3/6-\beta\theta^2+\beta^2\theta\right)}{(3\tau_c-2\beta+\theta)}}{\tau_I}-\frac{\left(3\tau_c^2-\theta^2/2+2\beta\theta-\beta^2\right)}{(3\tau_c-2\beta+\theta)}$	$\beta=\psi\left[1-\left(\left(1-\frac{\tau_c}{\tau}\right)^3 e^{-\theta/\psi}\right)^{1/2}\right]$
$\frac{Ke^{-\theta s}}{(\tau_1 s+1)(\tau_2 s+1)}$	$\frac{\tau_I}{K(4\tau_c-\beta_1+\theta)}$	$(\tau_1+\tau_2+\beta_1)-\frac{\left(6\tau_c^2-\theta^2/2+\theta\beta_1-\beta_2\right)}{(4\tau_c-\beta_1+\theta)}$	$\frac{(\tau_1\tau_2+(\tau_1+\tau_2)\beta_1+\beta_2)-\frac{\left(4\tau_c^3+\theta^3/6-\beta_1\theta^2/2+\theta\beta_2\right)}{(4\tau_c-\beta_1+\theta)}}{\tau_I}-\frac{\left(6\tau_c^2-\theta^2/2+\theta\beta_1-\beta_2\right)}{(4\tau_c-\beta_1+\theta)}$	$\beta_1=\frac{\tau_1^2\left[\left(1-\frac{\tau_c}{\tau_1}\right)^4 e^{-\theta/\tau_1}-1\right]-(\tau_2)^2\left[\left(1-\frac{\tau_c}{\tau_2}\right)^4 e^{-\theta/\tau_2}-1\right]}{(\tau_2-\tau_1)}$ $\beta_2=\tau_2^2\left[\left(1-\frac{\tau_c}{\tau_2}\right)^4 e^{-\theta/\tau_2}-1\right]+\tau_2\beta_1$
$\frac{Ke^{-\theta s}}{s(\tau s+1)}$	$\frac{\tau_I}{\psi K(4\tau_c-\beta_1+\theta)}$	$(\psi+\tau+\beta_1)-\frac{\left(6\tau_c^2-\theta^2/2+\theta\beta_1-\beta_2\right)}{(4\tau_c-\beta_1+\theta)}$	$\frac{(\psi\tau+(\tau+\psi)\beta_1+\beta_2)-\frac{\left(4\tau_c^3+\theta^3/6-\beta_1\theta^2/2+\theta\beta_2\right)}{(4\tau_c-\beta_1+\theta)}}{\tau_I}-\frac{\left(6\tau_c^2-\theta^2/2+\theta\beta_1-\beta_2\right)}{(4\tau_c-\beta_1+\theta)}$	$\beta_1=\frac{\psi^2\left[\left(1-\frac{\tau_c}{\psi}\right)^4 e^{-\theta/\psi}-1\right]-(\tau)^2\left[\left(1-\frac{\tau_c}{\tau}\right)^4 e^{-\theta/\tau}-1\right]}{(\tau-\psi)}$ $\beta_2=\tau^2\left[\left(1-\frac{\tau_c}{\tau}\right)^4 e^{-\theta/\tau}-1\right]+\tau\beta_1$
$\frac{Ke^{-\theta s}}{\tau s-1}$	$-\frac{\tau_I}{K(3\tau_c-2\beta+\theta)}$	$(-\tau+2\beta)-\frac{\left(3\tau_c^2-\theta^2/2+2\beta\theta-\beta^2\right)}{(3\tau_c-2\beta+\theta)}$	$\frac{(-2\tau\beta+\beta^2)-\frac{\left(\tau_c^3+\theta^3/6-\beta\theta^2+\beta^2\theta\right)}{(3\tau_c-2\beta+\theta)}}{\tau_I}-\frac{\left(3\tau_c^2-\theta^2/2+2\beta\theta-\beta^2\right)}{(3\tau_c-2\beta+\theta)}$	$\beta=\tau\left[\left(\left(1+\frac{\tau_c}{\tau}\right)^3 e^{\theta/\tau}\right)^{1/2}-1\right]$
$\frac{Ke^{-\theta s}}{(\tau s-1)(as+1)}$	$-\frac{\tau_I}{K(4\tau_c-2\beta+\theta)}$	$(a-\tau+2\beta)-\frac{\left(6\tau_c^2-\theta^2/2+2\beta\theta-\beta^2\right)}{(4\tau_c-2\beta+\theta)}$	$\frac{(-a\tau-2\beta(-a+\tau)+\beta^2)-\frac{\left(4\tau_c^3+\theta^3/6-\beta\theta^2+\beta^2\theta\right)}{(4\tau_c-2\beta+\theta)}}{\tau_I}-\frac{\left(6\tau_c^2-\theta^2/2+2\beta\theta-\beta^2\right)}{(4\tau_c-2\beta+\theta)}$	$\beta=\tau\left[\left\{\left(\frac{\tau_c}{\tau}+1\right)^4 e^{\theta/\tau}\right\}^{1/2}-1\right]$

Table 1. IMC-PID Controller Tuning Rules

3.3 Second-Order Plus Dead time Process (SOPDT)

Consider a stable SOPDT system:

$$G_p = G_d = \frac{Ke^{-\theta s}}{(\tau_1 s+1)(\tau_2 s+1)} \tag{19}$$

The optimum IMC filter structure is $f = \left(\beta_2 s^2 + \beta_1 s + 1\right) / (\tau_c s+1)^4$. The IMC controller becomes $q = (\tau_1 s+1)(\tau_2 s+1)\left(\beta_2 s^2 + \beta_1 s+1\right) / K(\tau_c s+1)^4$ and the ideal feedback controller equivalent to the IMC controller is $G_c = \dfrac{(\tau_1 s+1)(\tau_2 s+1)\left(\beta_2 s^2+\beta_1 s+1\right)}{K\left[(\tau_c s+1)^4 - e^{-\theta s}\left(\beta_2 s^2+\beta_1 s+1\right)\right]}$. The resulting PID tuning rules are listed in Table 1.

3.4 First-Order Delayed Integrating Process (FODIP)

Consider the following FODIP system:

$$G_p = G_d = \frac{Ke^{-\theta s}}{s(\tau s+1)} \tag{20}$$

It can be approximated as the SOPDT model, becoming:

$$G_p = G_d = \frac{Ke^{-\theta s}}{s(\tau s+1)} = \frac{\psi K e^{-\theta s}}{(\psi s+1)(\tau s+1)} \tag{21}$$

where Ψ is a sufficiently large arbitrary constant. Thus, the optimum IMC filter is same as that for the SOPDT, $f = \left(\beta_2 s^2 + \beta_1 s+1\right) / (\tau_c s+1)^4$ and the resulting IMC controller becomes $q = (\tau s+1)(\psi s+1)\left(\beta_2 s^2+\beta_1 s+1\right) / K\psi(\tau_c s+1)^4$. The resulting PID tuning rules are listed in Table 1.

3.5 First-Order Delayed Unstable Process (FODUP)

One of the most commonly considered unstable processes with time delay is the FODUP:

$$G_p = G_d = \frac{Ke^{-\theta s}}{\tau s-1} \tag{22}$$

The optimum IMC filter is $f = (\beta s+1)^2 / (\tau_c s+1)^3$. Therefore, the IMC controller becomes $q = (\tau s-1)(\beta s+1)^2 / K(\tau_c s+1)^3$ and the ideal feedback controller is $G_c = \dfrac{(\tau s-1)(\beta s+1)^2}{K\left[(\tau_c s+1)^3 - e^{-\theta s}(\beta s+1)^2\right]}$. The resulting PID tuning rules are listed in Table 1.

3.6 Second-Order Delayed Unstable Process (SODUP)

The process model is:

$$G_p = G_d = \frac{Ke^{-\theta s}}{(\tau s - 1)(as + 1)} \tag{23}$$

The optimum IMC filter is $f = (\beta s + 1)^2 / (\tau_c s + 1)^4$ and the IMC controller is $q = (\tau s - 1)(as + 1)(\beta s + 1)^2 / K(\tau_c s + 1)^4$. The resulting PID tuning rules are listed in Table 1.

4. Performance and robustness evaluation

The performance and robustness of a control system are evaluated by the following indices.

4.1 Integral error criteria

Three commonly used performance indices based on integral error can evaluate performance: the integral of the absolute error (IAE), the integral of the squared error (ISE), and the integral of the time-weighted absolute error (ITAE).

$$IAE = \int_0^\infty |e(t)| dt \tag{24}$$

$$ISE = \int_0^\infty e(t)^2 dt \tag{25}$$

$$ITAE = \int_0^\infty t|e(t)| dt \tag{26}$$

where the error signal $e(t)$ is the difference between the set-point and the measurement. The ISE criterion penalizes larger errors; the ITAE criterion, long-term errors. The IAE criterion tends to produce controller settings that are between those of the ITAE and ISE criteria.

4.2 Overshoot

Overshoot is a measure of by how much the response exceeds the ultimate value following a step change in set-point and/or disturbance.

4.3 Maximum sensitivity (M_s) to modeling error

A control system's maximum sensitivity, $M_s = \max\left|1/[1 + G_p G_c(i\omega)]\right|$, can be used to evaluate its robustness. Since M_s is the inverse of the shortest distance from the Nyquist curve of the loop transfer function to the critical point (-1, 0), a low M_s indicates that the control system has a large stability margin. M_s is typically between 1.2 and 2.0 (Åström et al., 1998). To ensure a fair comparison, all the controller simulation examples considered here are designed to have the same robustness level in terms of maximum sensitivity.

4.4 Total Variation (TV)

To evaluate the manipulated input usage the *TV* of the input u(t) is computed as the sum of all its moves up and down. Considering the the input signal as a discrete sequence $[u_1, u_2, u_3 ...., u_i ...]$, leads to $TV = \sum_{i=1}^{\infty} |u_{i+1} - u_i|$, which should be minimized. *TV* is a good measure of a signal's smoothness (Skogestad, 2003).

4.5 Set-point and derivative weighting

The conventional form of the PID controller used here for simulation is:

$$G_{PID} = K_c \left(1 + \frac{1}{\tau_I s} + \tau_D s \right) \tag{27}$$

A more widely accepted control structure that includes set-point weighting and derivative weighting is given by Åström and Hägglund (1995):

$$u(t) = K_c \left(\left[br(t) - y(t) \right] + \frac{1}{\tau_I} \int_0^t \left[r(t) - y(t) \right] d\tau + \tau_D \frac{d\left[cr(t) - y(t) \right]}{dt} \right) \tag{28}$$

where *b* and *c* are additional parameters. The integral term must be based on error feedback to ensure the desired steady state. The controller structure in Eq. (28) has two degrees of freedom. The set-point weighting coefficient *b* is bounded by $0 \le b \le 1$ and the derivative weighting coefficient *c* is bounded by $0 \le c \le 1$. The overshoot for set-point changes decreases with decreasing *b*.

The controllers obtained with different values of *b* and *c* respond to disturbances and measurement noise in the same way as a conventional PID controller, i.e. the values of *b* and *c* do not affect the closed-loop responses to disturbances (Chen and Seborg, 2002). Therefore, the PID tuning rules developed here are also applicable for the modified PID controller in Eq. (28). However, the set-point response does depend on the values of *b* and *c*. Throughout this study, *c=1*, while the set-point filter $f_R = (b\tau_I s + 1) / (\tau_I \tau_D s^2 + \tau_I s + 1)$ was used with $0 \le b \le 1$.

5. Simulation results

Six process models are simulated here. They are common processes widely used in the chemical industry and have been studied by other researchers. In each case, different performance and robustness matrices have been calculated and are compared with other existing methods. The method of Shamsuzzoha and Lee (2007a), the SL method for conciseness, is compared with other reported tuning methods.

5.1 Example 1. FOPDT

Consider the following FOPDT model (Chen and Seborg, 2002):

$$G_p = G_d = \frac{100e^{-s}}{100s+1} \tag{29}$$

A unit step disturbance acts at the plant input and the corresponding simulation result is shown in Figure 2a. All the tuning methods, except the ZN method, were adjusted to have the same robustness level of M_s=1.94 by varying τ_c. Comparison of the SL method with other existing IMC-PID methods (Horn et al., 1996; Lee et al., 1998; Rivera et al., 1986), the direct synthesis method (Chen and Seborg, 2002) and the ZN method in a performance matrix for disturbance rejection containing IAE, ISE and ITAE (Table 2) shows that the SL method performs better than the other IMC-PID methods. The ZN method gives the lowest IAE, ISE, ITAE and overshoot values, though it has an M_s of 2.29. The SL method shows better performance indices than the ZN method at M_s=2.29.

Tuning methods	τ_c	K_c	τ_I	τ_D	M_s	set-point TV	set-point IAE	set-point ISE	set-point ITAE	set-point Overshoot	disturbance TV	disturbance IAE	disturbance ISE	disturbance ITAE	disturbance Overshoot
SL (b=1.0)	1.51	0.827	3.489	0.356	1.94	1.67	3.08	1.86	8.22	1.45	1.96	4.30	3.74	15.91	1.26
SL (b=0.4)	1.51	0.827	3.489	0.356	1.94	0.79	2.37	1.79	3.72	1.03	1.96	4.30	3.74	15.91	1.26
DS-d	1.2	0.828	4.051	0.353	1.94	1.57	3.06	1.77	8.69	1.40	1.88	4.89	4.08	20.04	1.27
Lee et al.	1.32	0.810	3.928	0.307	1.94	1.54	3.09	1.83	8.47	1.44	1.87	4.85	4.26	19.29	1.31
Horn et al.	1.68	15.146	100.5	0.497	1.94	1.62	3.56	1.95	23.23	1.31	1.69	6.64	6.41	30.85	1.47
ZN	-	0.948	1.99	0.498	2.29	3.25	3.58	2.21	11.07	1.67	3.04	3.22	2.18	12.7	1.17
IMC	0.85	0.744	100.5	0.498	1.94	1.18	2.11	1.46	15.29	1.0	1.58	84.47	77.74	3634	1.29

Table 2. PID Controller Settings for Example 1 (θ/τ=0.01) (FOPDT) Horn et al.,

$$G_c = K_c\left(1+\frac{1}{\tau_I s}+\tau_D s\right)\frac{1+cs+ds^2}{1+as+bs^2} \text{ where } a = 100.2127;\ b = 21.2687;\ c = 4.2936, \text{ and } d = 0$$

The resulting output responses when unit step changes are introduced to the set-point are shown in Figure 2a. Under a 1DOF control structure, any controller with good disturbance rejection is essentially accompanied by excessive overshoot in the set-point response. A 2DOF control structure can be used to avoid this water-bed effect and the corresponding response is shown in the same figure. In the 1DOF case, only the ZN method shows a larger overshoot of the output response than the SL method, though the SL method shows the quickest settling of the considered methods. The overshoot of the SL controller can be eliminated without affecting its disturbance response by setting b=0.40 in the 2DOF controller, as suggested by Åström and Hägglund (1995) and Chen and Seborg (2002). The output responses and the performance matrices listed in Table 2 suggest that the SL controller performs best.

The controllers' robustness is evaluated by simultaneously inserting a perturbation uncertainty of 20% in all three parameters to obtain the worst case model mismatch, i.e. $G_p = G_d = 120e^{-1.2s}/(80s+1)$ as an actual process. The simulation results for both set-point and disturbance rejection are listed Table 3. The methods of SL and Chen & Seborg (2002) (DS-d) give similar error integral values for disturbance rejection. In terms of servo response, the IMC has clear advantages and the methods of Shams & Lee, DS-d, Lee et al. (1998) and Horn et al. (1996) show almost similar robustness.

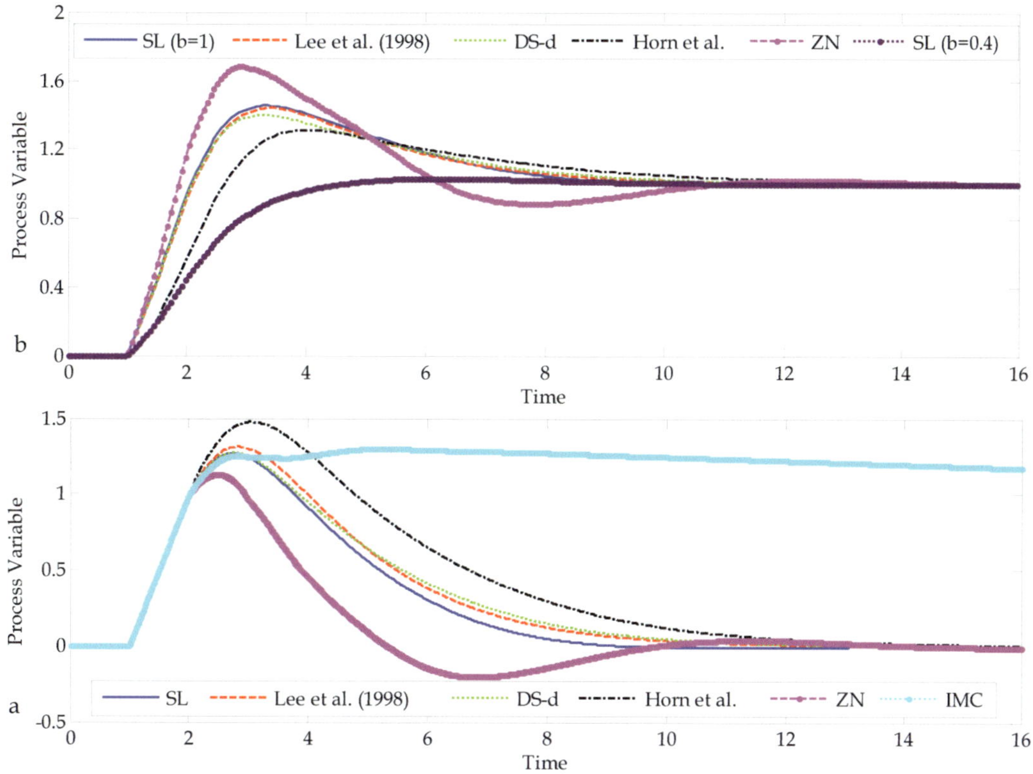

Fig. 2. Responses of the nominal system in Example 1.

5.2 Example 2. DIP (Distillation Column Model)

Consider next the distillation column model studied by Chien and Fruehauf (1990) and Chen & Seborg (2002) (DS-d). The column separates a small amount of low-boiling material from the final product. Its bottom level is controlled by adjusting the steam flow rate. The process model for the level control system is represented as the following DIP model, which can be approximated by the FOPDT model for design of the PID controller:

$$G_p = G_d = \frac{0.2e^{-7.4s}}{s} = \frac{20e^{-7.4s}}{100s+1} \tag{30}$$

Tuning methods	set-point					disturbance				
	TV	IAE	ISE	ITAE	Over shoot	TV	IAE	ISE	ITAE	Over shoot
SL (b=1.0)	6.36	5.47	3.50	27.77	2.12	7.33	6.31	6.39	36.03	1.95
DS-d	6.20	5.23	3.24	26.46	2.06	7.19	6.28	6.50	35.53	1.96
Lee et al.	5.93	5.61	3.49	30.0	2.09	7.00	6.80	7.07	40.39	2.1
Horn et al.	4.44	5.15	2.92	35.98	1.84	5.91	7.76	9.52	45.09	2.23
ZN	24.21	12.89	8.01	166.2	2.56	24.13	12.50	8.36	170.8	1.83
IMC	4.58	3.26	1.81	17.53	1.43	6.15	84.59	79.07	3616	1.92

Table 3. Robustness Analysis for Example 1 (FOPDT) $G_p = G_d = \dfrac{120e^{-1.2s}}{80s+1}$

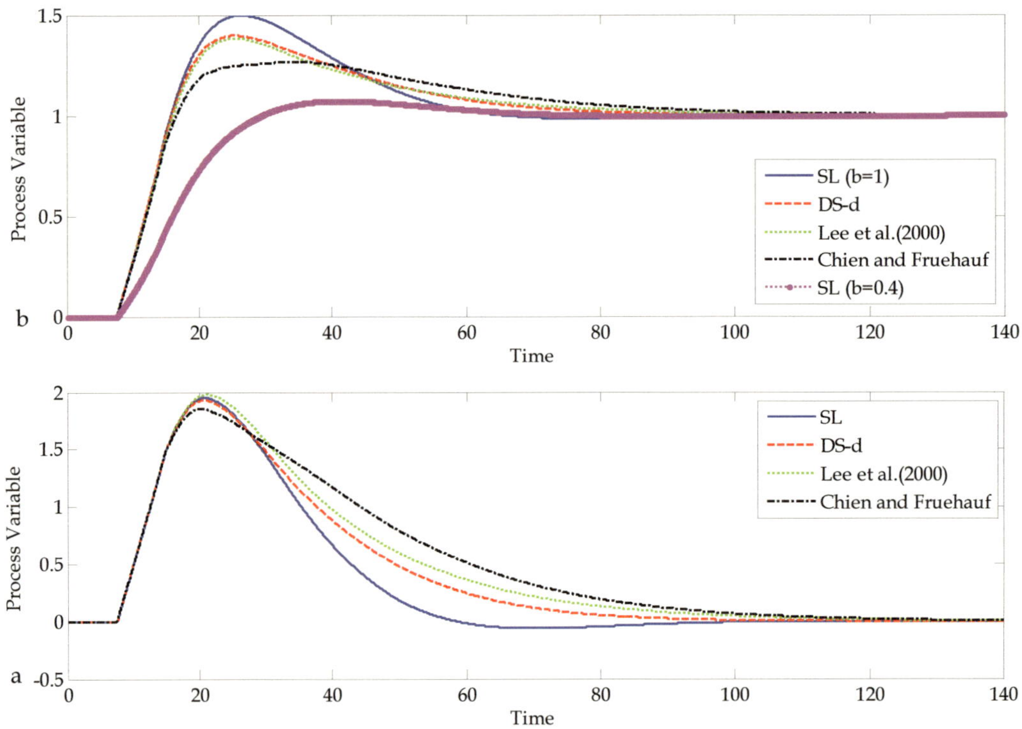

Fig. 3. Responses of the nominal system in Example 2.

The methods proposed by Chen and Seborg (2002), Lee et al. (2000) and Chien and Fruehauf (1990) are used to design PID controllers (Figure 3 and Table 4). $\tau_c = 11.3$ was chosen for the SL method, $\tau_c = 9.15$ for the method of Chen and Seborg (2002), $\tau_c = 11.0$ for that of Lee et al. (2000) and $\tau_c = 15.28$ for that of Chien and Fruehauf (1990), consistently resulting in M_s=1.90. The output responses for disturbance rejection (Figure 3a) show that the SL tuning rules result in the quickest settling. Chien and Fruehauf's method (1990) shows the slowest response and the longest settling time. The performance matrix (Table 4) shows that the SL method gives the lowest error integral value, while Chien and Fruehauf's (1990) method gives the highest at the same robustness level.

The simulation results for a unit set-point change (Figure 3b) show that the SL method gives a large overshoot but the shortest settling time of the considered 1DOF controllers. The overshoot can be minimized using a 2DOF controller with *b=0.4*, with the SL method performing better than the other conventional methods (Table 4 and Figure 3).

Tuning methods	τ_c	K_c	τ_I	τ_D	M_s	set-point					disturbance				
						TV	IAE	ISE	ITAE	Over shoot	TV	IAE	ISE	ITAE	Over shoot
SL (b=1.0)	11.3	0.531	24.533	2.467	1.90	1.128	24.04	14.65	495.4	1.49	1.99	49.19	66.86	1366	1.95
SL (b=0.4)	11.3	0.531	24.533	2.467	1.90	0.58	18.41	13.48	243.8	1.07	1.99	49.19	66.86	1366	1.95
DS-d	9.15	0.543	31.15	2.558	1.90	1.0	23.28	13.36	508.3	1.39	1.84	57.47	73.35	1794	1.93
Lee et al.	11	0.536	35.137	2.286	1.90	0.95	23.46	13.14	550.2	1.38	1.78	65.35	81.95	2292	1.98
Chien and Fruehauf	15.28	0.526	37.96	3.339	1.90	0.99	23.68	12.62	630	1.27	1.81	71.88	86.57	2716	1.86

Table 4. PID Controller Settings for Example 2 (Distillation Column Model)

The controllers' robustness is evaluated by simultaneously inserting a perturbation uncertainty of 20% in both of the process's parameters. The worst plant-model mismatch case after 20% perturbation is $G_p = G_d = 0.24e^{-8.88s}/s$. The simulation results (Table 5) show that disturbance rejection is best by the SL method, followed by the methods of DS-d and Lee et al. (2000). The set-point responses of the methods of DS-d, Lee et al. (2000) and Chien and Fruehauf (1990) are similar to those of the SL method.

Tuning methods	set-point					disturbance				
	TV	IAE	ISE	ITAE	Over shoot	TV	IAE	ISE	ITAE	Over shoot
SL (b=1.0)	1.89	27.81	20.30	565.4	1.88	3.27	49.86	88.05	1335	2.54
DS-d	1.80	25.38	17.79	499.5	1.77	3.09	57.47	90.24	1793	2.52
Lee et al.	1.74	25.85	17.56	549.3	1.75	3.06	65.34	99.4	2288	2.57
Chien and Fruehauf	1.62	25.14	15.23	580.9	1.58	2.90	71.69	98.48	2683	2.43

Table 5. Robustness Analysis for Example 2 (Distillation Column Model)

$$G_p = G_d = \frac{0.24e^{-8.88s}}{s}$$

5.3 Example 3. SOPDT

Consider the SOPDT process described by Chen and Seborg (2002):

$$G_p = G_d = \frac{2e^{-s}}{(10s+1)(5s+1)} \tag{31}$$

The SL, DS-d (Chen and Seborg, 2002), ZN and DS (Smith et al. 1975, Seborg et al. 2004) methods were used to design PID controllers. DS and IMC-PID (Rivera et al. 1986) give exactly similar tuning formulae for the SOPDT process. The parameters of the PID controller settings for the DS and ZN controllers are taken from Chen and Seborg (2002). All the other methods were adjusted to have M_s =1.87 to ensure a fair comparison. The output responses for disturbance rejection (Figure 4a) show that the SL method has a similar overshoot to the DS-d method, while the ZN tuning method gives the highest peak. The SL method shows the lowest settling time; the DS method shows the slowest response. Apart from the DS-d method, all the considered tuning methods have either a higher overshoot or a slower response for disturbance rejection.

Simulation results for a unit set-point change (Figure 4b) show that the overshoot of the ZN method is the largest, followed by that of the SL method. However, the SL method shows the shortest settling time of the considered 1DOF controllers. Overshoot can be minimized using a 2DOF controller with *b=0.4*. Comparison of the various methods' performances in the above figures and in the performance matrix (Table 6) shows that the SL method gives the best performance.

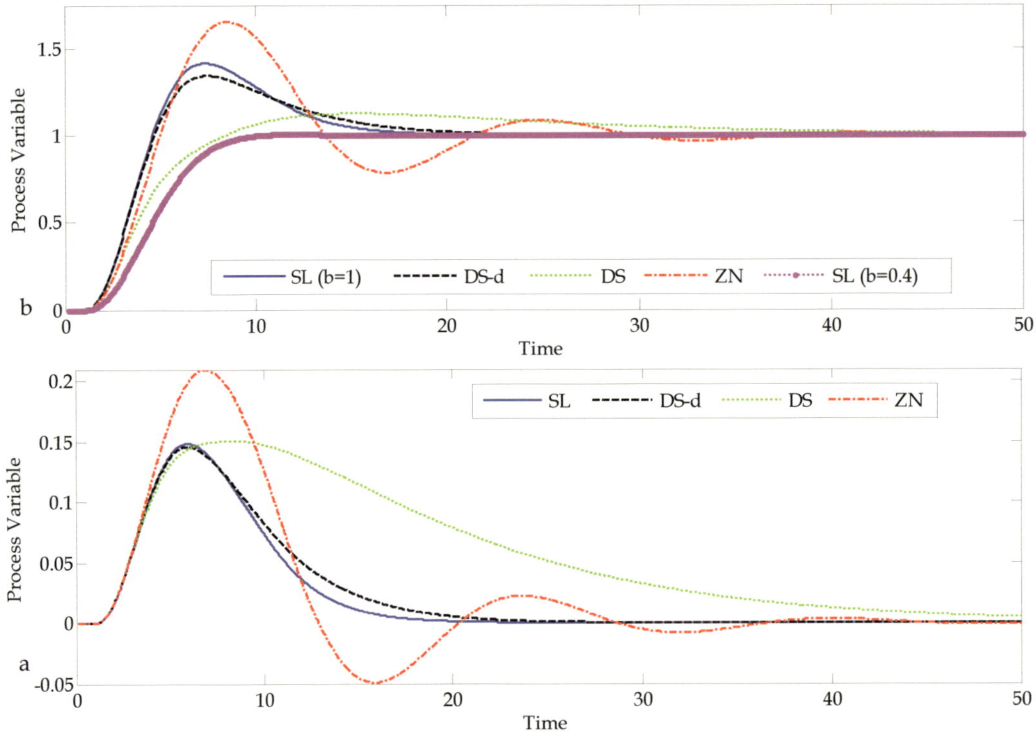

Fig. 4. Responses of the nominal system in Example 3.

The methods' robustness is evaluated by considering the worst plant-model mismatch case of $G_p = G_d = 2.4e^{-1.2s}/(8s+1)(4s+1)$ by simultaneously inserting a perturbation uncertainty of 20% in all three parameters (Table 7). Of the considered methods, the error integral values of the SL method are the best for both set-point and disturbance rejection. The overshoot of the SL method is similar to those of the DS-d and DS methods. The ZN method shows the largest overshoot of the considered methods.

Tuning methods	τ_c	K_c	τ_I	τ_D	M_s	set-point TV	set-point IAE	set-point ISE	set-point ITAE	set-point Over shoot	disturbance TV	disturbance IAE	disturbance ISE	disturbance ITAE	disturbance Over shoot
SL (b=1.0)	1.6	6.415	6.859	1.9798	1.87	11.59	5.66	3.36	28.50	1.41	1.83	1.06	0.11	7.90	0.14
SL (b=0.4)	1.6	6.415	6.859	1.9798	1.87	4.86	4.72	3.63	13.17	1.0	1.83	1.06	0.11	7.90	0.14
DS-d	2.4	6.384	7.604	2.0977	1.87	10.82	5.67	3.22	31.19	1.34	1.71	1.19	0.12	9.77	0.14
SIMC	0.43	3.496	5.72	5.0	1.87	5.514	10.08	4.77	133.6	1.36	1.47	2.50	0.26	38.50	0.16
DS	0.5	5.0	15.0	3.33	1.91	7.18	6.53	3.28	68.19	1.12	1.42	3.0	0.31	49.47	0.15
ZN	-	4.72	5.83	1.46	2.26	12.75	8.73	4.88	78.02	1.65	2.71	1.79	0.23	18.9	0.21

Table 6. PID Controller Settings for Example 3 (SOPDT)

Tuning methods	set-point TV	set-point IAE	set-point ISE	set-point ITAE	set-point Over shoot	disturbance TV	disturbance IAE	disturbance ISE	disturbance ITAE	disturbance Over shoot
SL (b=1.0)	41.88	5.11	2.95	29.45	1.46	6.41	1.09	0.11	8.61	0.17
DS-d	47.93	5.14	2.87	33.12	1.38	7.40	1.21	0.19	10.49	0.17
SIMC	50.36	8.71	3.99	104.9	1.37	14.19	2.31	0.25	31.78	0.18
DS	82.83	6.34	2.99	76.93	1.22	16.50	3.00	0.30	49.41	0.17
ZN	14.90	6.00	3.88	30.23	1.68	3.09	1.29	0.22	8.69	0.24

Table 7. Robustness Analysis for Example 3 (SOPDT) $G_p = G_d = \dfrac{2.4e^{-1.2s}}{(8s+1)(4s+1)}$

5.4 Example 4. FODIP (reboiler level model)

Consider the level control problem proposed by Chen and Seborg (2002). It is an approximate model of the liquid level in the reboiler of a steam heated distillation column, which is controlled by adjusting a control valve on the steam line. The process model is given by:

$$G_p = G_d = \frac{-1.6(-0.5s+1)}{s(3s+1)} \tag{32}$$

This kind of "inverse response time constant" (negative numerator time constant) can be approximated as a time delay such as $\left(-\theta^{inv}s+1\right) \approx e^{-\theta^{inv}}$

This is reasonable since an inverse response can deteriorate control similar to a time delay (Skogestad, 2003).

Therefore, the above model can be approximated as:

$$G_p = G_d = \frac{-1.6e^{-0.5s}}{s(3s+1)} \tag{33}$$

This process can be treated as a FODIP for PID controller design and the tuning parameters can be estimated by the analytical rules given in Table 1.

Disturbance rejection is compared by evaluating the output responses of the SL, the DS-d, the IMC and the ZN methods (Figure 5a, Table 8). The PID controller settings for all the other methods were taken from Chen and Seborg (2002). The SL method has τ_c=0.935, giving M_s=1.94. The SL output response shows the smallest overshoot and fastest settling, followed by the DS-d and IMC methods. The ZN method shows a very aggressive response with significant overshoot and subsequent oscillation that takes a long time to settle. The SL method clearly shows the best performance of the considered tuning rules.

The output responses for a unit set-point change in the reboiler level model (Figure 5b) show that the SL method with a 1DOF structure gives a large overshoot but shows fast settling to its final value. The overshoot can be greatly reduced using a 2DOF controller. The SL method shows superior performance over the other tuning methods for both set-point and disturbance rejection (Figure 5 and Table 8).

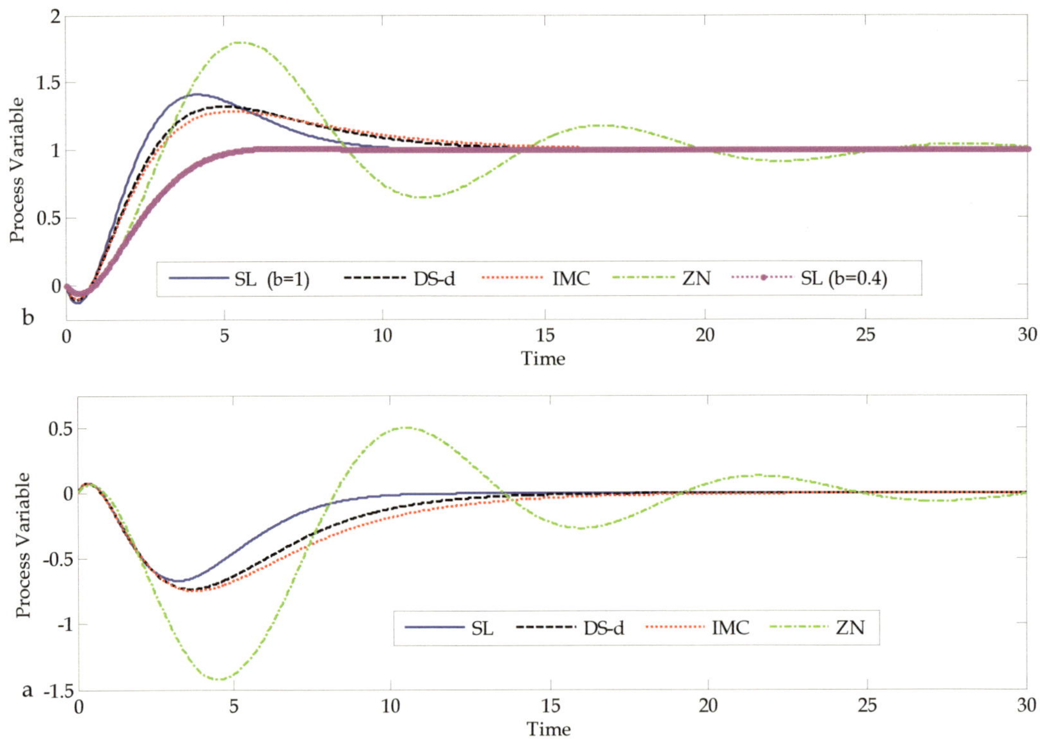

Fig. 5. Responses of the nominal system in Example 4.

Tuning methods	τ_c	K_c	τ_I	τ_D	M_s	set-point					disturbance				
						TV	IAE	ISE	ITAE	Over shoot	TV	IAE	ISE	ITAE	Over shoot
SL (b=1.0)	0.935	-1.456	4.195	1.250	1.94	4.70	3.087	1.90	9.64	1.40	3.43	2.96	1.38	12.11	-0.66
SL (b=0.4)	0.935	-1.456	4.195	1.250	1.94	1.49	2.536	1.96	4.14	1.01	3.43	2.96	1.38	12.11	-0.66
DS-d	1.6	-1.25	5.3	1.45	1.93	3.70	3.42	1.91	13.57	1.32	3.14	4.32	2.17	22.49	-0.72
IMC	1.25	-1.22	6.0	1.5	1.95	3.58	3.505	1.88	15.49	1.28	3.10	5.00	2.48	29.53	-0.74
ZN	-	-0.752	3.84	0.961	2.77	2.89	7.401	4.09	59.91	1.79	4.03	9.65	7.64	84.17	-1.42

Table 8. PID Controller Settings for Example 4 (Level Control Problem)

The controllers' robustness is evaluated considering the worst case under a 20% uncertainty in all three parameters: $G_p = G_d = -1.92(-0.6s+1)/s(2.4s+1)$. The SL method shows the best error integral and overshoot values (Table 9). The overshoot and IAE of the ZN method are the highest among the other tuning methods.

Tuning methods	set-point					disturbance				
	TV	IAE	ISE	ITAE	Over shoot	TV	IAE	ISE	ITAE	Over shoot
SL (b=1.0)	17.89	2.318	1.61	8.49	1.26	13.23	3.13	1.30	12.41	-0.59
DS-d	13.85	2.74	1.63	12.22	1.23	11.95	4.47	2.06	22.68	-0.67
IMC	14.70	2.85	1.63	14.04	1.21	12.93	5.14	2.39	29.55	-0.68
ZN	2.59	4.49	2.98	19.07	1.70	3.71	6.47	6.40	31.58	-1.49

Table 9. Robustness Analysis for Example 4 (Level Control Problem)

$$G_p = G_d = \frac{-1.92(-0.6s+1)}{s(2.4s+1)}$$

5.5 Example 5. FODUP

Consider the following FODUP (Huang & Chen, 1997 and Lee et al., 2000):

$$G_p = G_d = \frac{e^{-0.4s}}{s-1} \tag{34}$$

The closed-loop time constant $\tau_c = 0.63$ is selected for the SL tuning method to give $M_s = 3.08$. Setting $\tau_c = 0.5$ results in the same $M_s = 3.08$ for Lee et al.'s method, thus providing a fair comparison. The setting parameters of the other existing method are taken from Lee et al. (2000). The output responses for disturbance rejection by the various methods (Figure 6a) show that the SL method performs best. Comparison of performance and robustness matrices of the considered tuning rules (Table 10) also shows that the SL method performs best.

Responses to a unit step set-point change using 1DOF controllers (Figure 6b) show significant overshoots. Using a 2DOF controller can reduce the overshoot, as shown in the response of the SL method when *b=0.1*. Comparison of the controllers' performances (Figures 6 and Table 10) demonstrates the advantages of the SL method.

Tuning methods	τ_c	K_c	τ_I	τ_D	M_s	set-point					disturbance				
						TV	IAE	ISE	ITAE	Overshoot	TV	IAE	ISE	ITAE	Overshoot
SL (b=1.0)	0.63	2.573	2.042	0.207	3.08	9.58	2.12	1.56	3.17	2.06	3.71	0.92	0.37	1.55	0.65
SL (b=0.1)	0.63	2.573	2.042	0.207	3.08	2.44	1.30	0.91	1.27	1.07	3.71	0.92	0.37	1.55	0.65
Lee et al.	0.5	2.634	2.519	0.154	3.03	9.13	2.09	1.68	2.79	2.21	3.54	0.96	0.44	1.50	0.71
De Paor and O Malley	-	1.459	2.667	0.25	4.92	11.01	8.74	6.94	57.66	2.53	7.02	5.96	3.49	39.14	1.13
Rotstein and Lewin	-	2.250	5.760	0.20	2.48	6.32	3.74	2.28	11.24	1.82	3.03	2.56	1.18	8.19	0.72
Huang and Chen	-	2.636	5.673	0.118	3.21	9.14	3.28	1.96	9.86	2.19	3.62	2.15	0.80	7.57	0.76

Table 10. PID Controller Settings for Example 5 (FODUP)

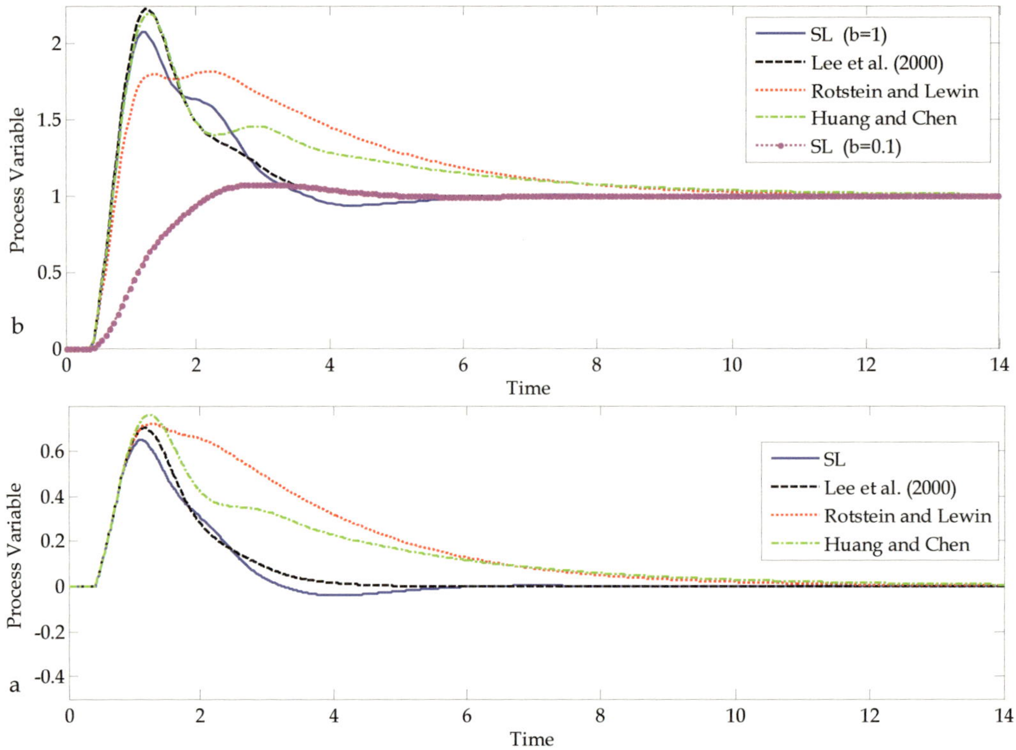

Fig. 6. Responses of the nominal system in Example 5.

To assess the controllers' robustness, perturbation uncertainties of 20% are introduced to all three parameters of the process models. The new process after 20% parameters uncertainty is defined as $G_p = G_d = 1.2e^{-0.48s}/(1.2s-1)$. The simulation results of model mismatch for the all compared tuning rules are given in Table 11. The SL method gives the best performance with model mismatch both for set-point and disturbance rejection.

Tuning methods	set-point					disturbance				
	TV	IAE	ISE	ITAE	Overshoot	TV	IAE	ISE	ITAE	Overshoot
SL (b=1.0)	14.90	2.11	1.89	2.95	2.45	5.63	0.86	0.42	1.42	0.76
Lee et al.	17.49	2.51	2.27	4.31	2.58	6.49	1.03	0.51	1.98	0.82
De Paor and O Malley	7.38	5.62	4.33	23.86	2.31	4.79	3.95	2.45	17.23	1.08
Rotstein and Lewin	7.97	3.48	2.01	10.89	2.05	3.69	2.56	1.09	9.28	0.83
Huang and Chen	18.29	3.24	2.41	9.71	2.49	6.90	2.15	0.84	8.35	0.86

Table 11. Robustness Analysis for Example 5 (FODUP) $G_p = G_d = \dfrac{1.2e^{-0.48s}}{1.2s-1}$

5.6 Example 6. SODUP

Consider the following unstable process (Huang & Chen, 1997, and Lee et al., 2000):

$$G_p = G_d = \frac{e^{-0.5s}}{(5s-1)(2s+1)(0.5s+1)} \tag{35}$$

It can be approximated (Huang & Chen, 1997; Lee et al., 2000) to a SODUP model as:

$$G_p = G_d = \frac{e^{-0.939s}}{(5s-1)(2.07s+1)} \tag{36}$$

$\tau_c = 1.2$ is used for the method of Lee et al. (2000) and $\tau_c = 0.938$ is used for the SL method, giving $M_s = 4.35$. The methods of Huang & Chen (1997) and Huang & Lin (1995) are also considered and the controller parameters are obtained from Lee et al. (2000). Comparison of the various tuning rules' output responses for disturbance rejection (Figure 7a) shows that the SL tuning method gives the fastest. Lee et al.'s (2000) method gives a smaller peak but is slower to converge and has greater oscillation. The listed controller setting parameters and performance indices (Table 12) show the advantages of the SL method.

Tuning methods	τ_c	K_c	τ_I	τ_D	M_s	set-point					disturbance				
						TV	IAE	ISE	ITAE	Over shoot	TV	IAE	ISE	ITAE	Over shoot
SL (b=1.0)	0.938	7.017	5.624	1.497	4.35	36.44	5.35	3.59	24.84	1.89	5.21	0.85	0.11	4.85	0.20
SL (b=0.3)	0.938	7.017	5.624	1.497	4.35	12.27	3.49	2.58	9.46	1.05	5.21	0.85	0.11	4.85	0.20
Lee et al.	1.20	7.144	6.696	1.655	4.34	36.45	5.19	3.03	24.93	1.71	5.12	0.95	0.10	5.76	0.19
Huang and Chen	-	6.186	7.17	1.472	3.63	26.04	5.57	3.70	25.95	1.85	4.25	1.16	0.17	7.04	0.23
Huang and Lin	-	3.954	4.958	2.074	2.18	11.99	8.99	5.55	73.31	1.86	3.16	2.19	0.38	20.13	0.29
Poulin and Pomerleau	-	3.050	7.557	2.07	1.86	8.71	11.0	6.98	105.8	1.88	3.00	3.81	0.97	40.02	0.40

Table 12. PID Controller Settings for Example 6 (SODUP)

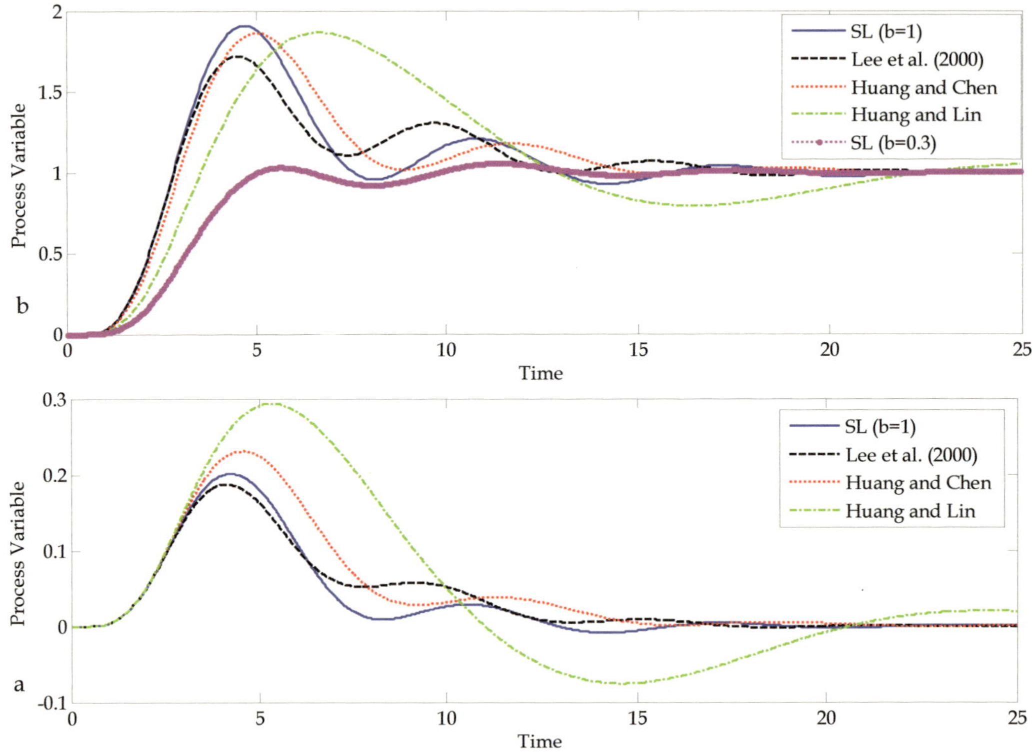

Fig. 7. Responses of the nominal system in Example 6.

Comparison of the output responses for a unit step set-point change (Figure 7b) shows that the 2DOF controller can improve the set-point response by eliminating the overshoot. The listed performance indices with 20% uncertainty in gain and dead time (Table 13) show that the SL method gives a better response both for set-point and disturbance rejection when compared with the method of Lee et al. (2000). Due to different M_s bases in the nominal case, it is difficult to achieve a fair comparison, Huang & Lin's (1995) method has more robust performance than Huang & Chen's (1997) method.

Tuning	set-point					disturbance				
methods	TV	IAE	ISE	ITAE	Overshoot	TV	IAE	ISE	ITAE	Overshoot
SL (b=1.0)	167.03	13.54	5.88	285.7	2.04	23.56	1.95	0.15	41.29	0.22
Lee et al.	248.09	15.19	5.45	446.6	1.84	34.50	2.26	0.15	63.86	0.21
Huang and Chen	65.05	7.915	4.32	75.7	1.96	10.48	1.44	0.18	14.24	0.25
Huang and Lin	13.34	7.303	4.55	46.38	1.82	3.43	1.78	0.34	13.33	0.31
Poulin and Pomerleau	8.25	8.69	5.45	64.61	1.79	2.90	3.12	0.84	26.52	0.41

Table 13. Robustness Analysis for Example 6 (SODUP) $G_p = G_d = \dfrac{1.2e^{-0.6s}}{(5s-1)(2s+1)(0.5s+1)}$

6. Discussions

6.1 Effects of τ_c on tuning parameters

The SL IMC-PID tuning method has a single tuning parameter, τ_c, that is related to the closed-loop performance and the robustness of the control system. It is important to analyze the effects of τ_c on the PID parameters, K_c, τ_I and τ_D. Consider the FOPDT model:

$$G_p = G_d = \frac{e^{-\theta s}}{s+1} \tag{37}$$

The PID parameters are calculated using the SL method for different values of the closed-loop time constant, τ_c, for each case of θ/τ= *0.25, 0.5, 0.75* and 1.0.

As the θ/τ ratio decreases, the effect of τ_c on K_c is increases (Figure 8), implying that K_c becomes less sensitive to τ_c with increasing θ/τ ratio. As τ_c increases, the variation of K_c decreases significantly.

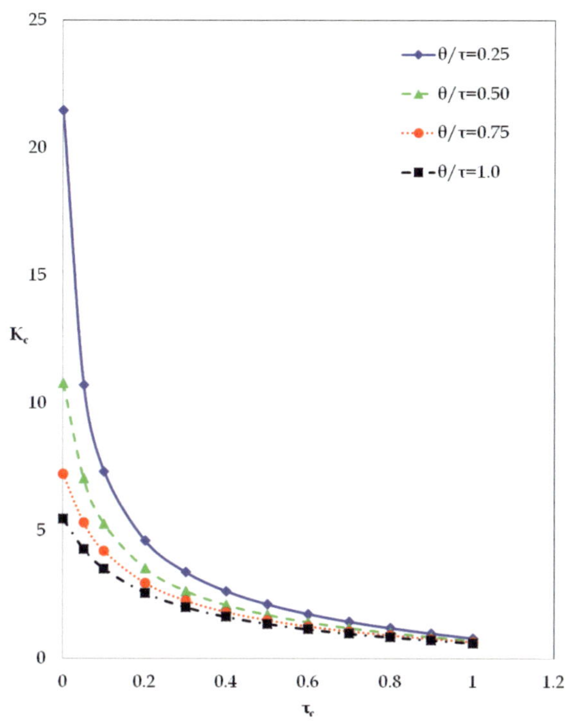

Fig. 8. Proportional gain (K_c) settings with respect to τ_c

τ_I increases initially with increasing τ_c for the different θ/τ ratios, but as the θ/τ ratio increases from 0.25 to 1.0, the variation of τ_I decreases, as clearly demonstrated at θ/τ =1.0 (Figure 9). The trend of increasing τ_I with τ_c reverses after a specific τ_c value for each θ/τ ratio. For the considered θ/τ ratios, τ_I decreases with increasing τ_c after a certain value of τ_c, which increases as θ/τ ratio increases. Note that for some large values of τ_c, τ_I is positive.

In Figure 10, the variation of τ_D with τ_c is shown and it is clear from figure that the τ_D remains positive with increasing τ_c at the various considered θ/τ ratios.

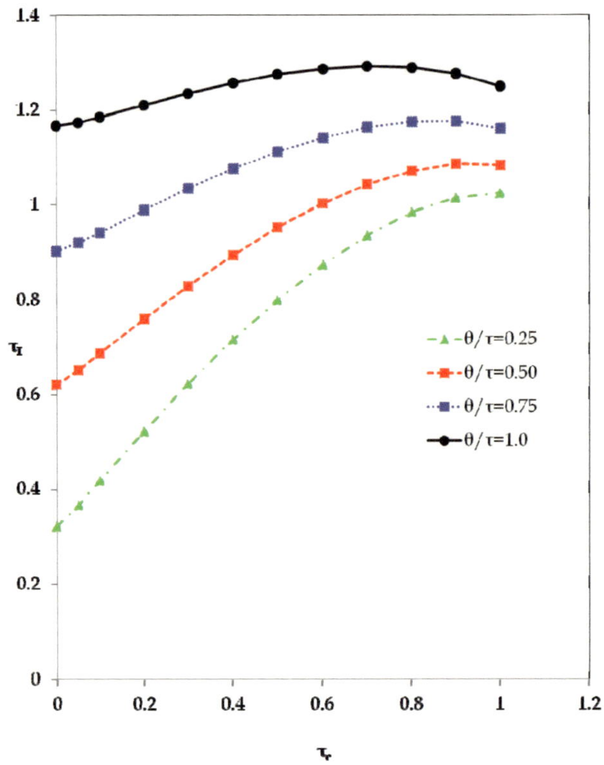

Fig. 9. Integral time constant (τ_I) settings at different τ_c values

6.2 τ_c guidelines for the IMC-PID parameter settings

Since the closed-loop time constant, τ_c, is the only user-defined tuning parameter in the SL tuning rules, it is important to have some τ_c guidelines to provide the best performance with a given robustness level. Figure 11 shows the plot of τ_c/τ vs. θ/τ ratios for the FOPDT model at different M_s values. Figure 12 shows the τ_c guideline plot for the DIP model, where τ_c can be calculated for a desired θ value at different M_s values. Figure 13 shows the τ_c guidelines for the SOPDT model; Skogestad's half rule is used to obtain this τ_c/τ *vs.* θ/τ plot. A SOPDT can be converted to a FOPDT model using the half rule. For any given θ/τ ratio of the converted FOPDT model, it is possible to obtain the τ_c/τ value from the plot in Figure 13 for the SOPDT. Although this model reduction technique introduces some approximation error, it is within an acceptable limit. Figures 14 and 15 show the τ_c guideline plots for the FODIP and FODUP models, respectively.

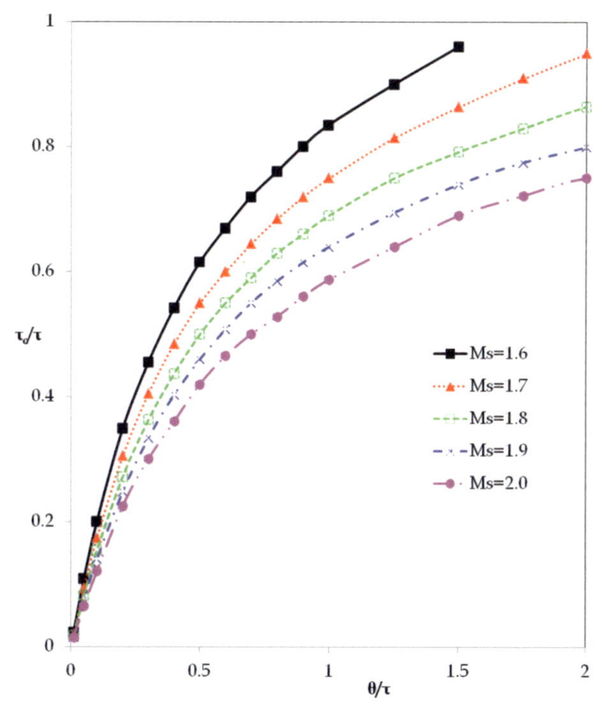

Fig. 10. Derivative time constant (τ_D) settings with respect to τ_c

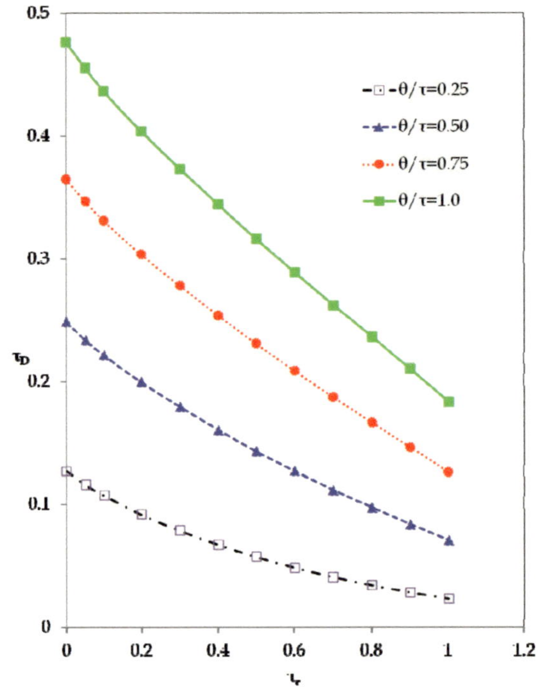

Fig. 11. τ_c guidelines for a FOPDT

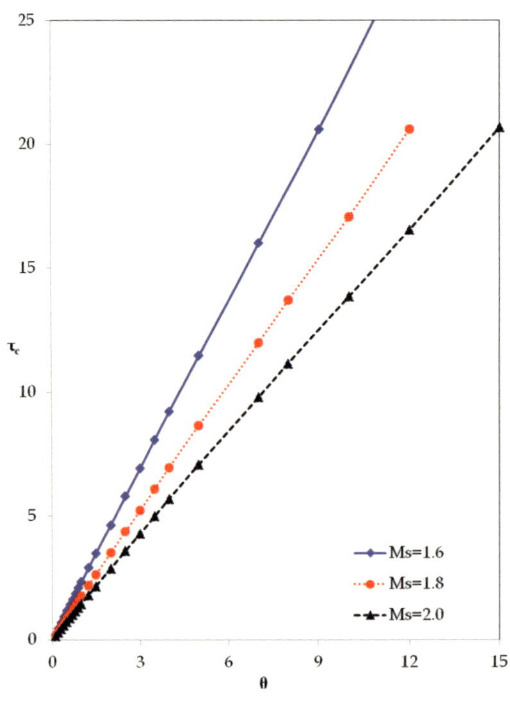

Fig. 12. τ_c guidelines for a DIP

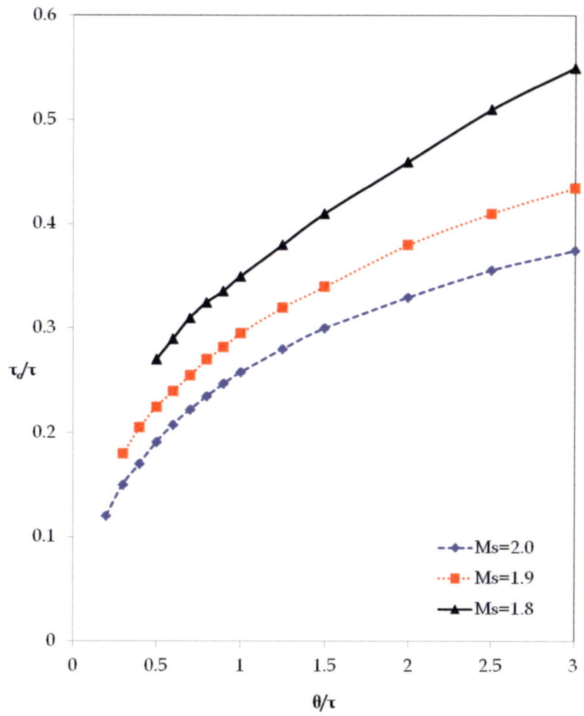

Fig. 13. τ_c guidelines for a SOPDT

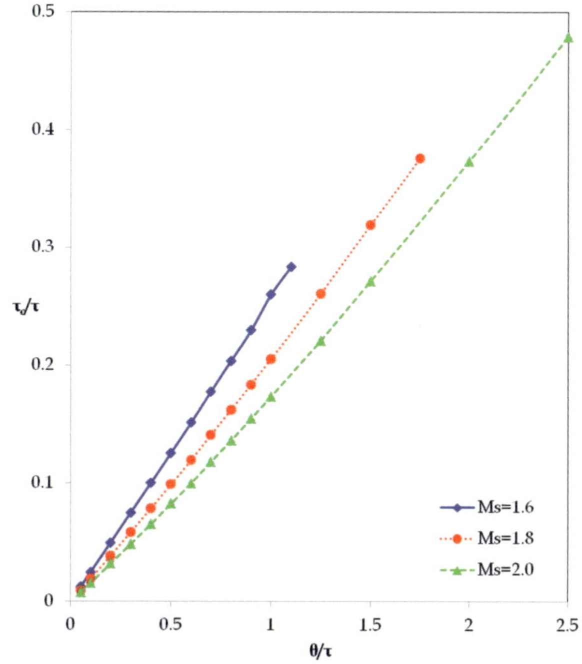

Fig. 14. τ_c guidelines for a FODIP

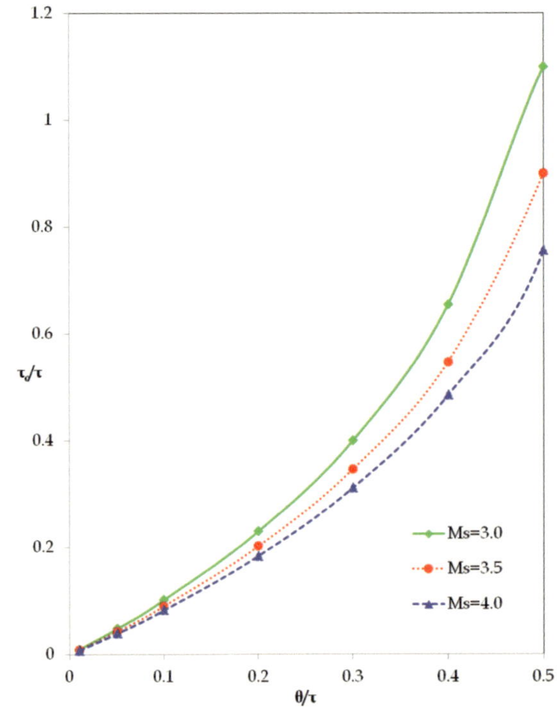

Fig. 15. τ_c guidelines for a FODUP

6.3 Beneficial range of the SL method

The load performance of the SL PID controller is superior as a lag time dominates but its superiority over a controller based on a conventional filter diminishes as a dead time dominates. In the case of a dead time-dominant process (*i.e.* $\theta/\tau >> 1$), the filter time constant should be chosen for stability as $\lambda \approx \theta >> \tau$. Therefore, the process pole at -1/τ is not a dominant pole in the closed-loop system. Instead, the pole at $-1/\tau_c$ determines the overall dynamics. Thus, introducing a lead term, $(\beta s + 1)$, to the filter to compensate the process pole at -1/τ has little impact on the speed of the disturbance rejection response. The lead term generally increases the complexity of the IMC controller, which in turn degrades the performance of the resulting controller by causing a large discrepancy between the ideal and the PID controllers. It is also important to note that as the order of the filter increases, the power of the denominator, $(\tau_c s+1)$, also increases, causing an unnecessarily slow output response. As a result in the case of a dead time-dominant process, a conventional filter without any lead term could be advantageous. Comparison of the IAE values of the load responses of the PID controllers based on the SL filter and on the conventional filter for the process model, $G_p = G_d = 10e^{-\theta s}/(s+1)$, clearly indicates that the IAE gap between the two filters decreases as the θ/τ ratio increases (Figure 16).

The robustness of controllers for dead time-dominant processes based on the conventional filter and the SL method show similar performances for perturbation uncertainties in the process parameters

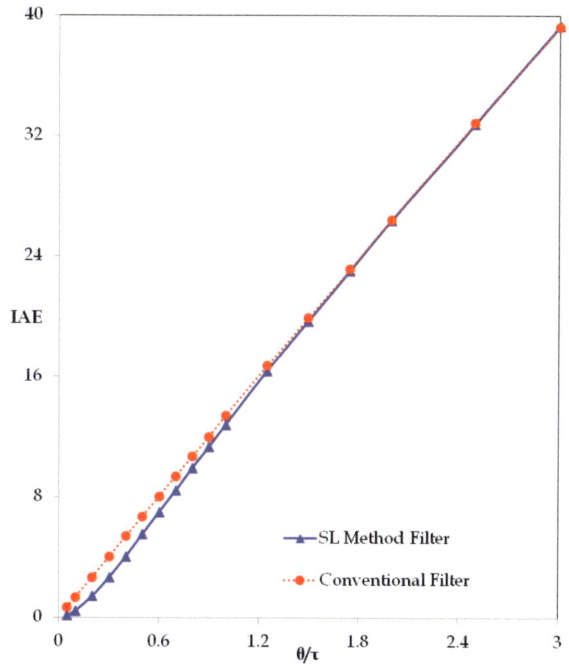

Fig. 16. Comparison of the performances of a filter based on the SL method and a conventional filter

6.4 Optimum filter structure for IMC-PID design

A common problem with conventional IMC-PID approaches is that the IMC filter is usually selected based on the performance of the resulting IMC, while the ultimate goal of IMC filter design is to obtain the best PID controller. The conventional approach for filter design assumes that the best IMC controller results in the best PID controller. However, since all IMC-PID approaches employ model reduction techniques to convert the IMC controller to the PID controller, approximation error necessarily arises. Therefore, if an IMC filter structure entails significant error in its conversion to a PID controller, the resulting PID controller could have poor control performance, despite being derived from the best IMC performance. The performance of the resulting PID controller depends on both the conversion error and the dead time approximation error, which is also directly related to the filter structure and the process model. Therefore, an optimum filter structure exists for each specific process model which gives the best PID performance. For a given filter structure, as τ_c decreases the discrepancy between the ideal and the PID controller increases, while the nominal IMC performance improves. This indicates that an optimal τ_c value also exists that can balance these two effects to give the best performance. Therefore, the best filter structure is defined here as that which gives the best PID performance for the optimal τ_c value.

To find the optimum filter structure, IMC filters with structures of $(\beta s+1)^r/(\tau_c s+1)^{r+n}$ for the first order models and of $(\beta_2 s^2+\beta_1 s+1)^r/(\tau_c s+1)^{2r+n}$ for the second order models are evaluated, where r and n are each varied from 0 to 2. A high order filter structure is then generally shown to give a better PID performance than a low order filter structure. For example, for an FOPDT model, the high order filter, $f(s)=(\beta s+1)^2/(\tau_c s+1)^3$, provides the best disturbance rejection in terms of IAE. Based on the optimum filter structures, PID controller tuning rules are derived for several representative process models (Table 1).

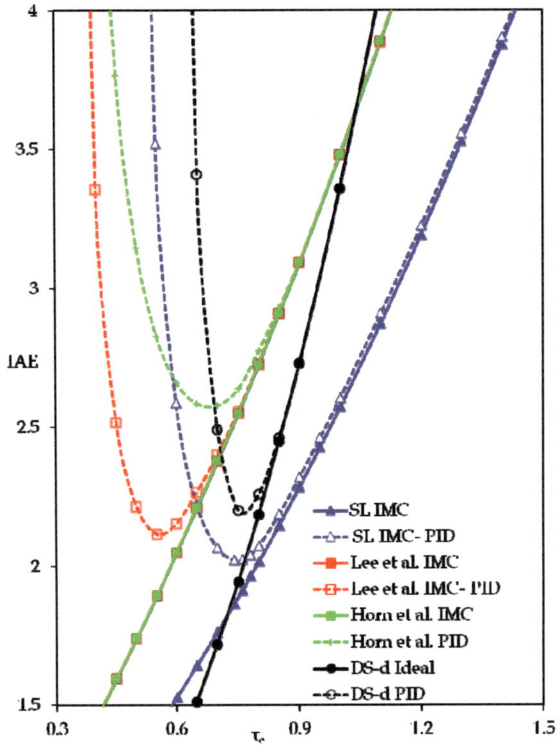

Fig. 17. τ_c vs. IAE for various tuning rules for a FOPDT

Figure 17 shows the variation of IAE with τ_c for the tuning methods of the FOPDT model considered in example 1. The tuning rules proposed by Horn et al. (1996) and Lee et al. (1998) are based on the same filter $f(s)=(\beta s+1)/(\tau_c s+1)^2$. Horn et al. (1996) use a 1/1

Pade approximation for the dead time when calculating both β and the PID parameters. Lee et al. (1998) obtained the PID parameters using a Maclaurin series approximation. Since both methods use the same IMC filter structure, the IMC controllers of Lee et al. (1998) and Horn et al. (1996) coincide with each other (Figure 17). Due to the approximation error in $e^{-\theta s}$ when calculating the PID, the performance of Lee et al.'s (1998) method is better than that of Horn et al. (1996). Down to some optimum τ_c value, the ideal (or IMC) and the PID controllers have no significant difference in performance; after some minimum IAE point, the gap increases sharply towards the limit of instability. The smallest IAE value can be achieved by the SL tuning method while that of Horn et al. (1996) shows the worst performance. The SL method also performs best in the case of model mismatch where a large τ_c value is required.

It is worthwhile to visualize the performance and robustness of the controller design. Plotting the M_s robustness index with respect to the IAE performance index for the different tuning methods for the FOPDT model of Example 1 (Figure 18) shows that at any given M_s, the PID controller developed by the SL method always produces a lower IAE than those by the other tuning rules.

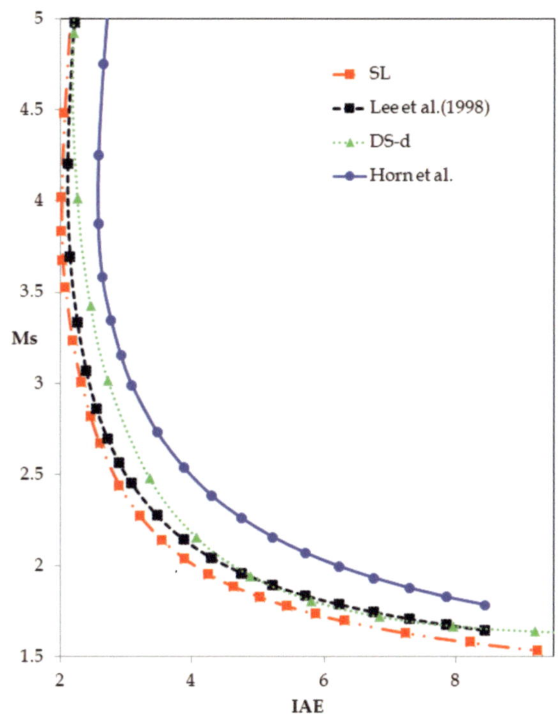

Fig. 18. M_s vs. IAE for different tuning rules for a FOPDT

7. Conclusions

The SL method could produce optimum IMC filter structures for several representative process models to improve the PID controllers' disturbance rejection. The method's filter structures could be used to derive tuning rules for the PID controllers using the generalized IMC-PID method. Simulation results demonstrate the superiority of the SL method when various controllers are compared by each being tuned to have the same degree of robustness in terms of maximum sensitivity. The SL method is more beneficial as the process is lag time-dominant. Robustness analysis, by inserting a 20% perturbation in each of the process parameters in the worst direction, demonstrates the robustness of the SL method against parameter uncertainty. Closed-loop time constant, τ_c, guidelines are also proposed by the SL method for several process models over a wide range of θ/τ ratios.

8. Acknowledgment

The authors gratefully acknowledge the funding support provided by the Deanship of Scientific Research at King Fahd University of Petroleum & Minerals (project No. **SB101016**).

9. References

Åström, K. J., Hägglund, T. (1984). Automatic tuning of simple regulators with specifications on phase and amplitude margins, Automatica, 20, pp.645–651.

Åström, K.J., Panagopoulos, H., Hägglund, T. (1998). Design of PI controllers based on non-convex optimization. Automatica, 34, pp.585-601.

Åström, K. J., Hägglund, T.(1995). PID controllers: Theory, Design, and Tuning, 2nd ed,; Instrument Society of America: Research Triangle Park, NC.

Chen, D., Seborg, D. E. (2002). PI/PID controller design based on direct synthesis and disturbance rejection. *Ind. Eng. Chem. Res.*, 41, pp. 4807-4822.

Chien, I.-L., Fruehauf, P. S. (1990). Consider IMC tuning to improve controller performance. *Chem. Eng. Prog.* 1990, *86*, pp.33.

De Paor, A. M., (1989). Controllers of Ziegler Nichols type for unstable process with time delay. *International Journal of Control, 49*, pp.1273.

Desborough, L. D., Miller, R.M. (2002). Increasing customer value of industrial control performance monitoring—Honeywell's experience. Chemical Process Control - VI (Tuscon, Arizona, Jan. 2001), AIChE Symposium Series No. 326. Volume 98, USA.

Horn, I. G., Arulandu, J. R., Christopher, J. G., VanAntwerp, J. G., Braatz, R. D.(1996). Improved filter design in internal model control. *Ind. Eng. Chem. Res.* 35, pp.3437.

Huang, C. T., Lin, Y. S. (1995). Tuning PID controller for open-loop unstable processes with time delay. *Chem. Eng. Communications, 133*, pp.11.

Huang, H. P., Chen, C. C. (1997). Control-system synthesis for open-loop unstable process with time delay. *IEE Process-Control Theory and Application, 144*, pp.334.

Kano, M., Ogawa, M. (2010). The state of the art in chemical process control in Japan: Good practice and questionnaire survey, Journal of Process Control, 20, pp. 968-982.

Lee, Y., Lee, J., Park, S. (2000). PID controller tuning for integrating and unstable processes with time delay. *Chem. Eng. Sci.*, 55, pp. 3481-3493.

Lee, Y., Park, S., Lee, M., Brosilow, C. (1998). PID controller tuning for desired closed-loop responses for SI/SO systems. *AIChE J.* 44, pp.106-115.

Morari, M., Zafiriou, E. (1989). *Robust Process Control;* Prentice-Hall: Englewood Cliffs, NJ.

Poulin, ED., Pomerleau, A. (1996). PID tuning for integrating and unstable processes. *IEE Process Control Theory and Application, 143*(5), pp.429.

Rivera, D. E., Morari, M., Skogestad, S.(1986). Internal model control. 4. PID controller design. *Ind. Eng. Chem. Process Des. Dev.*, 25, pp.252.

Rotstein, G. E., Lewin, D. R. (1992). Control of an unstable batch chemical reactor. *Computers in Chem. Eng.*, *16* (1), pp.27.

Schei, T. S.(1992). A method for closed loop automatic tuning of PID controllers, Automatica, 20, 3, pp. 587-591.

Seborg, D. E., Edgar, T. F., Mellichamp, D. A. (2004). Process dynamics and control, 2nd ed., John Wiley & Sons, New York, U.S.A.

Seki, H., Shigemasa, T. (2010). Retuning oscillatory PID control loops based on plant operation data, Journal of Process Control, 20, pp. 217-227.

Shamsuzzoha, M., Lee, M. (2007a). An enhanced performance PID.filter controller for first order time delay processes, Journal of Chemical Engineering of Japan, 40, No. 6, pp.501-510.

Shamsuzzoha, M., Lee, M. (2007b). IMC-PID controller design for improved disturbance rejection, Ind. Eng. Chem. Res., 46, No. 7, pp. 2077-2091.

Shamsuzzoha, M., Lee, M. (2008a). Analytical design of enhanced PID.filter controller for integrating and first order unstable processes with time delay, Chemical Engineering Science, 63, pp. 2717-2731.

Shamsuzzoha, M., Lee, M. (2008b). Design of advanced PID controller for enhanced disturbance rejection of second order process with time delay, AIChE, 54, No. 6, pp. 1526-1536.

Shamsuzzoha, M., Skogestad, S. (2010a). Internal report with additional data for the set-point overshoots method, Available at the home page of S. Skogestad. http://www.nt.ntnu.no/users/skoge/.

Shamsuzzoha, M., Skogestad, S. (2010b). The set-point overshoot method: A simple and fast closed-loop approch for PI tuning, Journal of Process Control 20, pp. 1220–1234.

Skogestad, S. (2003). Simple analytic rules for model reduction and PID controller tuning. J. Process Control, 13, pp.291-309.

Smith, C. L., Corripio, A. B., Martin, J. (1975). Controller tuning from simple process models. *Instrum. Technol.* 1975, 22 (12), 39.

Tyreus, B.D., Luyben, W.L. (1992). Tuning PI controllers for integrator/dead time processes, Ind. Eng. Chem. Res., pp.2625–2628.

Veronesi, M., Visioli, A. (2010). Performance assessment and retuning of PID controllers for integral processes, Journal of Process Control, 20, pp. 261-269.

Yuwana, M., Seborg, D. E. (1982). A new method for on-line controller tuning, AIChE Journal, 28 (3), pp. 434-440.

Ziegler, J. G., Nichols, N. B. (1942). Optimum settings for automatic controllers. *Trans. ASME*, 64, 759-768.